T0155897

ORGANO MAIN GROUP CHEMISTRY

ORGANO MAIN GROUP CHEMISTRY

KIN-YA AKIBA

A JOHN WILEY & SONS, INC., PUBLICATION

Published by John Wiley & Sons, Inc., Hoboken, New Jersey.
Published simultaneously in Canada.

Library of Congress Cataloging-in-Publication Data:

Akiba, Kin-ya, 1936–
Organo main group chemistry / Kin-ya Akiba.
p. cm.
ISBN 978-0-470-45033-8 (pbk.)
1. Organic compounds. 2. Organic compounds–Synthesis. 3. Carbon compounds.
4. Chemistry Organic. I. Title.
QD251.3.A35 2011
547–dc22

2010050401

oBook ISBN: 9781118025918
ePDF ISBN: 9781118025871
ePub ISBN: 9781118025888

10 9 8 7 6 5 4 3 2 1

CONTENTS

PREFACE

The fundamental and essential element of organic compounds is, without a doubt, carbon. There are excellent textbooks on organic chemistry, such as those written by McMurry, Jones, Jr., Morrison and Boyd, and Vollhardt and Schore. These are referred in undergraduate courses internationally and are available in almost all bookstores dealing with science. On the other hand, main group element chemistry has been traditionally described in inorganic chemistry textbooks before transition metal chemistry (such as in Cotton and Wilkinson, Huheey, Housecroft and Sharpe, and Schriver, Atkins and Langford).

Heteroatoms, that is, elements of groups 15, 16, and 17 bearing unshared electron pair (s) and elements of groups 1, 2, and 12–18 in which valence electrons reside in sp orbitals, are important elements contained in carbon skeletons; they modify the character of carbon compounds. The chemistry of the sp elements, namely, main group elements, has been investigated primarily to discover and to use their unique characteristics, as compared to carbon.

There are books and reviews on the chemistry of each main group element dealing with detailed and advanced researches. However, there is no concise book on organic chemistry focusing on the synthesis, structure, and reaction of main group element compounds.

This book, *Organo Main Group Chemistry*, consists of 12 chapters and 10 notes. Chapters 1–8 describe the fundamental and basic organic chemistry of main group elements that are classified according to their groups. These are appropriate for use as a textbook. Chapters 9–12 note the recent advances in the related field of hypervalent (higher coordinate) and hypovalent (lower coordinate) compounds of main group elements. The notes supplement the chapters by explaining basic ideas and also describing recent research in related fields.

Chapter 1 describes the fundamental properties of main group elements and explains the difference between a heteroatom and a main group element. Notes 1–3 remind readers of basic ideas of chemistry, emphasizing the importance of formal logic. Chapter 2 describes precisely the main group element effect in contrast to the heteroatom effect. Also, the effect of hypervalent bond, which does not appear in heteroatoms of a second period, is explained in considerable detail. Note 4 is a refresher on Hueckel molecular orbital (HMO) and electrocyclic reaction.

Chapters 3–8 discuss the synthesis, structure, and reaction of main group element compounds. Chapter 3 deals with groups 1 and 2, namely, lithium, magnesium, and copper compounds. Chapter 4 describes boron and aluminum compounds, and Chapter 5 explains silicon, tin, and lead compounds.

Chapter 6 describes compounds of group 15 elements of phosphorus, antimony, and bismuth. The synthesis of optically active phosphines, reaction of ylides, formation of phosphoranes, and effect of freezing pseudorotation are explained using phosphorus as a representative element. These are essentially common for compounds of groups 14, 15, 16, and 17, and the chemistry has been developed on phosphorus before that of sulfur because of its importance, and also for stability in handling the compounds. Note 6 mentions historical aspects of researches on hypervalent compounds, which stemmed from the dreams of Staudinger and Wittig. Note 7 explains possible mechanisms on nucleophilic substitution of MX_4-type compounds.

Chapter 7 describes compounds of group 16 elements of sulfur, selenium, and tellurium. Note 8 explains the mechanism and dynamic aspects of edge inversion. Chapter 8 deals with halogen compounds, with emphasis on fluorine and iodine compounds because organochlorine and bromine compounds are quite common and are treated well in textbooks for undergraduates.

Chapters 9, 10, and 11 describe recent advances in researches on main group elements of third period and heavier ones. Chapter 9 deals with the formation of hypervalent (higher coordinate) bonds, including silatrane and atrane. Chapter 10 explains the synthesis of hypovalent (lower coordinate) compounds of groups 14, 15, and also that of aromatic compounds. Chapter 11 illustrates ligand coupling reaction (LCR) on compounds of groups 15, 16, and iodine. Note 9 illustrates the synthesis of hexavalent tellurium compounds and that of a cation and an anion of pentacoordinate tellurium bearing the same kind of ligand.

Chapter 12 describes the synthesis of hypervalent carbon compounds, namely, pentacoordinate hypervalent carbon species (10-C-5), and attempts to hexacoordinate (hypervalent) carbon species (12-C-6), supported by theoretical calculations. Note 10 illustrates the synthesis of main group element porphyrins of Sb and P in which central atoms are hypervalent.

References throughout this book are based on arbitrary choices by the author, and not at all comprehensive. Particularly, the references in Chapters 1–8, except in Chapter 2, contain basic and fundamental references on the fields, accompanied by newer ones, where appropriate. Most consist of references that the author has read carefully before and found to contain basic experimental results. They were used in lectures for undergraduate and graduate courses in Japanese universities. Chapters 9–12 deal with recent advances in the related fields; therefore, references are cited to show basic ideas and recent researches. References for notes were chosen based on the same standpoint as for the chapters. Advanced series, such as "Comprehensives" of organic synthesis, organometallic, and heterocyclic chemistry and the "Patai's series" on functional groups were not cited as references.

In Japanese bookstores dealing with sciences, we find a variety of textbooks originally written in English that had been translated into Japanese. For instance, seven textbooks mentioned at the beginning of this preface, except the one by Housecroft and Sharpe, had been translated and are being used in Japanese universities.

There are similar books and countless review articles on specialized topics originally written in Japanese. The Chemical Society of Japan has been editing the *Encyclopedia of Experimental Chemistry* (Jikken Kagaku Kouza). The recent fifth edition consists of 30 volumes; it includes all fields of chemical sciences.

However, the articles written in Japanese can neither be referred in original research papers of international journals, nor in books written in English. It is a great pity, but it is the reality. This book was originally written in Japanese and published by Kodansha Scientific, Inc., in Tokyo under the title *Yuuki Tenkei Genso Kagaku* (translated into English as *Organo Main Group Chemistry*) in 2008. I am grateful to Kodansha Scientific, Inc., for allowing me to use all the equations and schemes in the Japanese edition, without any restriction, to write the present revised English version.

I am thankful to the international reviewers nominated by John Wiley & Sons, who heartily recommended to publish this book in English. Also, I wish to acknowledge the patience and support of my family, and the kind cooperation, skill, and good cheer of the editors at Wiley.

Professor Emeritus of Hiroshima University
Tokyo, Japan
November 2010

KIN-YA AKIBA

CHAPTER 1

MAIN GROUP ELEMENTS AND HETEROATOMS: SCOPE AND CHARACTERISTICS

1.1 AUFBAU PRINCIPLE AND SIGN OF ORBITALS

All the atoms consist of a dense nucleus surrounded by electrons. The nucleus consists of protons, which are positively charged, and neutrons, which are neutral. The number of electrons is equal to the number of protons; hence, the atom bears no charge and is neutral. The atomic number of an element is equal to the number of protons in the nucleus, and the mass number is the sum of the number of protons and the number of neutrons. The atomic weight is approximately the same as mass number, because an electron is very light in comparison (the ratio of mass, electron:proton = 1:1840; the mass of proton = 1.6726×10^{-24} g).

Electrons are concentrated in specific regions of space around the nucleus called *orbitals*. The orbitals are designated as s, p, d, and f according to their shape and energy (s orbital is spherical and the lowest in energy), and each orbital designated as above has 1, 3, 5, and 7 orbitals of equal energy. Furthermore, each electron rotates like the earth and there are two kinds (plus and minus) of rotations called *spin*. Each orbital can contain a maximum of two electrons with opposite spin (the Pauli exclusion principle).

Electrons are placed in orbitals starting from the lowest energy one (s<p<d<f). In case there are more than two (3, 5, or 7) orbitals of the same energy, electrons are first placed sequentially, one by one in different orbitals with parallel spin. Then, extra electrons are placed as the second electron with anti-parallel spin in the orbitals mentioned above. Thus, according to the atomic

Organo Main Group Chemistry, First Edition. Kin-ya Akiba.

TABLE 1.1 Periodic Table and Essential Characteristics of Main Group Elements

Group (n) / Period	1 1A	2 2A	12 2B	13 3B	14 4B	15 5B	16 6B	17 7B	18 0
1 1s	(0.37(H-H)) 1312 (0.32(C-H)) $^{1}_{1}$**H** ← (0.66 x 10^{-5}) (1.54 –) Hydrogen 2.1 435								(1.28°) 2372 $^{4}_{2}$**He** Helium [5.50]
2 [He] 2s2p	1.23 513 (0.78⁺) $^{7}_{3}$**Li** Lithium 1.0 248	0.89 899 (0.34^{2+}) $^{9}_{4}$**Be** Beryllium 1.5		0.88 801 $^{11}_{5}$**B** Boron 2.0 372	0.77 1086 $^{12}_{6}$**C** Carbon 2.5 368	0.70 1402 $^{14}_{7}$**N** Nitrogen 3.0 292	0.66 1314 (1.40^{2-}) $^{16}_{8}$**O** Oxygen 3.5 351	0.58 1681 (1.33^{-}) $^{19}_{9}$**F** Fluorine 4.0 441	2081 $^{20}_{10}$**Ne** Neon [4.84]
3 [Ne] 3s3p	(1.54°) 496 (0.98^{+}) $^{23}_{11}$**Na** Sodium 0.9	1.36 738 (0.79^{2+}) $^{24}_{12}$**Mg** Magnesium 1.2		1.25 577 (0.57^{3+}) $^{27}_{13}$**Al** Aluminium 1.5 255	1.17 787 $^{28}_{14}$**Si** Silicon 1.8 301	1.10 1012 $^{31}_{15}$**P** Phosphorus 2.1 264	1.04 1000 (1.74^{2-}) $^{32}_{16}$**S** Sulfur 2.5 272	0.99 1251 (1.81^{-}) $^{35}_{17}$**Cl** Chlorine 3.0 352	(1.74°) 1520 $^{40}_{18}$**Ar** Argon [3.20]
4 [Ar:$3d^{10}$] 4s4p	2.03 419 (1.33^{+}) $^{39}_{19}$**K** Potassium 0.8	1.74 590 (1.06^{2+}) $^{40}_{20}$**Ca** Calcium 1.0	1.25 906 (0.83^{2+}) $^{64}_{30}$**Zn** Zinc 1.6 176	1.25 579 (0.83^{3+}) $^{69}_{31}$**Ga** Gallium 1.6 247	1.22 762 $^{74}_{32}$**Ge** Germanium 1.8 237	1.21 947 $^{75}_{33}$**As** Arsenic 2.0 200	1.17 941 (1.91^{2}) $^{80}_{34}$**Se** Selenium 2.4 245	1.14 1140 (1.96^{-}) $^{79}_{35}$**Br** Bromine 2.8 293	1.89 1351 (1.69^{+}) $^{84}_{36}$**Kr** Krypton [2.94]
5 [Kr:$4d^{10}$] 5s5p	(2.48°) 403 (1.49^{+}) $^{85}_{37}$**Rb** Rubidium 0.8	1.92 550 (1.27^{2+}) $^{88}_{38}$**Sr** Strontium 1.0	1.41 867 (1.03^{2+}) $^{114}_{48}$**Cd** Cadmium 1.7 139	1.50 558 (0.92^{3+}) $^{116}_{49}$**In** Indium 1.7 165	1.40 709 $^{120}_{50}$**Sn** Tin 1.8 225	1.41 834 $^{121}_{51}$**Sb** Antimony 1.9 215	1.37 869 (2.24^{2-}) $^{130}_{52}$**Te** Tellurium 2.1	1.33 1008 (2.20^{-}) $^{127}_{53}$**I** Iodine 2.5 213	2.09 1170 (1.90^{+}) $^{132}_{54}$**Xe** Xenon [2.40]
6 [Xe:$4f^{14}5d^{10}$] 6s6p	2.35 376 (1.65^{+}) $^{133}_{55}$**Cs** Caesium 0.7	1.98 503 (1.43^{2+}) $^{138}_{56}$**Ba** Barium 0.9	1.44 1007 (1.12^{2+}) $^{202}_{80}$**Hg** Mercury 1.9 122	1.55 589 (1.05^{3+}) $^{205}_{81}$**Tl** Thallium 1.8 125	1.54 716 $^{208}_{82}$**Pb** Lead 1.8 130	1.52 703 $^{209}_{83}$**Bi** Bismuth 1.9 143	1.53 812 $^{209}_{84}$**Po*** Polonium 2.0	(0.57^{5+}) 930 (2.27^{-}) $^{216}_{85}$**At*** Astatine 2.2	1040 $^{222}_{86}$**Rn*** Radon [2.06]

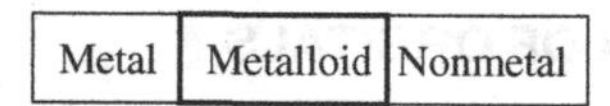

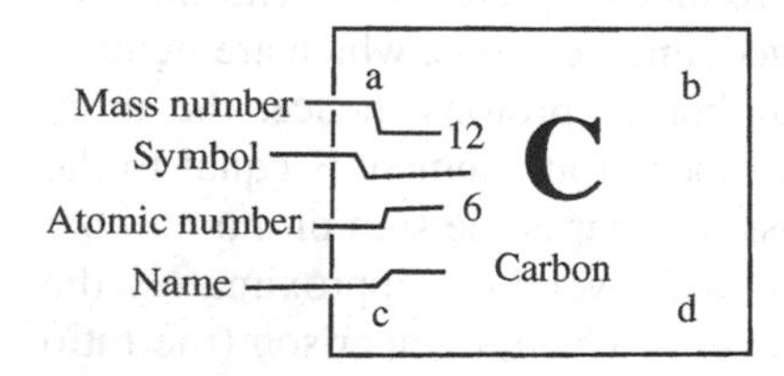

1 eV = 23.061 kcal/mol = 96.485 kJ/mol

a: Covalent radius (Å) of single bond (A–A) symbol in parentheses (o,+,–) shows atom, cation, and anion, respectively

b: First ionization energy (kilojoules per mole)

c: Pauling's electronegativiy (Allred-Rochow's number is in [ ])

d: Bond energy of C–E (kilojoules per mole)

*: Radioactive element (the most abundant isotope in nature)

number, the electronic configuration of an atom can be determined. This is called *aufbau principle*. For example, the atomic orbital of elements of first period (H and He) consists of 1s, the atomic orbitals of elements of second period consist of 2s and 2p in addition to 1s, and for those of third period, 3s and 3p are added to 2s, 2p, and 1s. For fourth period elements, 4s and 4p orbitals are added, in which orbitals of third period are filled and 3d orbitals are also filled from zinc (group 12) to krypton (group 18). For fifth period, [Kr] $4d^{10}$ are filled as kernel and for sixth period, [Xe] $4f^{14}$ $5d^{10}$ are filled as kernel (Table 1.1). The aufbau principle should be familiar to readers and, just to make sure, the shapes and signs of 1s, 2p, and 3d are shown in Fig. 1.1 [3, 4, 5].

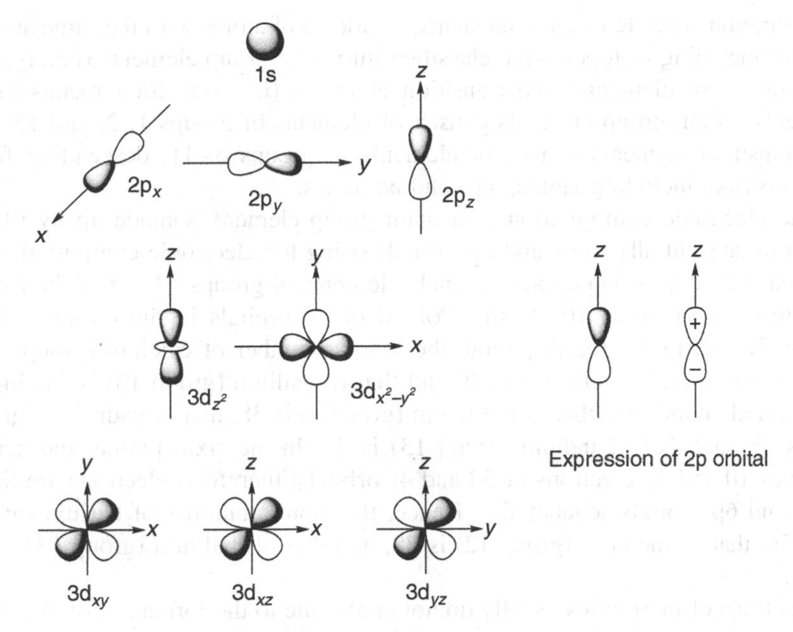

Figure 1.1 Atomic orbital of hydrogen atom, 1s, 2p, and 3d.

The orbitals of electrons are expressed by wave function (Φ) according to quantum mechanics. The probability of finding electrons is expressed as a square of wave function (Φ^2) and is thus positive, although wave function has a plus or minus sign according to symmetry. The orbital shapes in common text books of chemistry exhibit specific regions as contour to find electrons with 0.95 (or 0.90) probability, and the sign of wave function (Φ) is added to the contour.

The wave function of 1s orbital is spherical and thus has only one sign, either plus or minus. Usually, plus sign is used. The 2s orbital also has spherical symmetry but has one node. The 2p orbitals are dumbbell shaped and separated by an orthogonal plane (node) bisecting the dumbbell. The two lobes (parts of an orbital) of an orbital have opposite signs and this is shown by + or − or by different colors. The 3d orbitals consist of three cross shaped orbitals of *xy, yz*, and *zx* planes, in which two orthogonal axes are tilted by 45°; and the fourth orbital is aligned to the *xy* axis ($3dx^2 - y^2$) and the fifth is aligned to the *z* axis ($3dz^2$). Hence, there are five orbitals in total (Fig. 1.1) [6, 7].

1.2 ELECTRONIC CONFIGURATION OF AN ATOM: MAIN GROUP ELEMENTS AND HETEROATOMS

The electronic configuration of an atom is made up according to aufbau principle, and all the atoms are arranged in the periodic table based on the electronic configuration. The elements—substances made up of atoms with the same atomic number including isotopes—are classified into main group elements (i.e., typical elements or sp elements) and transition elements (i.e., transition metals or df elements). Main group elements consist of elements of groups 1, 2, and 12–18, and transition elements consist of elements of groups 3–11, descending from fourth period, including lanthanides and actinides.

The electronic configuration of a main group element is made up by filling electrons sequentially in ns and np orbitals using the electronic configuration of inert gas of the $(n-1)$ period as kernel. Elements of groups 12–18 of the fourth or fifth period contain 10 electrons of 3d or 4d orbitals in the corresponding kernel. Hence, in the fourth period, the atomic number of calcium (group 2) is 20 and that of zinc (group 12) is 30 and that of gallium (group 13) is 31. In the fifth period, atomic number of strontium (group 2) is 38, that of cadmium (group 12) is 48, and that of indium (group 13) is 49. In the sixth period, the kernel involves 10 and 14 electrons of 5d and 4f orbitals; therefore, electrons are filled in 6s and 6p orbitals sequentially. Hence, the atomic number of barium (group 2) is 56, that of mercury (group 12) is 80, and that of thallium (group 13) is 81 [6, 7].

Electrons of inert gases usually do not contribute to the formation of chemical bonds. It is convenient to show kernel, which consists of a nucleus and electrons of filled inner shells, as the symbol of the element. Residual electrons are contained in the outermost shell and contribute to the formation of chemical bonds; hence, they are valence electrons. It is clear that valence electrons of main group elements are accommodated in ns and np orbitals; thus, it is realized that main group elements are also called sp *elements*. The group 12 elements such as zinc, cadmium, and mercury are sometimes classified as transition elements; however, they are mainly considered as main group elements. This is because their outer shell electronic configurations are $(n-1)d^{10}ns^2$ and the ions bearing two positive charges are stable containing electrons of $(n-1)d^{10}$.

Organo-main-group chemistry (organic chemistry of main group elements) is a branch of organic chemistry dealing with structures, syntheses, and reactions of compounds bearing a carbon—main group element bond. This branch resides between traditional organic chemistry and organo-transition-metal (organometallic) chemistry and bridges the gap between them.

In Table 1.1, four fundamental characteristics of main group elements are summarized, that is, (i) covalent radius of single bond (A–A); (ii) the first ionization energy; (iii) electronegativity (Pauling); (iv) bond energy of carbon–element single bond (C–E) [1, 2].

To make the characteristics of these elements clear, a few bold lines are drawn in Table 1.1. A horizontal bold line between the second and third periods shows that there are distinct differences in these characteristics between the elements of the second period and those of the third period and heavier. There are two vertical bold lines, one between the groups 14 and 15 and the other between the groups 17 and 18. These show that the elements of groups 15, 16, and 17 are heteroatoms. Heteroatoms contain unshared electron pairs (lone pair of electrons), such as one pair of nitrogen, two pairs of oxygen, and three pairs of fluorine. A shorter vertical bold line between the groups 2 and 12 shows that transition elements exist between the two groups.

Six elements, which are often called *metalloids*, are depicted as surrounded by bold lines in the middle of Table 1.1; these are four diagonally located elements, boron, silicon, arsenic, and tellurium, and two accompanying elements, germanium and antimony. Elements lying on the right-hand side of these, including hydrogen, are nonmetals; on the other hand, those lying on the left-hand side of these are metals. There is ambiguity in this kind of classification, because it depends on the physical properties of these elements (metal, nonmetal, gas, liquid, solid, etc.) but not on the electronic configuration. Furthermore, there is an idea to classify the elements depending on electronegativity, that is, those with electronegativity >2.0 are nonmetals and those with electronegativity <2.0 are metals.

1.3 FUNDAMENTAL PROPERTIES OF MAIN GROUP ELEMENTS

The element is a group of atoms (including isotopes) with the same atomic number; however, element and atom are often used without clear distinction. As the fundamental characteristics of main group elements, the following four are summarized and illustrated in Table 1.1 [1, 2]:

1. covalent radius of single bond (A–A),
2. the first ionization energy (potential),
3. electronegativity (χ: Pauling), and Allred–Rochow (for inert gases)
4. bond energy of carbon–element single bond (C–E).

The corresponding numerical values of four characteristics are cited at the corners of a square in which an element symbol with mass number, atomic number, and the name of the element are written at the center.

At the upper left corner, covalent radius (Å) of the same atoms (A–A) is cited. Numerical values enclosed in parentheses and written below the covalent radius values, which bear one of signs, $^\circ$, $+$, or $-$, correspond to atomic, cationic, or anionic radius of the atom.

It is explicitly and numerically shown that cationic radius is considerably smaller than the covalent radius and anionic radius is considerably larger than the covalent radius. Covalent radius decreases according to the shift of the group to the right in the periodic table and this trend is the most prominent for elements of the second period. Hydrogen is special. Covalent radius of hydrogen (H–H) is 0.37 Å and that of C–H is 0.32 Å. Ionic radius of the cation of hydrogen (proton) is extremely small, 0.66×10^{-5} Å, whereas that of the anion of hydrogen (hydride) is as large as 1.54 Å, being larger than those of the fluoride ion (1.33 Å) and the helium atom (1.28 Å). For group 1 elements, covalent radius or atomic radius is shown with the radius of the monocation. For group 2 elements, covalent radius and the radius of the dication are shown.

At the upper right corner, the first ionization energy (in kilojoules per mole) is cited. It is the least energy required to detach an electron from a neutral atom to infinite distance. The energy to detach the second electron after the first is called the *second ionization energy* and that required to detach the third electron after the second one is the third ionization energy. Ionization energy is the same as ionization potential.

In each period of the periodic table, it is apparent that the first ionization energy is the least for group 1 element and the largest for inert gas. Within the elements of the same group, the first ionization energy decreases according to the increase of the period number. On the other hand, within the elements of the same period, ionization energy tends to increase with the increasing group number. It is larger, however, for group 15 elements than for group 16 elements in the second, third, and fourth periods. It is quite evident in the second period, that is, for nitrogen (1402 kJ/mol) and for oxygen (1314 kJ/mol). This is contrary to the general knowledge that an ammonium salt is prepared much more easily compared to an oxonium salt.

At the lower left corner, electronegativity (χ) proposed by Pauling is cited. Electronegativity is a numerical scale (0.0–4.0) to represent relative electron-withdrawing ability of an atom contained in a neutral molecule. Although there are a couple of expressions for electronegativity, that by Pauling is cited here to the first decimal place, which is accurate enough for use in organic chemistry. For inert gases, electronegativity by Allred–Rochow is quoted. Electronegativity increases according to the increase in the number of group within a period and decreases according to the increase of the period number.

At the lower right corner, a single bond energy (in kilojoules per mole) of carbon–element (C–E) bond is cited. This bond energy is much more useful in organic chemistry than that between the same elements (A–A). The bond energy (C–E) of the same group decreases according to the increase of the period number, and increases according to the group number within a period. However, among the second, third, and fourth periods, the bond energy between carbon and group 15 elements is the least among groups 13–17. This tendency is the most apparent for the nitrogen bond (C–N: 292 kJ/mol).

It is evident that there are considerable differences in the corresponding four characters between the elements of the second period and those of third period (see the horizontal bold line in Table 1.1).

The ionic character of C–E bond is also an important index. It is (i) directly related to the acidity of the C–H bond; (ii) the indication of stability of the corresponding carbanion; and (iii) fundamentally related to the order of nucleophilicity. Ionicity (C^-–E^+: %) of the C–E bond, in which carbon is sp^3, is cited with electronegativity of the element. Ionicity decreases when electronegativity of E increases.

	Ionicity of C–E bond (%)	Electronegativity of E (χ)
Li	43%	1.0
Na	47%	0.9
K	51%	0.8
Be	22%	1.5
Mg	35%	1.2
Ca	43%	1.0
B	6%	2.0
Al	22%	1.5
Tl	12%	1.6
Zn	18%	1.6
Cd	15%	1.7
Hg	9%	1.9
Sn	12%	1.8
Pb	12%	1.8

On the other hand, transition elements lie between groups 3 and 11 and from fourth to seventh periods. These are also classified as df elements because valence electrons are located in d and/or f orbitals. Ionicity of the C–E (E: transition metal) bond is in the range of 34–38% and electronegativity of E lies in the range 1.1–1.2. The values are quite close to one another and the df elements are metals.

1.4 ACIDITY OF CARBOXYLIC ACID AND SUBSTITUENT EFFECT

Among quantitatively measured characters of organic compounds, one of the most important is the strengths of acids and bases (acidity: pK_a) [8]. Substitution of a hydrogen of acetic acid (p$K_a = 4.76$) by a methyl group results in propanoic acid (p$K_a = 4.87$), which is weaker than acetic acid. Methyl (alkyl) group is, therefore, electron-donating by inductive effect. Substitution by a halogen instead of the methyl group renders carboxylic acid stronger because of the increased electronegativity of the halogen; thus, halogen is electron-withdrawing by inductive effect. Electronegativity of fluorine (4.0) is the largest and that of iodine (2.5) is the smallest among the four halogens, bearing the same electronegativity as carbon. But it should be noted that the acidity of iodoacetic acid (p$K_a = 3.15$) is much stronger than that of propanoic acid. Inductive effects of chlorine and bromine are almost the same and they are in between those of iodine and fluorine. Trifluoromethyl group is a stronger electron-withdrawing group than fluorine, and trichloro- and trifluoroacetic acids are usually classified as strong acids (p$K_a = 0.64$, 0.52).

Comparing the acidities of propanoic acid (4.87), chloroacetic acid (2.84), dichloroacetic acid (1.35), and trichloroacetic acid (0.64), the inductive effect of chlorine gets stronger with the increase in the number of chlorine atoms, but the effect is not linearly additive. This is due to steric effect, solvation effect, and so on, and detailed discussion is not useful here. On the other hand, let us take a look at the acidity of propanoic acid (4.87), α-bromo- (2.98), β-bromo- (4.02), and α,β-dibromopropanoic acid (2.20). We recognize that the additivity of inductive effect of bromine is valid. (cf.: $\{4.87 - [(4.87 - 2.98) + (4.87 - 4.02)] = 2.13 = 2.20\}$). It is therefore believed that the additivity of inductive effect is valid, in case other factors are not influential.

Here, let us think about the effect of substituent attached to the benzene ring on the rates and equilibria of many reactions. It was noted by Hammett in 1930s that there is a linear relationship between the acid strengths of substituted benzoic acids and the rates of many other reactions. The correlation is valid and a linear free-energy relationship [9] between two reaction series is maintained based on one of the three conditions : (i) ΔH terms are proportional and ΔS is constant for the series; (ii) ΔS terms are proportional and ΔH is constant; (iii) ΔH and ΔS are linearly related, which is often shown to be the case. Hammett equations are expressed for equilibria and rates as follows:

$$\log(K_{\mathrm{X}}/K_{\mathrm{H}}) = \sigma\rho \tag{1.1}$$

$$\log(k_{\mathrm{X}}/k_{\mathrm{H}}) = \sigma\rho \tag{1.2}$$

where σ is the substituent constant and ρ is the reaction coefficient.

There is an excellent linear correlation between $\mathrm{p}K_{\mathrm{a}}$ of substituted benzoic acids and rates of basic hydrolysis of ethyl benzoates.

The substituent constant (σ) is calculated as the ratio of acidity constant ($\mathrm{p}K_{\mathrm{a}}$) of a substituted benzoic acid and the parent benzoic acid in water at 25°C, assuming that $\rho = 1.00$ (ρ: sensitivity of a particular reaction to substituent effect, which is determined experimentally for each reaction). The σ value is applicable to a variety of equilibria and rates of substituted benzenes with the corresponding value of ρ for each system.

Every substituent effect is a combination of resonance and inductive effects and each substituent can be electron-donating and/or electron-withdrawing to influence the electron density of 1-position of a benzene ring, which is connected to the reaction center. The substituent effect has been determined for *meta* (σ_m) and *para* (σ_p) substituents, where steric effect can be free. In Table 1.3, $\mathrm{p}K_{\mathrm{a}}$ of substituted benzoic acids and substitutent constants (σ_m and σ_p) are listed.

When they are negative in value, they are electron-donating; on the other hand, they are positive when they are electron-withdrawing. Both values are positive for chlorine; however, the difference ($\Delta = \sigma_p - \sigma_m$: $-$ 0.14) is negative. This is consistent with the fact that the rate of electrophilic substitution of chloro(halogeno)benzene is slower than that of benzene itself but the orientation of substituent is *ortho–para*. For anisole, the methoxy group is electron-withdrawing inductively but electron-donating by resonance, and the latter effect overwhelms the former ($\Delta = \sigma_{\mathrm{p}} - \sigma_{\mathrm{m}}$: $-$ 0.50). The rate of electrophilic substitution is much greater than that of benzene and the orientation is *ortho–para*.

TABLE 1.2 pK_a of Substituted Acetic Acid

1)	H–CH_2–CO_2H	4.76
2)	CH_3–CH_2–CO_2H	4.87
3)	CH_3O–CH_2–CO_2H	3.57
4)	I–CH_2–CO_2H	3.15
5)	Br–CH_2–CO_2H	2.91
6)	Cl–CH_2–CO_2H	2.84
7)	F–CH_2–CO_2H	2.54
8)	CF_3–CH_2–CO_2H	2.02
9)	O_2N–CH_2–CO_2H	1.68
10)	Cl_2–CH–CO_2H	1.35
11)	F_2–CH–CO_2H	1.24
12)	Cl_3–C–CO_2H	0.64
13)	F_3–C–CO_2H	0.52

TABLE 1.3 Acidity of Substituted Benzoic Acid, Substituent Constant (σ), and Polarizability of Element (E)

	Substituent (X)	pK_a		σ constant		χ of E (Pauling)	Polarizability of E
		m	p	σ_m	σ_p		
(a)	NH_2	4.37	4.87	–0.16	–0.66	3.0	8.7
(b)	OH	4.08	4.58	0.12	–0.37	3.5	5.9
(c)	OMe	4.09	4.49	0.12	–0.28	3.5	5.9
(d)	CH_3	4.28	4.36	–0.07	–0.17	2.5	9.3
(e)	H	4.21	4.21	0.00	0.00	2.1	6.7
(f)	I	—	—	0.35	0.18	2.5	51.1
(g)	Br	3.81	4.00	0.39	0.23	2.8	33.4
(h)	Cl	3.82	3.99	0.37	0.23	3.0	22.8
(i)	F	3.87	4.14	0.34	0.06	4.0	3.8
(j)	CO_2Et	3.82	3.63	0.37	0.45	—	—
(k)	COMe	3.89	3.68	0.38	0.50	—	—
(l)	CN	3.60	3.53	0.56	0.66	—	—
(m)	NO_2	3.49	3.42	0.71	0.78	—	—

For nitrobenzene, the nitro group is electron-withdrawing both by inductive and by resonance effects. Hence, the rate is very much slower than that of benzene and the orientation is *meta* ($\Delta = \sigma_p - \sigma_m$: $+0.07$). Electronegativity (χ: Pauling) and polarizability of element (E) are also listed in Table 1.3. Characteristics of fluorine are apparently shown due to the fact that the difference ($\Delta = -0.28$) is quite large and negative, C–F single bond energy (441 kJ/mol) is the largest, polarizability (3.8) is the smallest, and electronegativity (4.0) is the largest. On the other hand, characteristics of iodine are contrary to those of fluorine. C–I single bond energy (213 kJ/mol) is the smallest among halogens and the difference ($\Delta = -0.17$) is almost the same with halogens except fluorine, polarizability

(51.1) is the largest, and electronegativity (2.5) is the same as that of carbon. Electron-withdrawing effect of trifluoromethyl group is different from that of fluorine but rather similar to that of cyano group. The experimental and theoretical researches on electronic effect of substituent and linear free-energy relationship have advanced quite far and a variety of substituent constants (σ°, σ^+, σ^-, σ_I, σ_R, etc.) are proposed and used [9]. The concept and unique idea drawn out of them have led to understand structure–reactivity relationship of carbocation, carbanion, radical, and functional group. These fundamental and quantitative researches formed the basis of organic chemistry of carbon compounds and are useful and important both for the interpretation of reaction mechanisms and for the prediction of rates and equilibria of many reactions.

1.5 HETEROATOM EFFECT

Heteroatoms are elements of groups 15, 16, and 17 which bear unshared electron pair(s) as their valence electrons. The characteristics of C–E (E: heteroatom) bond that are compared to those of C–C bond are (i) the presence of polarity of bond ($C^{\delta+}$–$E^{\delta-}$) due to the difference of electronegativity of the two atoms and (ii) the presence of unshared electron pair(s).

Heteroatom effect consists of the combined effect of (i) and (ii) on structure and reactivity. The representatives of heteroatoms are intuitively nitrogen and oxygen. The major effects of unshared electron pair (condition ii above) are (a) to stabilize α-carbocation by resonance and (b) to coordinate with Lewis acids including metallic ions. Usually, effect (ii) appears to exceed effect (i).

1.5.1 Stabilization of α-Carbocation by Resonance: Stereoelectronic Effect

Vilsmeier reagent (**1a**), known as a *formylating reagent*, is formed when *N*,*N*-dimethylformamide is treated with thionyl chloride or oxyphosphoric acid. This is an example of stabilization of α-carbocation by unshared electron pair of nitrogen [10–12].

$$\mathrm{Me_2N{-}CH({=}O) + SOCl_2 \longrightarrow [Me_2\overset{+}{N}{=}\underset{}{\overset{OS({=}O)Cl}{\overset{|}{C}}}{-}H]Cl^-} \longrightarrow \underset{\mathbf{1a}}{\mathrm{[Me_2\overset{+}{N}{=}CHCl]Cl^-}} + \mathrm{SO_2} \qquad (1.3)$$

On the other hand, Meerwein reagent (**1b**) is commercially available. Dimethyl ether is activated by protonation with a strong acid and the methyl group of the resulting oxonium ion is attacked by unshared electron pair of oxygen of a second molecule to afford **1b**.

$$\mathrm{Me_2O + Me_2\overset{+}{O}{-}H\overset{-}{B}F_4 \longrightarrow \underset{\mathbf{1b}}{Me_3\overset{+}{O}\overset{-}{B}F_4} + MeOH} \tag{1.4}$$

Moreover, the effect of unshared electron pairs on stabilization of α-carbocation is evident from the formation of reactive intermediates for hydrolysis from α-chloroether and α-chloroamine.

$$\mathrm{MeOCH_2Cl \longrightarrow [Me\overset{+}{O}{=}CH_2]\overset{-}{Cl} \xrightarrow{H_2O} MeOH + CH_2{=}O + HCl} \tag{1.5}$$

$$\mathrm{Me_2NCH_2Cl \longrightarrow [Me_2\overset{+}{N}{=}CH_2]\overset{-}{Cl} \xrightarrow{H_2O} Me_2NH + CH_2{=}O + HCl} \tag{1.6}$$

Until Eq. (1.6), there is no concern regarding stereochemistry, that is, angle dependence, of unshared electron pair and α-carbocation. It is stereoelectronic effect due to which spatial alignment of two components (functional group, unshared electron pair, vacant orbital, etc.) affects the structure and reactivity of a molecule. To elucidate the control of stereochemistry by two components, cyclic compounds are necessarily employed. Hence, monosaccharides are the compounds of choice to investigate the effect of relative stereochemistry of the two groups.

Mutarotation of glucose is well known, and here it is explained as one of examples of stereoelectronic effect. Pure β-D-glucose is obtained from dilute acetic acid by recrystallization, with specific rotation of $+19^\circ$ and melting point of $\sim 150^\circ$C. Pure α-D-glucose is obtained with specific rotation of $+112^\circ$ and melting point of $\sim 146^\circ$C (from water and successive drying over 50°C). When they are dissolved in water separately, specific rotation changes gradually and settles at the same value of $+52^\circ$. This is mutarotation and equilibrium is attained in water when a mixture of β-D-glucose and α-D-glucose is in the ratio of 64:36. This is due to the fact that both β- and α-isomers open the ring at 1,6 bond to yield a ring-opened aldehyde (**A**). The aldehyde recyclizes to give both β and α-isomers in the ratio of 64:36. Actually, aldehyde (**A**) is detected at a very low concentration of 0.003% in water (Eq. 1.7). However, the content of α-isomer is larger than that expected from relative conformational stability of tetrahydropyran (**4**) and the effect is called *anomeric effect*. The reason for the special attention is explained below.

2β ⇌ [**A**] ⇌ **2α** (1.7)

β-D-glucose — **2β**:**2α** = 64:36 — α-D-glucose

In chair conformation of cyclohexane, substituents are preferred to be equatorial in order to avoid steric repulsion between 1,3-diaxial substituents (Eq. 1.8). For methylcyclohexane (**3**, X = Me), the ratio of **3e**:**3a** = 95:5 with $\Delta\Delta G = 1.8$ kcal/mol and for hydroxycyclohexane (**3**, X = OH), the ratio of **3e**:**3a** = 2.3:1 with $\Delta\Delta G = 0.52$ kcal/mol.

In tetrahydropyran (**4**), 1,3-diaxial repulsion is larger than that of cyclohexane, because the bond length of carbon–oxygen is shorter than that of carbon–carbon (Eq. 1.9). With tetrahydropyran bearing alkyl groups at 2-position (**4**, X = methyl and ethyl), the proportion of equatorial isomer (**4e**) is larger by ~50% compared to that of cyclohexane (**3e**).

3e ⇌ 3a (1.8)

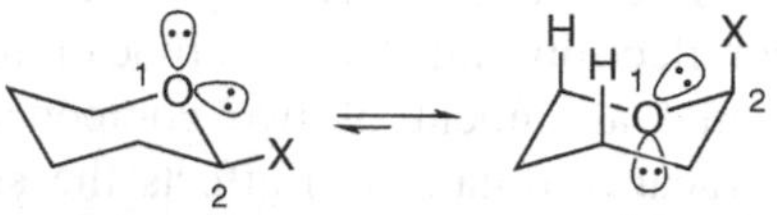

4e ⇌ 4a (1.9)

With these facts in mind, we notice that the proportion of **2**α in equilibrium is much larger than expected (Eq. 1.7). Moreover, the ratio becomes β-isomer:α-isomer = 14:86, when all the hydroxy groups are acetylated (**5**) and the proportion of **6** becomes pronounced as 6:94 when chlorine substitutes 1-acetoxy group of **5**. The fact shows that α-isomer (axial isomer) becomes preferred (predominant) when a stronger electron-withdrawing group is placed adjacent to the ring oxygen in pyranose. The effect is called *anomeric effect* (Fig. 1.2).

The effect of unshared electron pair of α-position becomes more evident when relative stereochemistry is fixed in bicyclic systems. The effect of α-oxygen on S_N1 reaction at tertiary carbon is described. The rate of solvolysis of **7** is $k_1 = 2.6 \times 10^{-10}$/s at 30°C in 50% dioxane–water. On the other hand, the corresponding rate for **8** is $k_1 = 6.0 \times 10^{2}$/s. Surprisingly, the relative ratio (**8/7**) is as large as 10^{12} (Eqs. 1.10 and 1.11).

There is one unshared electron pair of the ring oxygen in **8** but none in **7**, which is aligned anti-periplanar to the leaving group (ArO^-). The difference in relative stereochemistry exerted quite a profound accelerating effect of 10^{12} on **8** compared to **7** [10, 11].

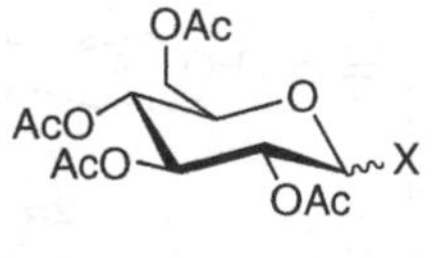

	β-Isomer α-Isomer (Equatorial):(Axial)
5: X = OAc	14:86
6: X = Cl	6:94

Figure 1.2 Ratio of β-isomer and α-isomer of acetylglucose.

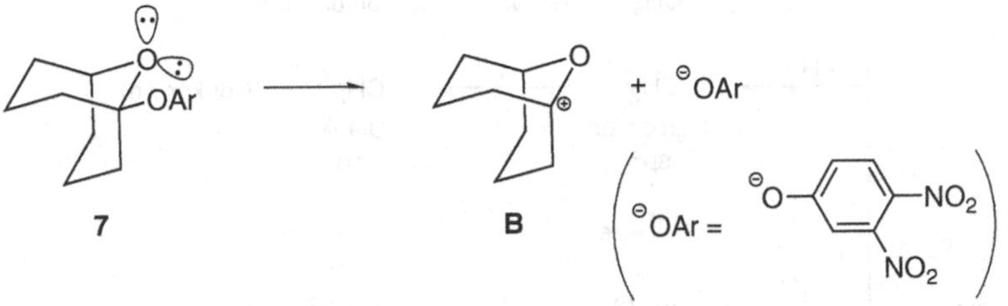

(1.10)

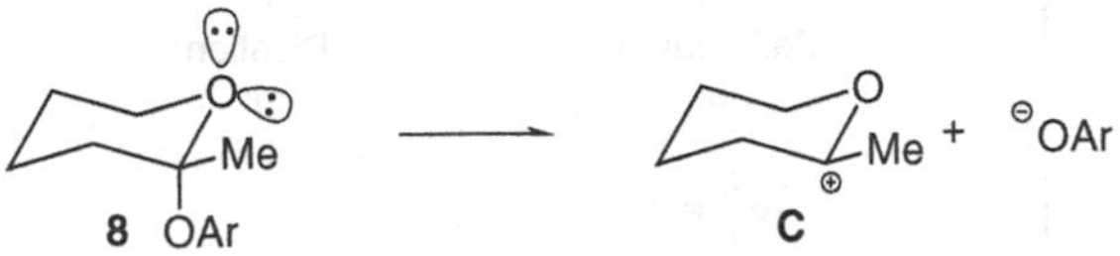

(1.11)

Photoirradiation of oxadiridine (**9**) affords lactam (**10**) completely but not a mirror image isomer (**10′**) at all (Eq. 1.12). During the reaction, carbon–carbon bond "a" which is anti-periplanar to the unshared electron pair of the nitrogen migrates solely but carbon–carbon bond "b" does not.

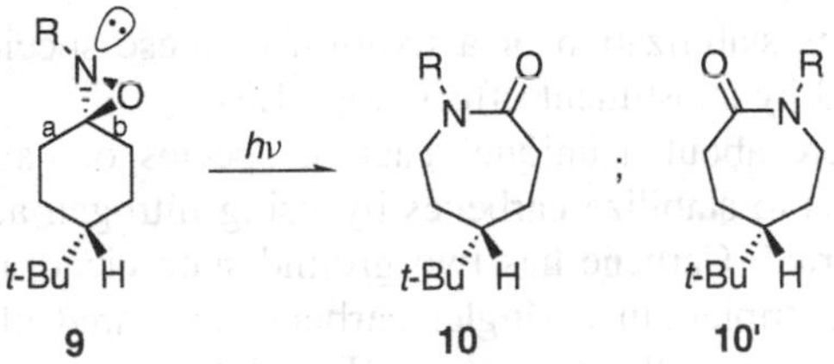

(1.12)

Anomeric effect and profound difference in reactivity mentioned here originate from donation of unshared electron pair (n_Y) of an adjacent heteroatom Y (O, N) to σ^*_{C-X} from anti-periplanar direction of C–X bond (X = O, N), that is, $n \rightarrow \sigma^*$ interaction. The direction of the shift of an unshared electron pair is evidently controlled, that is, stereoelectronic effect. The effect should be quite general, applying to all carbon atoms having two substituents connected by heteroatoms. This is generally illustrated by molecular orbital term (**D**) and by resonance term for α-haloether (**E**) [13, 14].

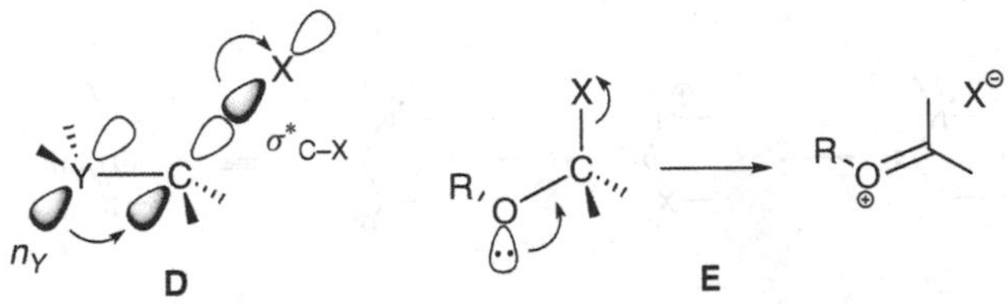

(1.13)

Here, it is appropriate to mention the formal relation of possible reactive species of carbon which can be derived from methane, because it can be taken for granted that every effect in organic chemistry (electronic, steric, heteroatom,

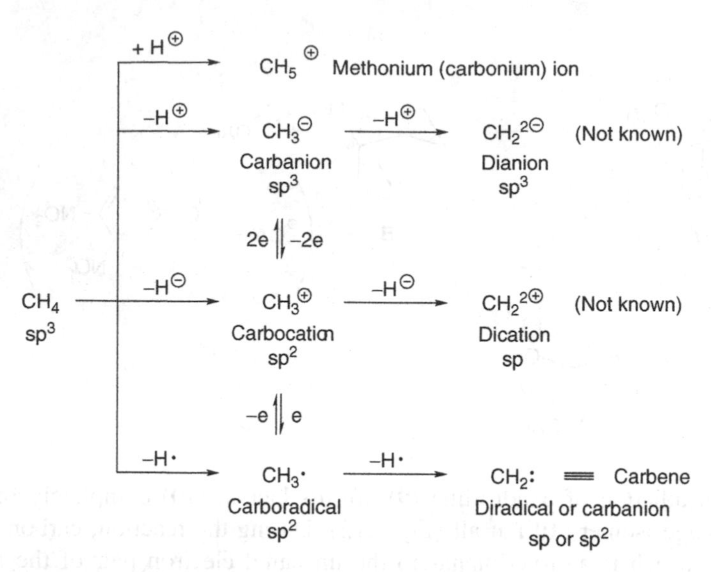

Figure 1.3 Relation of active species derived from methane.

and so on) concerns stabilization or activation of these species. Reactivities of these are controlled by substituent effect (Fig. 1.3).

Let us think here about a unique reactive species of carbene. Recently, a remarkable progress to stabilize carbenes by using nitrogen adjacent to the carbon has been reported. Carbene has two ground-state electronic configurations, that is, singlet and triplet. In a singlet carbene, unshared electron pair and a vacant 2p orbital reside on the sp^2 carbon (Fig. 1.4).

sp:triplet sp^2:singlet

Figure 1.4 Electronic configurations of ground state of carbene.

(1.14)

Aminocarbene:resonance structures

Stable carbenes (**11** and **12**) are prepared and the structures are determined by X-ray analysis and are called *Arduengo carbene* [12].

11 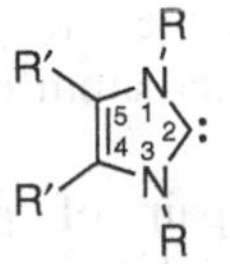 Imidazol-2-ylidene
R = Ad = adamantyl, R′ = H
stable carbene, mp 240 °C

12 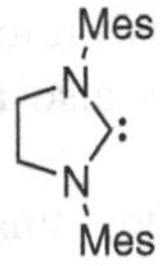 Imidazolin-2-ylidene
stable
Mes = mesityl

The role of nitrogen as a heteroatom is illustrated in Eq. (1.14), that is, (i) nitrogen donates unshared electron pair to the vacant 2p orbital (resonance effect) and (ii) nitrogen withdraws electrons by inductive effect. Compound **11** is aromatic but **12** is not; hence it is realized that aromaticity is not essentially required to stabilize carbene. Several types of new stable carbenes are reported and described in more detail as Note 5.

1.5.2 Coordination with Lewis Acids

Unshared electron pair of nitrogen coordinates with a proton to afford an ammonium salt, that is, acid–base reaction.

$$R_3\ddot{N} + HX \longrightarrow R_3\overset{\oplus}{N}\text{–H}\ \overset{\ominus}{X} \quad (1.15)$$

$$R_3\ddot{N} + R'X \longrightarrow R_3\overset{\oplus}{N}\text{–R}'\ \overset{\ominus}{X} \quad (1.16)$$

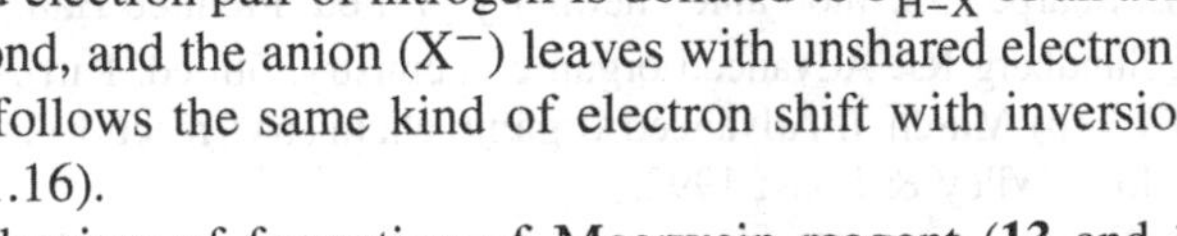

$$R_3N: + \text{H–X} \ (\sigma^*_{\text{H–X}}) \longrightarrow R_3\overset{\oplus}{N}\text{–H}\quad \ddot{X}^{\ominus} \quad (1.17)$$

F

Unshared electron pair of nitrogen is donated to $\sigma^*_{\text{H–X}}$ of an acid linearly along the H–X bond, and the anion (X^-) leaves with unshared electron pair (Eq. 1.15). Alkylation follows the same kind of electron shift with inversion of the carbon of R′ (Eq. 1.16).

The mechanism of formation of Meerwein reagent (**13** and **14**) is quite the same as given in Eq. (1.16). Proton of a strong acid coordinates to unshared electron pair of the oxygen of an ether to activate the C–O bond (the energy of $\sigma^*_{\text{C–O}}$ lowers), and unshared electron pair of a free ether attacks the activated C–O bond linearly from the rear side to cleave the bond. Strong methylating reagents such as **15** and **16** are obtained similarly.

$Me_3O^+ SbF_6^-$	$Et_3O^+ BF_4^-$	FSO_3Me	CF_3SO_3Me
13	**14**	**15**	**16**

Unshared electron pairs of nitrogen and oxygen also coordinate to lithium and magnesium cations preferably and control the site and stereochemistry of reactions. The effect is also an example of heteroatom effect and is explained in Chapter 2.

On the other hand, unshared electron pair of phosphorus preferably coordinates to transition metals (ions) to afford transition metal complexes. The different characteristics of phosphorus from those of nitrogen and oxygen can be understood by SHAB (soft and hard and acid and base) theory. The complexes are particularly useful for catalytic synthetic reactions, and quite active researches have been going on recently. A couple of examples employed as ligands for transition metals are shown and bidentate compounds (**19** and **20**) are used in catalytic asymmetric synthesis. This is shown quite briefly later; however, the theme is beyond the scope of this book (Figs. 6.1 and 6.2).

P 17; Cy–P(Cy)–Cy 18; PPh2/PPh2 19 Chiraphos; PPh2/PPh2 20 Binap

REFERENCES

1. Emsley J. The elements. 3rd ed. Oxford University Press; 1998.
2. Pauling L. The nature of the chemical bond. 3rd ed. Cornell University Press; 1960.
3. Henderson W. Main group chemistry. Royal Society of Chemistry; 2000.
4. Norman NC. Periodicity and the s- and p-block elements. Oxford University Press; 1997.
5. Barrett J. Structure and bonding. Royal Society of Chemistry; 2001.
6. Shriver DF, Atkins PW, Langford CH. Inorganic chemistry. 2nd ed. Oxford University Press; 1994.
7. Housecroft CE, Sharpe AG. Inorganic chemistry. 3rd ed. Prentice Hall; 2008.
8. (a) Carey FA, Sundberg RJ. Advanced organic chemistry. 5th ed. Part A, Chapter 3, Springer; 2007. (b) March J. Advanced organic chemistry: reactions, mechanisms, and structure. John Wiley & Sons; 1992.
9. (a) Hnasch C, Leo A, Taft RW. Chem Rev 1991;91:165. (b) Shorter J. Pure & Appl Chem 1997;69:2497. (c) Tsuno Y, Fujio M. Chem Soc Rev 1996;25:129. (d) Inamoto N. Hammett relationship: structure & reactivity. Maruzen (Japanese); 1983.
10. Deslongchamps P. Stereoelectronic effects in organic chemistry. Pergamon Press; 1983.
11. Kirby AJ. Stereoelectronic effects. Oxford University Press; 1996.
12. Arduengo III, AJ. Acc Chem Res 1999;32:913.
13. McMurry J. Organic chemistry. 5th ed. Pacific Grove, Brooks/Cole; 2000.
14. Jones Jr., M. Organic chemistry. 2nd ed. W.W. Norton & Co.; 2000.

NOTES 1

ELECTRONEGATIVITY

Electronegativity is a relative scale (numerically calculated) of electron-withdrawing ability of an atom in a neutral molecule. This is one of the important and fundamental characters of an atom and was first devised and introduced by Pauling.

Pauling determined electronegativity for each atom (χ_A, χ_B) (Eq. N1.3) on the basis of difference (Δ: Eq. N1.2) between bond dissociation energy of homolysis (Eq. N1.1a; of Table 1.1(d)) of bond A–B [D(A–B)] and one half of the sum of bond energy of A–A and B–B [D(A–A); D(B–B)]. The difference should be a positive value. The bond formed between different atoms is polarized to a certain degree; therefore, it is stronger than that between two same atoms because ionic resonance contribution of the bond and Coulombic interaction between the two should exist to attract each other.

$$\text{A–B} \begin{cases} \xrightarrow{a} \text{A}^{\bullet} + \text{B}^{\bullet} \\ \xrightarrow{b} \text{A}^{\oplus} + \text{B}{:}^{\ominus} \end{cases} \tag{N1.1}$$

$$\Delta = D(\text{A–B}) - \frac{D(\text{A–A}) + D(\text{B–B})}{2} \tag{N1.2}$$

$$\Delta = k\,(\chi_A - \chi_B)^2 \tag{N1.3}$$

where k is determined to calculate χ_B in which χ_A of fluorine is set as 4.0

Organo Main Group Chemistry, First Edition. Kin-ya Akiba.

Pauling calculated χ_B of each atom (B), setting $\chi_F = 4.0$ for the most electronegative atom of fluorine. This is the electronegativity of Pauling which is listed in Table 1.1 in Chapter 1. The energy of ionic cleavage (Eq. N1.1b) is quite large (about 400–1300 kJ/mol) compared to homolysis.

Electronegativity of Allred–Rochow (χ_{A-R}) is also frequently used. This is based on the theory that electronegativity is determined by the amount of the charge of atomic surface. Electrons of an atom are under the influence of effective nuclear charge (Z_{eff}) all the time. Coulomb energy on the surface of an atom is proportional to effective nuclear charge and is inversely proportional to atomic radius (r); hence the strength of the resulting electric field is proportional to Z_{eff}/r^2 (Eq. N1.4). Therefore, electronegativity of each atom can be obtained by calculation. Coefficients in Eq. (N1.4) are determined so that the obtained value is close to the electronegativity of Pauling.

$$\chi_{A-R} = 0.744 + \frac{0.3590 Z_{eff}}{r^2} \qquad \text{(N1.4)}$$

where r is the covalent radius of atom (Å), Z_{eff} is the effective nuclear charge. Coefficients are determined so that χ_{A-R} becomes close to Pauling's χ

Mulliken defined the electronegativity as an average of ionization energy (I) and electron affinity (A) and proposed numerical values converted to become close to Pauling's (Eq. N1.5). According to Eq. (N1.5), it is deduced that electronegativity becomes large when I and A are large and it becomes small when they are small. It is apparent that the values depend on the degree of ionization of an atom.

$$\chi_M = \frac{1}{2}(I + A) \qquad \text{(N1.5)}$$

Electronegativities of Mulliken and Pauling are almost in proportion. On the basis of three definitions mentioned above, it is realized that electronegativity of an atom depends on the electronic state (valence) of an atom. These numerical values are recorded to the second decimal place, but it is enough to recognize the values to the first decimal place in order to qualitatively estimate (make image of) polarity and reactivity of a single bond.

NOTES 2

IMPORTANCE OF FORMAL LOGIC-I: OXIDATION NUMBER AND FORMAL CHARGE

Oxidation number is defined as a number of charges in an atom of simple substance, ion, or molecule, which is assigned according to the following rule:

1. Oxidation number of simple substance is set as zero.
2. For atoms bound by ionic bond, oxidation number in each atom is the same as the charge assigned for the ionic bond.
3. For atoms bound by a covalent bond, the oxidation number is the number of charges in each atom: (i) for the bond consisting of the same kind of atoms, a pair of covalent electrons are assigned 1:1 for each atom; (ii) for the bond consisting of different atoms, two electrons are assigned to the atom with larger electronegativity.

Formal charge is defined as a number of charges of the atom in a molecule which is assigned according to the following rule:

1. Nonbonding electrons (lone pair and radical) are assigned to the atom.
2. For atoms bound by a covalent bond, a pair of covalent electrons are assigned 1:1 for each atom.
3. Formal charge is obtained by subtracting the sum of electrons of (1) and (2) above from the number of valence electrons in a free atom.

Organo Main Group Chemistry, First Edition. Kin-ya Akiba.

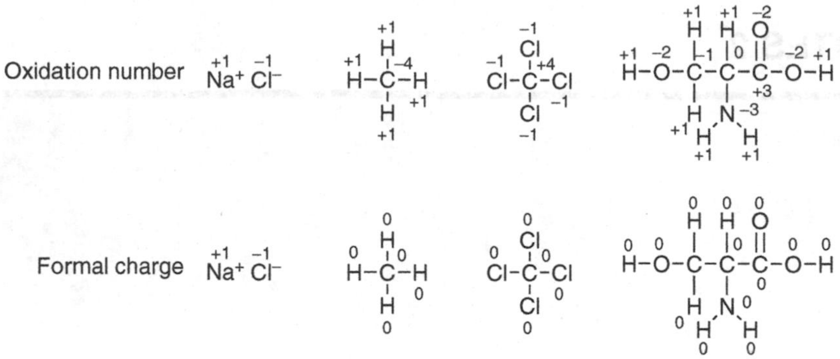

Oxidation number $\overset{+1}{Na^+}\ \overset{-1}{Cl^-}$

Formal charge $\overset{+1}{Na^+}\ \overset{-1}{Cl^-}$

Figure N2.1 Oxidation number and formal charge of each atom.

TABLE N2.1 Oxidation Number and Formal Charge of the Central Atom

	CH_4	CCl_4	$\bar{C}H_3$	$\bar{C}Cl_3$	$\overset{+}{C}H_3$	$\dot{C}H_3$	$:CH_2$	NH_3	$\overset{+}{N}H_4$	$\bar{N}H_2$	H^+	H^-	Li^+
Valence	4	4	3	3	4	3	2	3	4	2	0	0	0
Oxidation number	−4	+4	−3	+3	−2	−3	−2	−3	−4	−2	+1	−1	+1
Formal charge	0	0	−1	−1	1	0	0	0	+1	−1	+1	−1	+1

$$\text{Formal charge} = \text{number of valence electrons in a free atom}$$
$$- \text{number of bonds} - \text{number of nonbonding electrons}$$

Based on these definitions, the oxidation number and formal charge of each atom of simple molecules are illustrated in Fig. N2.1. Also the numbers of the central atoms of simple species are given in Table N2.1.

The oxidation number of the carbon of methane and tetrachloromethane is −4 and +4, respectively. For carbon atom of neutral molecules, these values are extraordinary and out of sense. On the other hand, the formal charge of the carbon of these molecules is zero. Formal charge cannot describe the electronic effect of substituents of organic molecules.

Oxidation number, however, is essential and fundamental to assign charges of inorganic compounds. It is a basic idea of oxidation and reduction between ions and of reactions of electrode and battery. It is one of the fundamentals of inorganic chemistry, which is often based on one-electron transfer.

On the other hand, the formal charge of every atom of covalent compounds is essentially zero. It is one of the important fundamentals in organic chemistry, which is based on slight shift of a pair of electrons (two electrons) induced by a bias of charge effected by difference of electronegativity of atoms. In order to comply with this, $+\delta$ or $-\delta$ is used to describe the partial charge.

It is apparent and well accepted in both cases that the calculated number of electrons does not localize on the corresponding atom. But both ideas are fundamental to chemistry and characters of compounds are understood based on them, as a fist approximation. The formalism is the basis of these and is fundamental to science (chemistry) in general.

REFERENCES

1. Parkin G. J Chem Educ 2006;83:791.
2. Smith DW. J Chem Educ 2005;82:1202.
3. McMurry J. Organic chemistry. 5th edn. Pacific Grove, Brooks/Cole; 2000, Chapter 2.

NOTES 3

IMPORTANCE OF FORMAL LOGIC-II: OCTET RULE, EIGHTEEN-ELECTRON RULE, HYPERVALENCE

Valence of carbon is four, and carbon binds substituents (atoms) with four covalent bonds (σ bond) (sp^3 hybridization). Unsaturated bonds of carbon [double bond (sp^2) and triple bond (sp)] consist of one σ bond and one or two π bond(s). When there is no substituent (atom) to bind, an unshared electron(s) results and two electrons become paired as lone pair electrons, belonging to the central atom. Therefore, the number of valence electrons (s^2p^6) is eight for main group elements (sp element) and octet rule is dominating. On the other hand, d electrons are involved as valence electrons ($s^2p^6d^{10}$) for transition metals (df element) and stable electronic configuration is attained with d^{10}. Hence, eighteen-electron rule is established fundamentally (f^{14} is added when f electrons are involved as valence electrons).

There are two kinds of substituents, that is, electron-withdrawing and electron-donating, and the situation is the same for ligands. Therefore, it is apparent that electron density on a central atom cannot be an integer of 8 or 18. Electron density of a certain atom should be different for each compound. Octet rule and eighteen-electron rule are accepted as fundamentals of chemistry, taking the above consideration for granted. In other words, formal logical rule based on "aufbau principle" (the Pauli exclusion principle) constitutes the backbone of science (chemistry) of materials.

By the way, it is well known and established that many compounds exist stable, bearing five or six bonds to combine substituents (ligands: L) to the central

Organo Main Group Chemistry, First Edition. Kin-ya Akiba.

atom (X) of the main group elements of third period and heavier ones. The central atom contains 10 or 12 valence electrons (N) formerly. The main group element compounds bearing formal valence electrons over the octet are classified as hypervalent compounds and are designated as N–X–L. This designation depends on formal logic but is convenient to express structure and character of hypervalent compounds in general and is usually employed. Molecular orbital of three-center four-electron bond (3c-4e bond) was devised in 1951 to hold extra electrons in nonbonding orbital(s), resulting in the formation of a polar, longer, and weaker bond compared to regular σ bond, thus avoiding the violation of the Pauli principle (Fig. 2.8 in Chapter 2). Actually, electron density on the central atom does not exceed eight and electrons are delocalized to the substituents (ligands).

REFERENCES

1. Akiba K.-y. Chemistry of hypervalent compounds. Wiley-VCH, Weinheim; 1999, Chapter 1.
2. Musher JI. Angew Chem Int Ed Engl 1969;8:54.
3. Weinhold F, Landis C. Valency and bonding. Cambridge: Cambridge University Press; 2005, Chapters 3.5 and 4.4–4.5.

CHAPTER 2

MAIN GROUP ELEMENT EFFECT

2.1 WHAT IS MAIN GROUP ELEMENT EFFECT?

Characteristics of heteroatoms (E: N, O, F) of the second period elements are (i) polarity of C–E bond due to the difference in electronegativity between the two atoms, (ii) electron-donating ability of an unshared electron pair to donate electrons to an electron-deficient center by resonance, and (iii) coordination of the unshared electron pair to Lewis acid (proton, metallic ion, etc.). These are in essence the heteroatom effect explained in Chapter 1 [1, 2]. In main group elements of the third period and heavier, the heteroatom effect continues to be effective but gets gradually weaker descending the period; on the other hand, "abnormal facts" compared to carbon compounds have been accumulated for these compounds. They are summarized as follows:

1. Long-chain compounds made of one element are rare.
2. Double bond made of main group elements is unstable.
3. A variety of pentacoordinate (pentavalent) and hexacoordinate (hexavalent) hypervalent compounds exist and are stable.

Organo Main Group Chemistry, First Edition. Kin-ya Akiba.

4. Structure of pentacoordinate compounds (10-X-5 and 10-X-4 or 10-X-3 bearing one or two unshared electron pair(s)) is fundamentally trigonal bipyramidal (tbp).
5. Structure of hexacoordinate compounds (12-X-6 and 12-X-5 or 12-X-4 bearing one or two unshared electron pair(s)) is fundamentally octahedral (Oh).
6. Unimolecular positional isomerization of pentacoordinate compounds (Berry pseudorotation) is very fast; however, that of hexacoordinate compounds (Bailar twist) is quite slow.
7. Acidity of onium salts is high and the corresponding ylides are easily formed and stable.
8. Different from S_N2, nucleophilic substitution proceeds via hypervalent intermediates, and stereochemistry at the central atom is retained and/or inverted competitively.
9. Ligand coupling reaction (LCR) takes place concertedly with hypervalent compounds.

On the basis of these facts, it is apparent that there is a large gap between the characteristics of the second period elements and the third period and heavier elements. The situation is summarized as the fundamental characteristics of the main group elements shown in Table 1.1 of Chapter 1. When all of these were combined and analyzed, the "heteroatom effect" was illustrated clearly for the second period elements and it became apparent that the "main group element effect" was established in a much broader sense, and is described in Table 2.1 [1, 2].

Characteristics of the "main group element effect" are explained below; however, it is noteworthy that the "heteroatom effect" is valid and is contained as one of the effective factors of "main group element effect".

TABLE 2.1 Main Group Element Effect

Second peroid elements	Heteroatom effect (unshared electron pair effect)	Electron-donating ability coordination to metal ion
	Hypervalence	Three-center four-electron bond (3c–4e)
Third to sixth period elements	Effect of $\sigma_{X\text{–}C}$	Electron-donating ability of HOMO
	Effect of $\sigma^*_{X\text{–}C}$	Electron-accepting (withdrawing) ability of LUMO

2.2 SINGLE BOND ENERGY AND π-BOND ENERGY

On the basis of the criteria of structure and reactivity of carbon compounds, there are nine "abnormal" facts with organic compounds of the main group elements of the third period and heavier as summarized in Section 2.1. The most fundamental of the 9 are 1 and 2, which are concerned with the stability of long chains and unsaturated bonds. These characteristics become obvious when the single bond and π-bond energies are compared with those of the second period elements and the third period elements and heavier [3, 4]. In Tables 2.2 and 2.3, the corresponding values are summarized, in which numerical values in Table 2.2 are experimental and those in Table 2.3 are calculated theoretically.

In Table 2.2, the single bond energy (E–X, kJ/mol) is classified (I) between the second period elements and (II) between the third period and the second period elements. Based on these values, the following facts are recognized.

1. The E(element)–H bond is stronger than E(element)–C bond in all the elements.
2. With the exception of the C–C bond (I-i: 346 kJ/mol), the bond between the same element (E–E) is the weakest in all groups (from I-ii to II-iii).
3. The trend in item 2 is prominent for the second period elements (I) as compared to those of the third period and the second period elements (II). This

TABLE 2.2 Bond Energy of Single Bonds (E–X) (kJ/mol)

	I			II		
	E	X	kJ/mol	E	X	kJ/mol
(i)	C	H	411	Si	H	318
	C	C	346	Si	C	301
	C	N	305	Si	O	458
	C	O	358	Si	F	565
	C	F	441	Si	Si	222
(ii)	N	H	386	P	H	322
	N	C	305	P	C	264
	N	N	167	P	O	407
	N	O	200	P	F	490
	N	F	283	P	P	201
(iii)	O	H	459	S	H	363
	O	C	358	S	C	272
	O	N	200	S	O	265
	O	O	142	S	F	284
				S	S	226

TABLE 2.3 π-Bond Energy of Double Bonds (E = X) (kJ/mol)

	I			II		
	E	X	kJ/mol	E	X	kJ/mol
(i)	C	C	272	Si	Si	105
	C	N	263	Si	P	121
	C	O	322	Si	S	209
	N	N	251	P	P	142
	N	O	259	P	S	167
(ii)	Si	C	159	P	N	184
	P	C	180	S	N	176
	S	C	217	Si	O	209
	Si	N	150	P	O	222

reflects that the polarity of the bond and the repulsion between unshared electron pairs are important in determining the bond energy.

4. The E–X (X = N, O, F) bond energy gets larger according to increasing electronegativity of X. The fact that an O–O bond (142 kJ/mol) is weaker than an N–O bond (200 kJ/mol) is the only one exception and the fact is in agreement with item 3.
5. E–H and E–C bonds of the second period elements are stronger than the corresponding bonds of the third period by 50–100 kJ/mol.
6. E–O and E–F bonds of the third period elements are stronger than the corresponding bonds of the second period by 100–200 kJ/mol.

These items can be explained qualitatively on the basis of the repulsion between unshared electron pairs (prominent in the second period elements due to the shorter bond length) and stabilization by polarity and polarizability (prominent in the third period elements due to the longer bond length).

In Table 2.3, the π-bond energy of E=X is shown as (I-i) between the second period elements, (II-i) between the third period elements, and (I-ii and II-ii) between the third and the second period elements. Based on these values, the following can be recognized.

1. There is more than 100 kJ/mol difference between the corresponding π-bond energy of I-i and II-i and that the energy of I-i is higher than that of II-i. It is understandable from this that unsaturated bonds between the third period elements and heavier should be unstable (or difficult to prepare). Especially, it is remarkable that the difference of the π-bond energy of the same elements of the second period and the third period is so prominent as C=C (272 kJ/mol) and Si = Si (105 kJ/mol), and N=N (251 kJ/mol) and P=P (142 kJ/mol).
2. The π-bond energy between the third period and the second period elements (I-ii and II-ii) are intermediate between I-i and II-i.

Combining all these items, it is obvious that there is a large gap in characteristics between the second period elements and the third period elements and heavier. In other words, it can be pointed out clearly here that the second period elements, especially carbon, are exceptional elements among main group elements.

It is interesting to compare the calculated size of orbitals of group 14 elements in relation to the bond energy. The size of ns and np orbitals becomes larger as the period decreases, and the difference in size between both orbitals also gets larger. Hence, the hybridization between ns and np orbitals becomes more difficult to take place. Comparing carbon and silicon, it is noticed that the difference in size of ns and np orbitals is the highest between two neighboring elements. For carbon, the size of 2s and 2p orbitals is quite close (almost the same), hence sp hybridization can take place easily (Fig. 2.1).

The calculated energy of the s orbital shows a large difference between carbon and silicon; otherwise the tendency looks the same as the size. In case relativistic effects are considered with heavy elements like tin and lead, the energy of orbital is shown to get lower (Fig. 2.2) [5a, b].

The double bond between group 14 elements is made of coordination of an unshared electron pair of a singlet carbene in sp^2 orbital of one atom to a vacant p orbital of the other. Therefore, two substituents on each atom cannot be in the plane of the double bond formed but maintain a certain angle to the plane of double bond (Fig. 2.3). On the other hand, the double bond between group 15 elements is made of the overlap of two p orbitals in the same plane. Thus two substituents lie in a plane but are orthogonal to the double bond formed and they are in opposite directions and the unshared electron pair should be in the s orbital (Fig. 2.4).

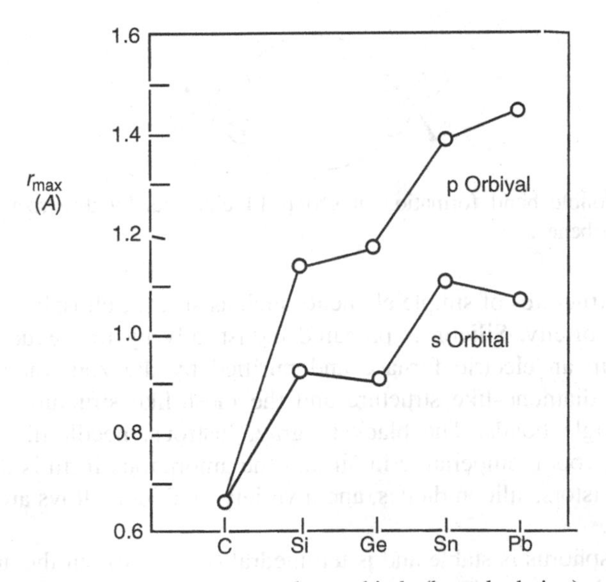

Figure 2.1 Size of ns and np orbitals (by calculation).

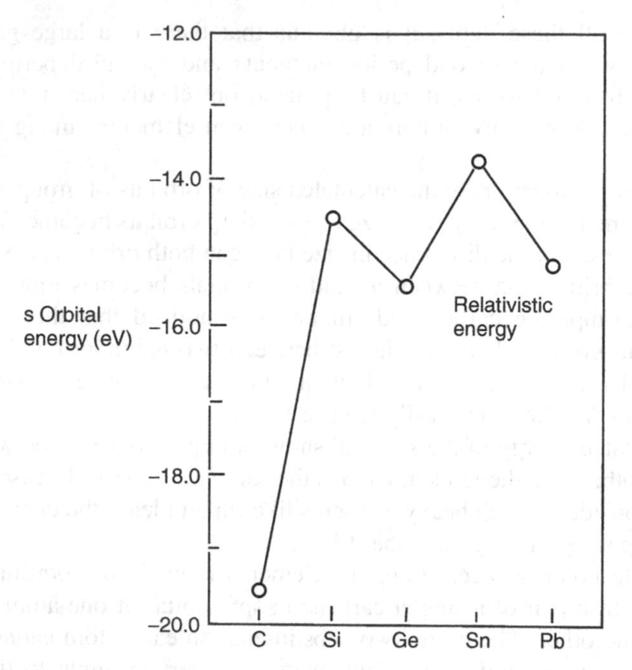

Figure 2.2 Energy of *ns* orbitals (by calculation).

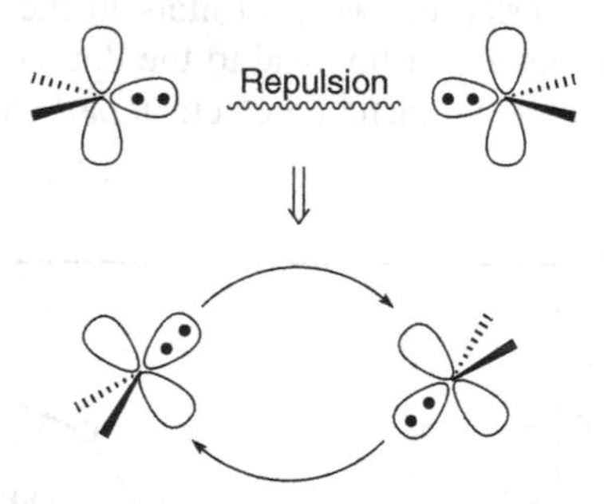

Figure 2.3 Double bond formation of group 14 elements by dimerization of singlet carbenes.

Here, the structures of simple elements such as silicon, phosphorus, and sulfur are explained briefly. Silicon is prepared industrially by the reduction of silica with carbon in an electric furnace and purified by the zone-melting method. Silicon has a diamond-like structure and the mesh-like structure is formed by long-chain single bonds. The black to gray, lustrous, needle-like crystals are quite stable at room temperature in air and the amorphous form is a dark brown powder. Transistors, silicon diodes, and a variety of silicon alloys are industrially manufactured.

White phosphorus is stable and is tetrahedral (P_4: toxic) but the angle P–P–P is 60°, which is very strained compared to a regular tetrahedron. It is sometimes

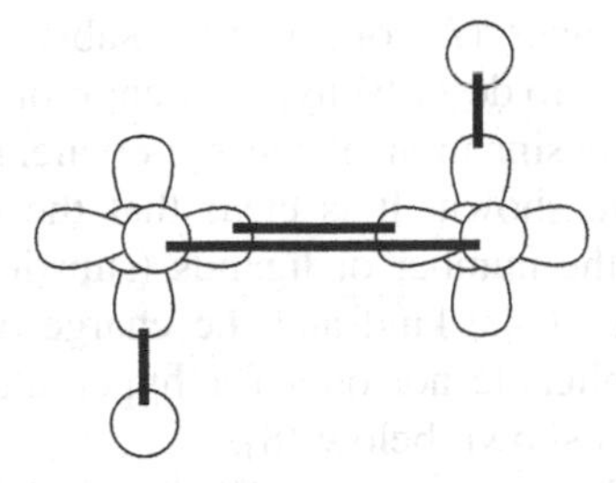

Figure 2.4 Double bond formation of group 15 elements by dimerization of p orbitals.

called *yellow phosphorus*, its color being due to impurities. When white phosphorus is heated (270–300°C), red phosphorus (red to violet powder) is obtained, which is amorphous. When heated under high pressure, white phosphorus is converted to black phosphorus which has a widespread planar structure resembling that of graphite. Sulfur has a variety of allotropes and polymorphisms, and S_8 is the stable amber colored crystals with a crown type structure. When it is heated above the m.p. (about 115°C), a complex mixture of chains and cyclic structures (S_4–S_8, etc.) results, but none of them is a long chain. Three forms of sulfur, that is, precipitated, sublimed, and washed sulfur, are recognized and used in pharmaceuticals.

2.3 HYPERVALENT COMPOUND

The number of valence electrons of the main group element (sp element) is 8 (ns^2, np^6) and the element can hold essentially four valence bonds (octet rule). Therefore, it is fundamental that eight valence electrons are the maximum that can be contained in a valence shell of the main group elements. However, the phosphorus atom of pentaphenylphosphorane (Ph_5P) has 10 valence electrons formally and the compound is pentacoordinate, pentavalent, neutral, and stable. Pentaphenylantimony and bismuth are also known to be stable. For group 15 elements of the third period and heavier, a variety of pentavalent compounds are known to exist as neutral and stable. In addition, pentacoordinate silicates (e.g., $[Ph_3SiF_2]^-Li^+$) and tetracoordinate sulfurane (e.g., $[(Me_2N)_2SF_2]$) have 10 valence electrons formally. The structures of all these compounds are essentially tbp. Moreover, a variety of hexacoordinate (hexavalent) silicon dianions, hexacoordinate phosphorus mono anions, and hexacoordinate neutral sulfur compounds are known to be stable. All of these bear 12 valence electrons formally and are octahedral molecules (Oh) [6, 7].

It is apparent that there exist main group element compounds bearing nine and more valence electrons formally for a variety of elements. These are called *hypervalent compounds*. Thus hypervalent compounds are defined as radicals, ions, and neutral compounds of a main group element (sp element: groups 1, 2, 13–18) that contain a number (N) of formally assignable electrons of more than the octet in a valence shell directly associated with the central atom (X) in directly bonding a number (L) of ligands (substituents). The designation N-X-L is conveniently used to describe hypervalent molecules [6–8]. In Fig. 2.5, the structure and N-X-L designation of the hypervalent compounds of silicon, phosphorus, and sulfur are shown. It is clear that the structure of these compounds is determined by the number of ligands (substituent including unshared electron pair), irrespective of the kind and the charge of the central atom. The designation N-X-L is applicable not only for hypervalent compounds but also for general species such as shown below [8]:

Carbene (CH_2), 6-C-2; carbocation (CH_3^+), 6-C-3; carboanion (CH_3^-), 8-C-3; methane (CH_4), 8-C-4; borane (BH_3), 6-B-3; transition state of S_N2, 10-C-5; pentaphenylphosphorane (Ph_5P), 10-P-5; and so on.

The structures of hypervalent compounds are described basically as tbp and Oh (Fig. 2.5). These are well known as the structures of transition metal compounds. In transition metals, the (n − 1)d orbitals are used to hybridize with ns and np orbitals to form tbp (dsp^3) and Oh (d^2sp^3) structures. In modern chemistry, where theoretical calculations have advanced very far because of the advances in computers, it is established that the contribution of the d orbitals for hybridization with s and p orbitals is small and can be negligible for the main group elements of third period and heavier. Instead, the hypervalent bond (three-center four-electron (3c–4e) bond) [9] has been invoked to contain formal valence electrons more than the octet, and the theory is well established [10]. According to the nature of the hypervalent bond, the structure and unique character of hypervalent compounds are realized and this is explained in Sections 2.4–2.6. The fundamental structure and hybridization of the orbitals are shown for organic and inorganic compounds in Fig. 2.6. Based on the structural similarity between the

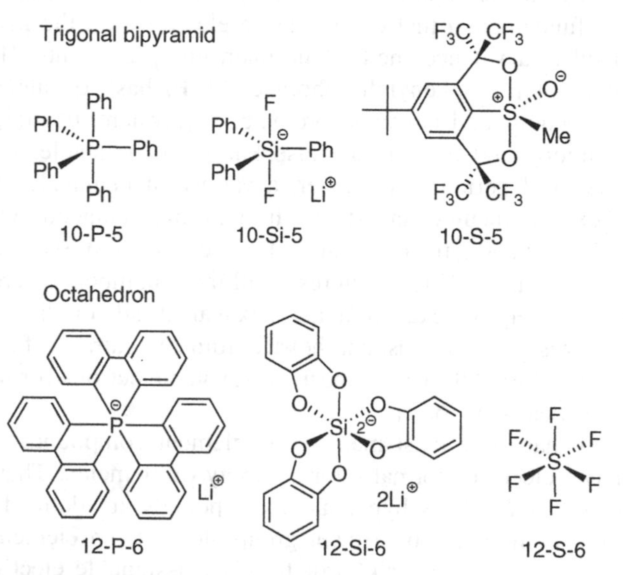

Figure 2.5 Hypervalent compounds of phosphorus, silicon, and sulfur.

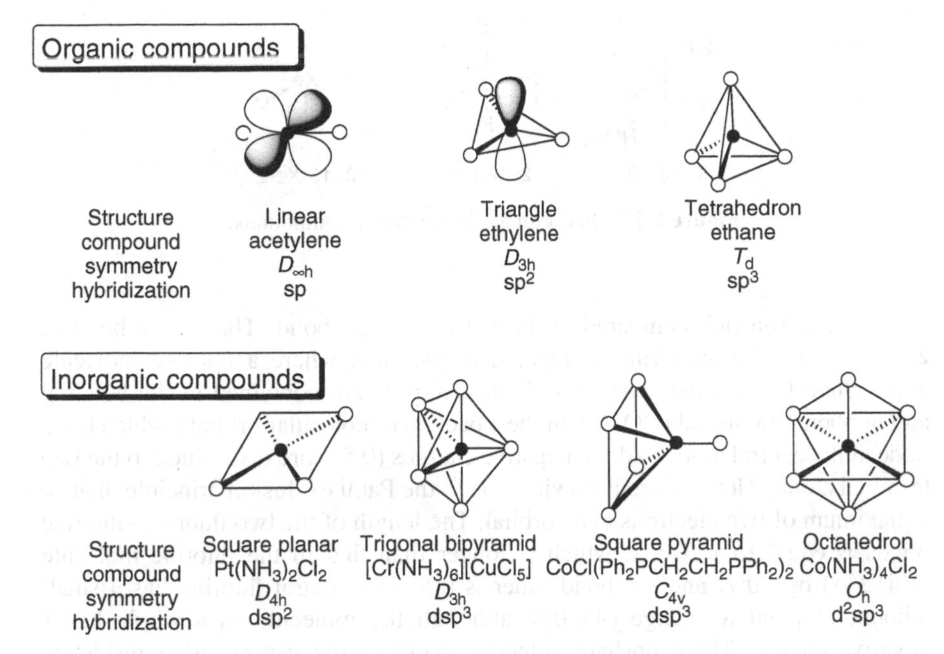

Figure 2.6 Shape of hybridized orbitals and their fundamental compounds.

transition metal compounds and hypervalent compounds, it is understood that the chemistry of hypervalent compounds lies in and bridges the gap between traditional organic and inorganic chemistry (Fig. 2.6).

2.4 EFFECT OF HYPERVALENT BOND (1): 3c–4e BOND AND STRUCTURE

The hypervalent bond is essentially described by a three-center and four-electron (3c–4e) bond. One of the most familiar and typical hypervalent compounds may be potassium triiodie $[I–I–I]^- K^+$ (**1**) which is deep blue colored, stable at room temperature, and is obtained by the reaction of iodine and potassium iodide in aqueous solution. The triiodie is linear and the central iodine atom has two bonds and three unshared electron pairs, thus 10 valence electrons formally. The lengths of the two iodine–iodine bonds are 2.82 and 3.10 Å. The average of the two is 11% longer than the length of iodine molecule (2.67 Å).

Lithium trifluoride $[F–F–F]^- Li^+$ (**2**) is made of a fluorine molecule and lithium fluoride and is stable at very low temperature (about −150°C). The fluorine nucleus is small and has the largest electronegativity; thus it is the most reluctant heteroatom for valence expansion. The trifluoride (**2**) is also linear and the central fluorine atom has two bonds and three unshared electron pairs like the iodine of **1**, thus having 10 valence electrons formally.

Molecular orbital of the anion is made of σ-type overlap of three 2p orbitals of fluorine to form three-center bond, which contains four electrons. Thus the 3c–4e

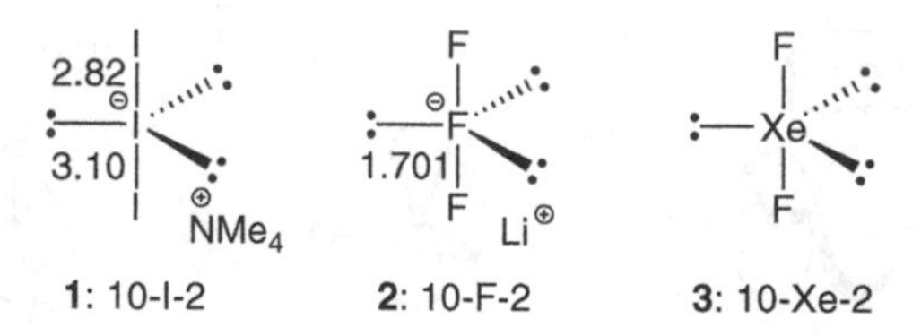

Figure 2.7 Dicoordinate hypervalent compounds.

bond is electron-rich compared to the normal 2c–2e bond. The 3c–4e bond of **2** is stabilized by 46 kJ/mol compared to the state where a fluorine molecule and a fluoride ion exist separately. Four electrons are contained in the bonding and nonbonding orbital (HOMO, highest occupied molecular orbital), which has a node at the central atom, and the negative charges (0.51) are distributed to the two terminal atoms. Hence there is no violation to the Pauli exclusion principle: that is, a maximum of two electrons in an orbital. The length of the two fluorine–fluorine bonds is equal to 1.701 Å, which is longer than that in the fluorine molecule (1.412 Å) by 20%, and the bond order is 0.5. The central fluorine has a small amount of positive charge (+0.03), although the molecule as a whole has 1 negative charge. Three unshared electron pairs of the central atom reside on a plane perpendicular to the linear 3c–4e bond. Regarding the three unshared electron pairs as substituents, the molecule is taken to be tbp. The bond in the axial direction of tbp is called the *apical bond* and that in the perpendicular plane is called the *equatorial bond*. The structure of **1** is essentially the same as **2** [11].

By the way, an allyl anion is a representative for a π-type 3c–4e bond, thus the allyl radical is 3c–3e bonded and the allyl cation is 3c–2e bonded. In Fig. 2.8, molecular orbitals of the σ-type and π-type 3c–4e bond are shown.

In both molecular orbitals, nonbonding is the HOMO and has a nodal plane on the central atom; thus the negative charge (−0.5) is distributed to both terminal atoms with bond order of 0.5 and the molecule is linear.

Examples of thermally stable hypervalent compounds are shown in Fig. 2.9 for tricoordinate, Fig. 2.10 for tetracoordinate, and Fig. 2.11 for pentacoordinate compounds [12].

The most fundamental class of hypervalent compounds is phosphorane (10-P-5) [13] which bears five substituents and is neutral and thermally stable. Let us explain structural characteristics of hypervalent compounds based on phosphorane because quite a number of structures of phosphoranes have been determined by X-ray crystallography (Fig. 2.11).

The fundamental structure of phosphorane is clearly shown by compounds **4** and **5**, which have five identical substituents. These are tbp and consist of three equatorial bonds (sp^2) which are in the plane and two apical bonds (hypervalent bond) which are perpendicular to the plane [14]. Actually, the bond lengths of **4** are Fa–P = 1.577 Å and Fe–P = 1.534 Å, and thus the apical bond is longer than the equatorial bond by 3%. Bond angles of **4** are $\angle$Fe–P–Fe = 120°, $\angle$Fa–P–Fe = 90°, $\angle$Fa–P–Fa = 180°; thus **4** is a typical tbp. The essential

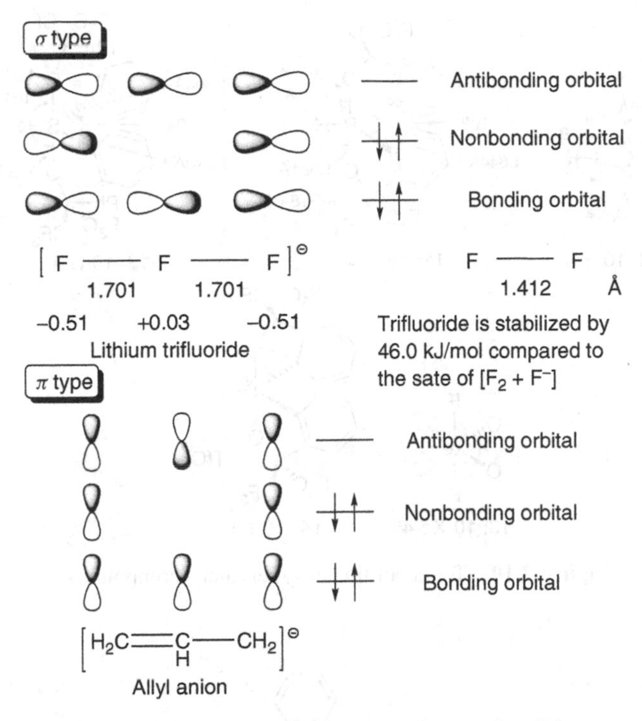

Figure 2.8 Molecular orbitals of three-center four-electron bond (σ-type and π-type).

15: 10-S-3

16: 10-M-3
(a) M = P (b) M = As (c) M = Sb

17: 10-Br-3

18: $(Ph\text{-}I{=}O)_n$ 10-I-3

19: 10-I-3

20: 10-I-3

21: 10-I-3

Figure 2.9 Tricoordinate hypervalent compounds.

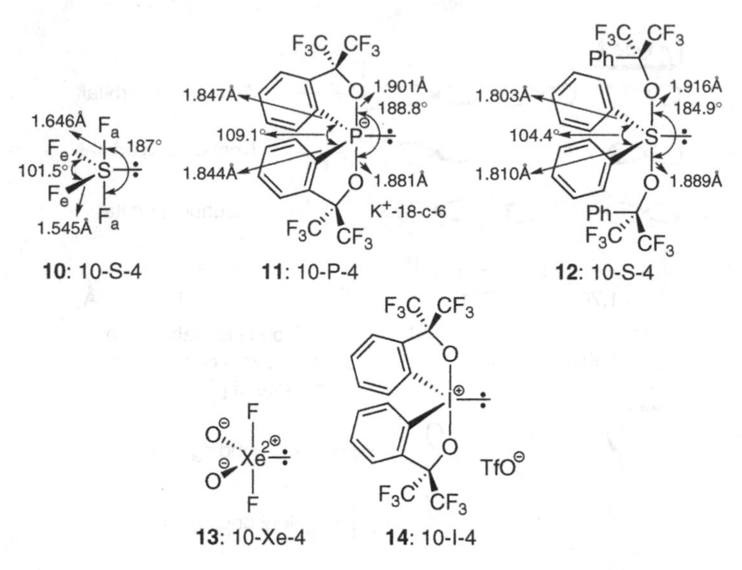

Figure 2.10 Tetracoordinate hypervalent compounds.

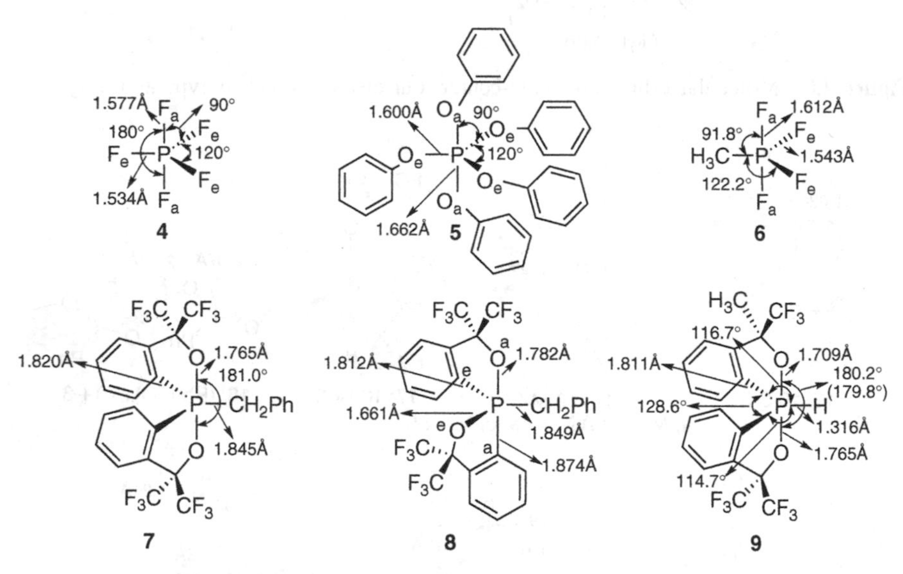

Figure 2.11 Pentacoordinate hypervalent compounds.

features of **5** are the same as those of **4** and **5**, that is, tbp with Oa–P = 1.662 Å and Oe–P = 1.600 Å.

Compound **6** has a methyl group in the equatorial position and is strained slightly because of steric repulsion of the methyl group.

Compounds **7** and **8** are positional isomers and are isolated separately as stable compounds. When **8** is heated, it isomerizes to **7**, which is thermally more stable. Compound **7** is symmetric, with two oxygens at the apical positions (P–O = 1.765 Å). This means that a hypervalent molecule becomes more stable when more electronegative substituents (oxygen compared to carbon) lie at the apex (apicophilicity, see Section 2.5). In compound **8**, the apical P–Oa bond (1.728 Å) is longer than the equatorial P–Oe bond (1.661 Å) and it is also the case for carbon; that is, the apical P–Ca bond (1.874 Å) is longer than the equatorial P–Ce (1.812 Å). In compound **9**, the apical P–O bond (1.765 Å) with two trifluoromethyl groups is longer and more polarized than that (1.709 Å) with a trifluoromethyl and a methyl group. The apical bond is linear (∠O–P–O = 180.2°), but in the equatorial plane ∠C–P–C (128.6°) is larger than 120° and ∠C–P–H (114.7°) is compressed from 120° [15].

Examples of tetracoordinate compounds are shown in Fig. 2.10. SF_4 (**10**) is essentially tbp with S–Fa = 1.646 Å and S–Fe = 1.545 Å. The apical bond ∠Fa–S–Fa = 187° is tilted to Fe and the equatorial bond ∠Fe–S–Fe = 101.5° is considerably compressed from 120°. The deviation from ideal angles is due to repulsion by an unshared electron pair which lies at the equatorial position.

The effect of unshared electron pair becomes evident when lithium phosphoranide (**11**, 10-P-4) is compared with phosphorane (**7**, 10-P-5). The bond angle (∠O–P–O) is 188.8° in **11** and 181.0° in **7** and ∠C–P–C is 109.1° in **11** and 120° in **7**. The bond length P–O is 1.90 Å in **11** and 1.77 Å in **7**. It is clear that unshared electron pair donates electrons to elongate the P–O bond and repels carbon and oxygen ligands. The structure of a neutral molecule of sulfurane **12** with an unshared electron pair resembles the anion **11** rather than the neutral molecules **7** and **9**.

The xenon derivative (**13**) has two positive charges formally on the central xenon, and the iodine derivative (**14**) has one positive charge on the iodine, and both are essentially tbp.

A planar bicyclic molecule of thiathiophthene (**15**; 10-S-3) has long been known as a tricoordinated sulfur compound and is named 1,6,6a-trithia(6a-S^{IV})pentalene in Chemical Abstracts (Fig. 2.9) [16]. Compound **15** is symmetric about the S(6a)–C(3a) bond. The bond distance S–S (2.363 Å) is larger than the sum of the covalent radii of sulfur (2.08 Å) by 14% and the length of S(6a)–C(3a) (1.78 Å) is more than S(1)–C(2) (i.e., S(6)–C(1); 1.684 Å). Based on these facts, the structure (**15b**) featuring a double bond between S(6a)–C(3a) cannot be correct. Therefore, the idea of no bond resonance was proposed to explain the presence of a weak bond (bond order of 0.5) between S(1)–S(6a) (i.e., S(6a)–S(6); Eq. (2.1)).

At present, it is well known that the S–S–S bond is a 3c–4e bond and S(6a)–C(3a) bond is an ylide type bond and the central sulfur has two unshared electron pairs, one above and the other below the molecular plane. Thus **15** is a hypervalent compound and is designated as 10-S-3.

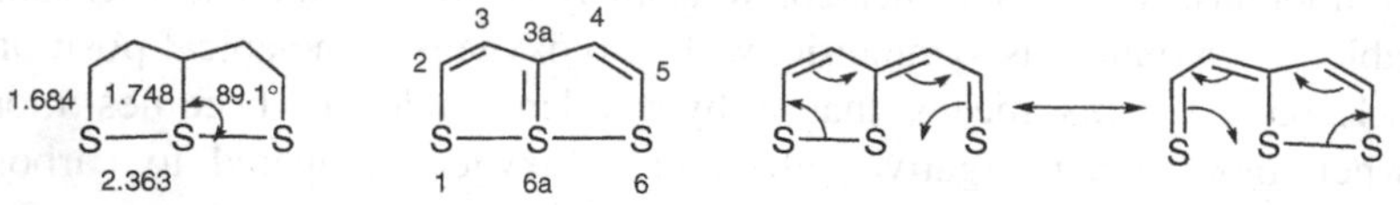

1.684 1.748 89.1° 2.363 **15a** **15b** **15c** **15d** (2.1)

Tricoordinate hypervalent compounds of phosphorus (**16a**), arsenic (**16b**), and antimony (**16c**) have been synthesized and these are isoelectronic to **15**. Compound **16** is planar and there are two positive charges on the ligand formally; thus the electron-withdrawing ability of the apical oxygen is enhanced to stabilize the molecule (Eq. 2.2) [17].

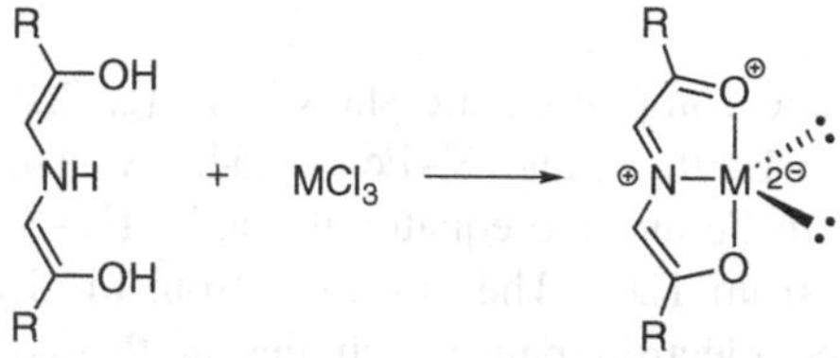

$$+ \ MCl_3 \longrightarrow \qquad (2.2)$$

16 (10-M-3)

16a, M=P; 16b, M=As; 16c, M=Sb

As tricoordinate halogen compounds (Fig. 2.7), a variety of iodine compounds are known including the bromine one such as **17** (Fig. 2.9).

Iodosobenzene (**18**: Ph–I=O) and dichloroiodobenzene ($PhICl_2$) are basic compounds for hypervalent organoiodanes. Structures of a variety of iodane compounds such as alkynyl triflate (**19**), alkenyl tetrafluoroborate (**20**), and azide (**21**) have been determined by X-ray analysis. All tricoordinate hypervalent compounds are usually called *T-shape compounds*; they are actually tbp(s) bearing two unshared electron pairs as equatorial substituents.

Many hexacoordinate phosphorus compounds (phosphoranate: 12-P-6) are well characterized [18] but it may not be easy to understand their stability as anions intuitively. In order to realize this, the results of theoretical calculations on PH_5, PCl_5, and so on, are described here. The electronic charges and bond lengths of PH_5 are shown in Fig. 2.12. P–Hap (1.478 Å) is longer than P–Heq (1.418 Å) and the negative charges (−0.21) are concentrated on Hap and the central P(V) has a positive charge (+0.66). The result illustrates the characteristics of Hap–P–Hap bond as a 3c–4e bond.

The calculated electronic charges of related phosphorus compounds are summarized in Table 2.4. For tetrahedral (PH_4^+, PCl_3, PCl_4^+) and octahedral (PH_6^-, PCl_6^-) compounds, all the substituents (H or Cl) have equal charges. For tbp compounds PH_4^-, PH_5, PCl_4^-, and PCl_5, the apical substituents bear larger negative charges compared to the equatorial ones. Here, one of the important results is that all the phosphorus atoms have positive charges, even those with a negative charge as a whole [19].

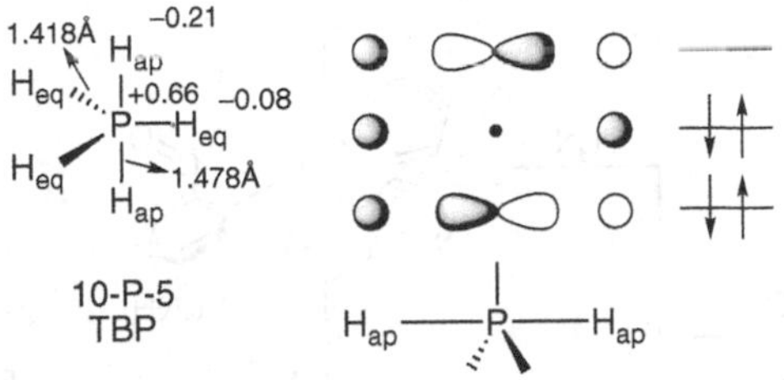

Figure 2.12 Structure of PH_5 and three-center four-electron bond consisting of apical bonds.

TABLE 2.4 Calculated Charges of Higher Coordinate Phosphorus Compounds

		Substituent (H or Cl)		
Compound	Charge on P	Apical	Equatorial	Equivalent
PH_4^+	+0.68	—	—	+0.08
PH_4^-	+0.20	−0.60	0.00	—
PH_5	+0.66	−0.21	−0.08	—
PH_6^-	+0.68	—	—	−0.28
PCl_3	+0.39	—	—	−0.13
PCl_4^+	−0.12	—	—	+0.28
PCl_4^-	+0.58	−0.63	−0.16	—
PCl_5	+0.03	−0.21	+0.13	—
PCl_6^-	+0.14	—	—	−0.19

Based on the facts above mentioned, it is understood that hexacoordinate monoanionic organophosphorus compounds (12-P-6) would be easily formed and are stable. Actually, **22** (10-P-5), where the carbonyl oxygen of a carboxyl group coordinates weakly to phosphorus, generates a stable isolable phosphoranate anion (**23**, 12-P-6) (Fig. 2.13). When 2 moles of triethylamine are added to 2 moles of **22**, the equilibrium shifts completely to give a cyclic dimer **24** (12-P-6). Equilibrium is seen in solution between the trivalent phosphorus compound **25** and the hexacordinate phosphoranate **26**, and the latter is the major component. All the hexacoordinate phosphorus anions show the very high chemical shifts of phosphorus (δ^{31}P: −124, −135, −93) [18a].

Examples of hexacoordinate hypervalent compounds (**27–32**) are shown in Fig. 2.14. They are essentially octahedral, irrespective of the kind of central atoms or the presence or absence of negative charges. Their bonds are fundamentally made of three 3c–4e bonds and six substituents consisting of the same elements and are equivalent. Hence, intramolecular positional isomerization does not take place by Berry mechanism at all. When it does, it is quite slow and the mechanism is said to be a Bailar twist. Silicon compounds are dianions (**27, 28**), phosphorus ones are monoanions (**29, 30**), and selenium and tellurium ones are neutral (**31, 32**).

Et$_3$N

2Et$_2$NH

22
^{31}P(ppm) δ −56.9

23
^{31}P(ppm) δ −124

24
^{31}P(ppm) δ −135.5

25
^{31}P(CH$_2$Cl$_2$) = 191.3ppm

26
^{31}P(CH$_2$Cl$_2$) = −93.3ppm

Figure 2.13 Formation of phosphoranate anions (12-P-6).

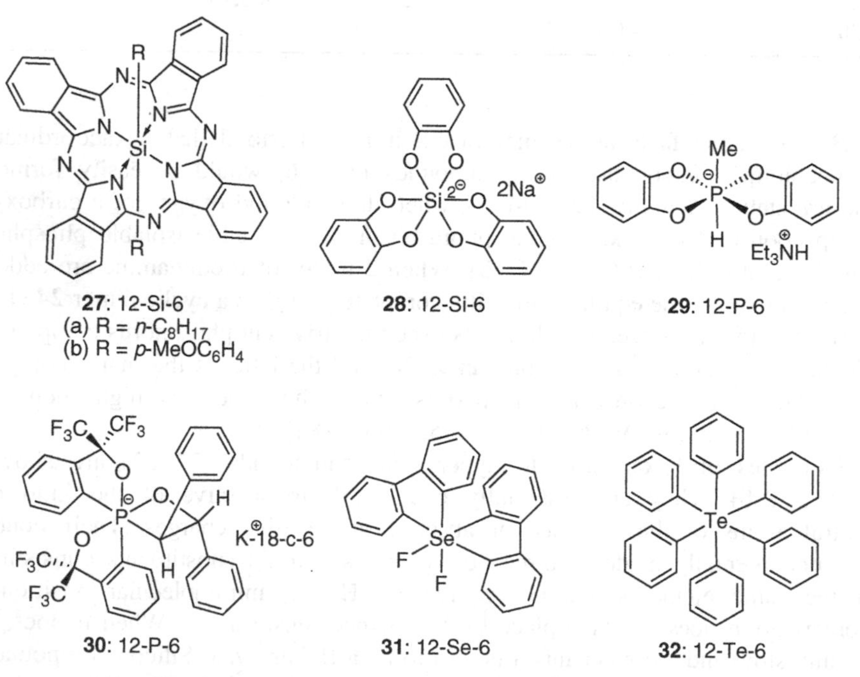

Figure 2.14 Hexacoordinate hypervalent compounds.

2.5 EFFECT OF HYPERVALENT BOND (2): APICOPHILICITY AND PSEUDOROTATION

In the preceding section, structures from 2–6 coordinate hypervalent compounds and the fundamental characteristics of hypervalent bond were explained. What kinds of unique and general phenomena would be expected from them? Firstly, they are apicophilicity and pseudorotation (positional isomerization). These are common to pentacoordinate hypervalent compounds including tetracoordinate compounds with an unshared electron pair (sulfur, selenium, and tellurium) and monoanions of silicon (silicate, etc.).

In hypervalent trigonal bipyramidal molecules, the apical bonds are the electron-rich 3c–4e bond (hypervalent bond) and the apical substituents hold excess electrons. Therefore, substituents that are more electron-withdrawing and smaller in size (steric hindrance is more effective at apical than equatorial position) reside preferentially at the apical positions. This is apicophilicity (Fig. 2.15) [20].

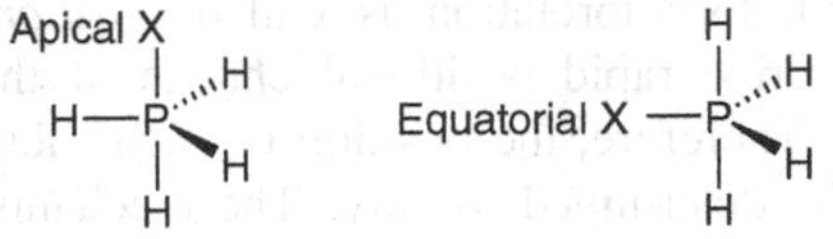

X (apicophilicity): F > H > OH (OR) ≈ Cl > NR_2 > Ph > H ≈ O^- > Me

Figure 2.15 Order of apicophilicity of substituents (X).

The order of apicophilicity of substituents has been examined experimentally and theoretically.

The theoretically calculated order of apicophilicity of substituents [21] is

$$Cl > F > H > OH \sim SH \sim CH_3 > PH_2 > NH_2 > SiH_3 > BH_2.$$

The experimentally determined order of apicophilicity of substituents [22] is

$$F > H > OR(OH) \sim Cl > NR_2 > Ph > H \sim O^- > CH_3.$$

It is apparent that at least two factors, that is, electronegativity and steric hindrance, influence the apicophilicity of substituents because the apicophilicity of the methyl group is smaller than hydrogen or the oxide anion.

It is important in organic chemistry to determine the order of apicophilicity of carbon substituents; however, it is considerably difficult because the difference in electronegativity among them is very small.

By determining the equilibrium ratio of **33a** and **33b** in solution, the apicophilicity of the R group against the methyl group was obtained. This was possible because the rate of pseudorotation is rendered low enough due to intramolecular hydrogen bonding between the hydroxyl group and the apical oxygen (Eq. 2.3). The apicophilicity order of the carbon substituents (R) thus obtained is as follows [23]:

$CDCl_3$

33a **33b** (2.3)

R: Ph $\gg$ CH_2OMe $\gg$ Me > $CH_2C_6H_4$-*p*-F > CH_2Ph > $CH_2C_6H_4$-*p*-Me $\sim$ Et > *n*-Pr $\sim$ *n*-Bu.

Pseudorotation occurs to equalize the character of the five σ bonds connected to the central atom, which consists of two apical bonds (3c–4e) and three equatorial bonds (sp^2). Pseudorotation is called *positional isomerization*: in another words, to describe rapid positional change of the substituents at the five apices of the tbp. Therefore, the position of each substituent bonded to the central atom cannot be determined usually. The mechanism of pseudorotation is clearly explained by Berry pseudorotation (Fig. 2.16) [24].

First, place a number for each substituent according to the sequence rule (the general rule which puts a number to each substituent to determine absolute configuration). Then, place substituents 1 and 2 (1 > 2) at the apical positions and remaining three substituents (3 > 4 > 5) in the equatorial plane so that the three substituents are placed clockwise when looked down from substituent 1. This positional isomer is depicted as **12**. One of the equatorial substituents (here 5 is picked up arbitrarily) is used as a pivot, then the bond angle of the apical substituents (1, 2) is decreased to 120° from 180° and that of equatorial ones (3, 4) is increased to 180° from 120° simultaneously. Finally, the obtained isomer (tbp **43**) is rotated 90° counterclockwise along the pivot (5), and then the new isomer **43** is obtained. Here, equatorial substituents (1, 2, 5) are placed clockwise when they are looked down from the apical substituent 4. This isomer can also be depicted as $\overline{\mathbf{34}}$. In $\overline{\mathbf{34}}$, the equatorial substituents (1, 2, 5) are placed counterclockwise when looked down from 3. Both are identical; however, we use this way to get **43** in this book.

TBP **12** $\Psi 5$ C_{4v} $\Psi 5$ TBP **43** = **43** = $\overline{\mathbf{34}}$

Figure 2.16 Positional isomerization by Berry pseudorotation (BPR).

At the transition state (C_{4v}) of pseudorotation, the five substituents become equivalent because of the adjusting bond lengths. The movement of the substituents is essentially a bending motion and does not contain rotational motion at all. However, the new positional isomer **43** is finally obtained by an imaginary counterclockwise rotation of 90° along the axis of the pivot; thus the total movement is called *pseudorotation*.

Twenty positional isomers exist for a pentacoordinate compound bearing five different kinds of univalent substituents. It is explained by the Desargues–Levi diagram, illustrated at the center of Fig. 2.17 [25]. There are 12 apices of two hexagons and six apices connecting the two apices of each hexagon and two apices (**13:31**) lying inside the two hexagons. Each apex corresponds to one positional isomer; thus 20 isomers are correlated and identified. It is quite difficult (actually impossible) to determine the position of each substituent because pseudorotaion takes place quite rapidly. By employing suitable bidentate (bivalent) ligands to form heterocycles with the central atom, a number of possible

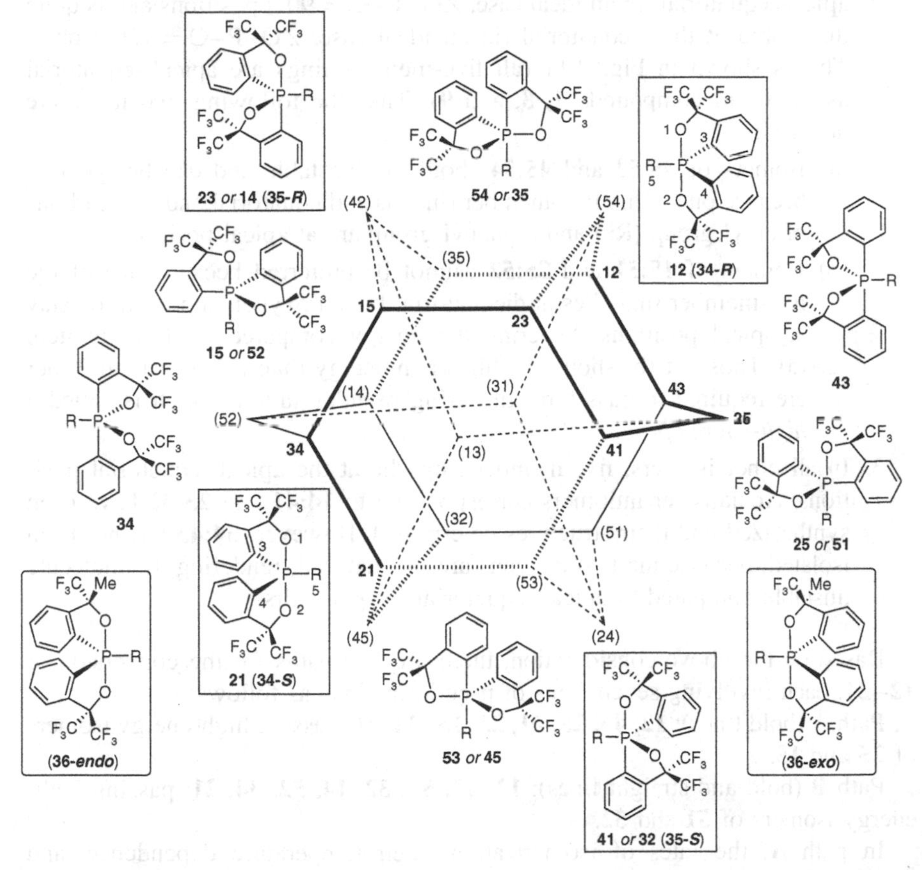

Figure 2.17 Desargues–Levi diagram and seven step BPR for inversion (**12·21**) of spirophosphorane.

isomers can be omitted and the rate of pseudorotation can also be slowed. By this idea, the position of substituents can be determined according to relative apicophilicity of the substituents. Among the several attempts, a recent successful example is explained below.

By using the Martin ligand, that is, [bis(trifluoromethyl)benzyl alcohol], stable enantiomers of phosphorane (**34**) and metastable phosphorane (**35**), and a diastereomer **36**, having one methyl group substituted for a trifluoromethyl group of **34**, were prepared and their structures determined by X-ray analysis (when R = $PhCH_2$, compound **34** is the same as compound **7** and **35** is **8** in Fig. 2.11. R = Bu was employed for the measurement described in Fig. 2.17). In Fig. 2.17, the numbering of substituents is shown for enantiomers of **12** (**34-*R***) and **21** (**34-*S***). Thus we can understand the following [22d, 26]:

1. Positional isomers of **13:31** and **24:42** can be omitted, because cyclic ligands (1–3; 2–4) cannot occupy apical positions at the same time.
2. Owing to the ring strain, the five-member ring prefers to be at apical–equatorial (in an ideal case, $\angle$O–P–C = 90°) positions and is quite disfavored at the diequatorial (in an ideal case, $\angle$O–P–C = 120°) ones. This is shown in Fig. 2.11 (all five-member rings are apical–equatorial as shown in compounds **7, 8**, and **9**). Thus the following two items are deduced:
   (a) Isomers of **35:53** and **45:54** should be unstable and can be ignored, because one of the five-member rings is at diequatorial and two carbons of alkyl group (R^5) and a phenyl group are at apical positions.
   (b) Isomers of **15:51** and **25:52** cannot be preferred because one of the five-member rings lies at diequatorial but an oxygen and a carbon stay at apical positions, lowering the energy compared to those in item (a). Those of (b) should be higher in energy than those of item 3, but are required to pass through during pseudorotation, which is called a *high-energy pass*.
3. In all other isomers, five-member rings lie at the apical–equatorial positions. Actually, enantiomers corresponding to **14:41** and **23:32** have been synthesized and their structures determined. However, **34:43** has not been isolated, because the two oxygens are diequatorial, rendering the molecule unstable compared to apical–equatorial oxygen pairs.

Based on the above consideration, there are two paths for the conversion of **12–21**, each involving seven steps of pseudorotations as follows:

Path **A** (bold lines): **12, 43, 25, 41, 23, 15, 34, 21**; passing high-energy isomers of **25** and **15**.

Path **B** (bold and straight lines): **12, 43, 51, 32, 14, 52, 34, 21**; passing high-energy isomers of **51** and **52**.

In path **A**, the rates of isomerization, their temperature dependence, and activation energy have been obtained for **41** → **23**, **41** → **21**, and **12** → **21**. Four (two pairs) trifluoromethyl groups are observed at low temperature

(−80°C) for **41** (**35**-S) and **23** (**35**-R) because they are not symmetric. At higher temperature (16°C), four peaks yielded to two and the coalescence temperature for each was $T_c = 235$ K (−38°C) (R = n-Bu; in toluene). When R = n-Bu, kinetic parameters of $\Delta H^{\ddagger} = 10.0$ kcal/mol, $\Delta S^{\ddagger} = -3.4$ eu are obtained for the one-step pseudorotation (**41** → **23**).

41 (**35**-S) isomerizes to **21** (**34**-S) through four-step pseudorotations when heated (30–50°C). The rate of this isomerization did not change in toluene, pyridine, and ethanol. Hence the isomerization proceeds intramolecularly without opening of the five-member ring. The kinetic parameters are obtained as $\Delta H^{\ddagger} =$ 22.0 kcal/mol, $\Delta S^{\ddagger} = -10$ eu.

12 (**34**-R) and **21** (**34**-S) are quite stable and the rate of their conversion could not be observed. **36**-*exo* and **36**-*endo*, which are diastereomers having one methyl group substituted for a trifluoromethyl group of **34**, were prepared and separated. The rate of equilibration of both diastereomers were measured separately in t-butyltoluene (170–200°C) and the kinetic parameters are obtained as $\Delta H^{\ddagger} = 34.0$ kcal/mol, $\Delta S^{\ddagger} = -8.8$ eu. Therefore, it is realized that **34** (O-trans) is more stable than **35** (O-cis) by 12 (34–22) kcal/mol.

These results obtained for Path **A** correspond to isomerizations of **32** → **14**, **32** → **21**, and **12** → **21** for Path **B**. According to experiments mentioned above, it is not possible to decide whether the isomerization follows Path **A** or Path **B** [27].

One of the reasons why important reactions of phosphorylation and hydrolysis of phosphoric esters can take place smoothly at ambient temperatures is that the intermediate pentavalent phosphoranes are stable enough for pseudorotaion to lower the activation energy of those reactions. In Chapter 6, more discussions on this will be presented.

2.6 EFFECT OF HYPERVALENT BOND (3): LIGAND COUPLING REACTION (LCR) AND EDGE INVERSION

Two fundamentally and essentially new reactions of heavier main group elements are introduced briefly here in advance, that is, LCR and edge inversion. LCR is intramolecular and is the concerted coupling of two ligands (substituents) that are bonded directly in σ-type to a hypervalent main group element. During the coupling reaction, valence of the central atom (M) decreases by 2, that is, $[M^{(N+2)} \rightarrow M^{(N)}]$, and stereochemistry of the coupling substituents are retained. Transition metals can catalyze LCR but it is not essential. The scope and variety of LCRs will be described in Chapter 11.

$$L_nM^{(N+2)}\begin{matrix} X \\ Y \end{matrix} \longrightarrow X{-}Y + L_nM^{(N)} \tag{2.4}$$

A fundamental and important possibility of a new inversion mechanism was recently proposed and experimentally and theoretically investigated to show its validity. It is edge inversion and the process is illustrated in Eq. (2.5).

Vacant p orbital

A

Edge inversion process

(2.5)

A heavier group 14 element compound bearing four σ-bonds can invert when the bond angle of A–B and C–D enlarges to 180° perpendicularly and come on a plane at transition state **A** and continues to go to its mirror image. During the process, four bonds are kept at the central atom M by two 3c–4e bonds and a vacant p orbital appears at M. This was theoretically proposed and supported by experiments. Here one of four substituents can be the unshared electron pair and the stereochemical identity is not lost, different from vertex inversion (cf. NH_3). The theoretical aspects and experimental examples are explained in Notes 8.

2.7 EFFECT OF σ_{X-C}

The σ-bond energy of the main group element and carbon decreases when the former descends the period of the periodic table. The energy level of the HOMO increases and that of the LUMO (lowest unoccupied molecular orbital) decreases for the corresponding σ-bond, while they lie between the energy levels of σ-bond of the second period element and carbon. Hence, the HOMO of main group element–carbon bond can behave as an electron-donating species when it is arranged to interact with an electron-deficient species (for instance, carbocation) in space. This can be realized as the same kind of effect of the σ-bond as "hetero atom effect" of the unshared electron pair, although this is not so prominent as the latter (cf. Section 1.4, of Chapter 1). Well-known examples enabled by such interaction are pinacol–pinacolone, Wagner–Meerwein, Beckmann and Hofmann rearrangements, and so on. As a reminder, two examples are cited here.

When pinacol (**37**) is heated in an acidic solution, carbocation (**B**) is generated by the dehydration of the protonated **37** and a methyl group migrates to the cation keeping the electron pair of the carbon–carbon bond. The newly generated carbocation at the hydroxyl group is the same as a protonated carbonyl group (**C**), and thus *pinacolone* (**38**) is obtained. In this example, stereochemistry of the migrating group and the carbocation cannot be seen because of the free rotation of the C–C bond.

$$\text{(37)} \rightleftharpoons \text{H}_3\text{C}-\text{C}(\text{CH}_3)(\text{OH})-\text{C}(\text{CH}_3)(\text{OH}_2^+)-\text{CH}_3 \rightleftharpoons \text{(B)} \rightarrow \text{(C)} \rightarrow \text{(38)} \tag{2.6}$$

H+, $-H_2O$, $-H+$

In Beckmann rearrangement, the group anti to the hydroxyl group specifically rearranges to the nitrogen when oxime (**39**) is treated with PCl_5 in ether or benzene (**D**). After hydrolysis, amide (**40**) is obtained but the isomer cannot be obtained at all. By the cleavage of $O{=}PCl_3$ and Cl^- keeping the electron pair of the N–O bond, the *p*-tolyl group migrates bearing the electron pair of the C–C bond to the vacant σ^* orbital appearing anti to the N–O bond (**D**) [28].

$$\text{(39)} \xrightarrow[-\text{HCl}]{\text{PCl}_5} [\text{D}] \rightarrow [\text{E}: \text{O}{=}\text{PCl}_3 + \text{Cl}^-] \xrightarrow{\text{H}_2\text{O}} \text{(40)} \tag{2.7}$$

These interactions are shown below.

F G

In Chapter 5, the reactions of vinylsilanes and allylsilanes with electrophiles are explained, where the electron-donating interaction from the HOMO of the silicon–carbon bond to the β-carbocation controls regioselectivity [29]. Two examples are illustrated here in advance. In these, orientation of the electrophile is also controlled because the interaction becomes most effective when the carbocation and the silicon–carbon bond lie in the same plane. The energy level of the HOMO of silicon–carbon bond is generally lower than that of element–carbon bond of lower periods (for instance, Ge–C, Sn–C, etc.). Therefore, this kind of effect appears more prominent for the latter. The donating effect of σ_{X-C} to electrophilic species (e.g., carbocation) here described is called *σ–π conjugation*, to use another expression.

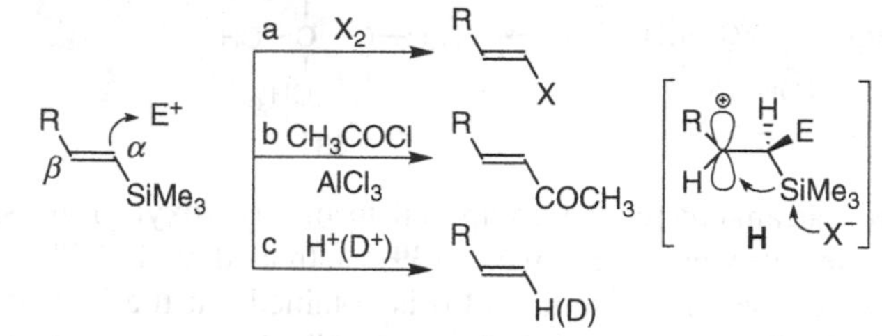

(2.8)

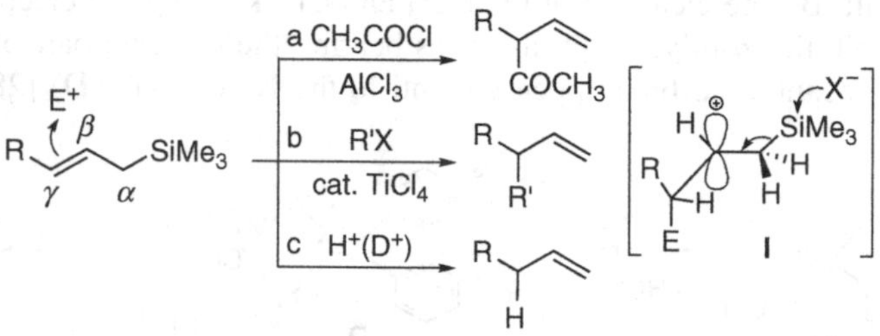

(2.9)

2.8 EFFECT OF σ^*_{X-C}

The effect of σ^*_{X-C}, that is, the electron-accepting (withdrawing) ability of the LUMO of a single bond of X–C, is usually not so apparent as the electron-donating ability of σ_{X-C} [30]. However, as early as the 1960s, the role of the X–C bond for stabilizing α-carbanion was investigated.

A mixture of dodecyldimethylamine (**41**), dodecyldimethylphosphine (**42**), and *t*-butyllithium (1:1:1 ratio) in hexane was kept room temperature for four days. The mixture was quenched with D_2O (Eq. 2.10). Deuterated compounds were obtained, that is, **41**-d_1 5%; **42**-d_1 51%; **42**-d_2 13%. The ratio of reactivity of **41:42** (the ratio of stability of α-carbanion **J**) is approximately 1:15 (cf. 5:77). The stability of the α-carbanion **J** of **41** is expected to be higher than that of **42** because the electronegativity of nitrogen (3.0) is larger than that of phosphorus (2.1). But the result was contrary to expectations [31].

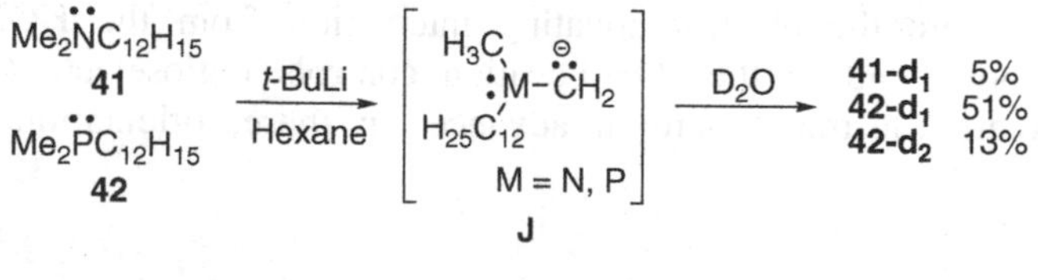

(2.10)

Another method to generate a carbanion α to a heteroatom is to add alkyllithium to a double bond. By the addition of alkyllithum to vinylphosphine and vinylsulfide, the corresponding carboxylic acid was obtained in about 50% yield after the addition of carbon dioxide to the reaction mixture (Eq. 2.11). In the case of vinyl phenyl ether, the ether was recovered intact (more than 90%), accompanied by a small amount of the decomposition product (phenol) [32]. It is apparently shown by the result that even an element bearing unshared electron pair such as phosphorus and sulfur can stabilize the α-carbanion (**K,L**) but oxygen cannot. The stabilizing effect is ascribed to the effect of σ^*_{X-C}.

$$Ph_2\ddot{P}-CH=CH_2,\ Ph\ddot{S}-CH=CH_2,\ Ph\ddot{O}-CH=CH_2 \xrightarrow[Et_2O]{RLi} Ph_2\ddot{P}-\overset{\ominus}{C}HCH_2R\ (\mathbf{K}),\ Ph\ddot{S}-\overset{\ominus}{C}HCH_2R\ (\mathbf{L}) \xrightarrow{CO_2} Ph_2\ddot{P}-CH(CO_2H)CH_2R,\ Ph\ddot{S}-CH(CO_2H)CH_2R \tag{2.11}$$

(for PhO–CH=CH₂: Recovery + decomposition (PhOLi))

Ylide (**M**) can be generated by deprotonation from the methyl group of the onium salt (**43**) (Eq. 2.12). When the reaction was carried out in a deuterated solvent (water or alcohol), H–D exchange took place to give **43-D**. The rate should be controlled by the stability of ylide (**M**). This is a bimolecular reaction of an onium salt (**43**) and hydroxide ion ([OD$^-$]) and the results are shown in Table 2.5. The experiments were carried out only at two temperatures for **43**, and the activation enthalpy and entropy were calculated. The reaction can be said to be enthalpy-controlled and the activation entropy was almost constant for each **43** [33].

$$\underset{\mathbf{43}}{\overset{\oplus}{E}-CH_3} + \overset{\ominus}{O}D/D_2O \longrightarrow \underset{\mathbf{M}}{\left[\overset{\oplus}{E}-\overset{\ominus}{C}H_2 \longleftrightarrow E=CH_2\right]} \xrightarrow{D_2O} \underset{\mathbf{43\text{-}D}}{\overset{\oplus}{E}-CH_2D} \tag{2.12}$$

E : Me_3N, Me_3P, Me_2S

First, it is impressive to see that the rate of H-D exchange of ammonium salt is very slow. The rate of sulfonium salt is faster than that of phosphonium salt by about 10 times, but there is quite a large gap between the latter two and that of ammonium salt (about 10^4 times). On the basis of the fundamental factors of electronegativity and Coulomb force, the ammonium salt is expected to be the fastest. The activation enthalpy of phosphonium or sulfonium salt can be obtained by multiplication of that of the ammonium salt (134.6 kJ/mol) by the ratio of the bond length of the salt (E^+–C) to the ammonium salt (N^+–C) (Eq. 2.11). Then, the activation enthalpy is calculated for the phosphonium salt (171.2 kJ/mol) and for the sulfonium salt (166.6 kJ/mol).

TABLE 2.5 H–D Exchange Reaction of Onium Salts by DO^- in D_2O

Onium salt	Temperature (°C)	k_2 (L/mol/sec)	$\Delta H^{\ddagger}$(kJ/mol)
$Me_3\overset{+}{N}-CH_3$	83.6	3.11×10^{-9}	134.6 ± 2.5
$Me_3\overset{+}{P}-CH_3$	26.8	9.85×10^{-6}	107.0 ± 0.8
$Me_2\overset{+}{S}-CH_3$	26.8	1.15×10^{-4}	93.6 ± 2.1

* $[OD^-] = 0.26 \sim 0.39$ mol/L
* Rates were measured only at two temperatures for each onium salt. Numerical values of $\Delta H^{\ddagger}$ and $\Delta S^{\ddagger}$ were calculated from these data. Numerical values of $\Delta S^{\ddagger}$ were almost constant and omitted for clarity.

$$\Delta H^{\ddagger}_{E^+} = (\Delta H^{\ddagger}_{N^+}) \times r_{E^+}/r_{N^+} \qquad (2.13)$$

Bond distance : $r_{E^+} = E^+\text{–}C$ Ar $_{N^+} = N^+\text{–}C$
($N^+\text{–}C = 1.47$, $P^+\text{–}C = 1.87$, $S^+\text{–}C = 1.81$ Å)

The difference between these calculated values and those in Table 2.5 ($\Delta = 64.4$ and 71.9 kJ/mol) would certainly be due to the extra ability to stabilize α-carbanion by the phosphonium and sulfonium groups. The major reason is now ascribed to the effect of σ^*_{X-C} (electron-accepting ability) with partial contribution of the d orbital, which had been believed to be the sole contribution of d orbital without any concrete evidence for a long time (disclosed by the advance in recent computer chemistry). According to this investigation, it is concluded that the extra stability of ylides of the group 15 (P^+, As^+, Sb^+) and the group 16 (S^+, Se^+, Te^+) compared to the ammonium ylide is in the range of 55–66 kJ/mol, which can be taken as constant values of ~60 kJ/mol.

Naturally, the effect of σ^*_{X-C} depends on relative direction to the carbanion. The rate of H–D exchange of cyclic sulfonium salt (**44**) was measured in D_2O–NaOD (Fig. 2.18). The rate for the external methyl group was the fastest at 1.4×10^{-4} L/mol/s (35°C) and that for Hd was 4×10^{-6} L/mol/s (35°C), but the rate for Hu could not be observed (apparently too slow). The rate of deuteration of the methyl group was higher than that of the methylene group (Hd) by 23 times, showing the importance of stereochemistry and the destabilizing effect of antiparallel dianions [34].

The relative rate of deprotonation of bicyclic sulfonium salt (**45**) by alkyllithium was H_{eq}:H_{ax} = 35:1 [35]. The effect is due to the stabilization of the carbanion generated at H_{eq} by interaction with $\sigma^*_{S^+-C}$ shown as a bold line at the center of the bicycle (Fig. 2.19). The carbanion at Heq and $\sigma^*_{S^+-C}$ lie antiparallel

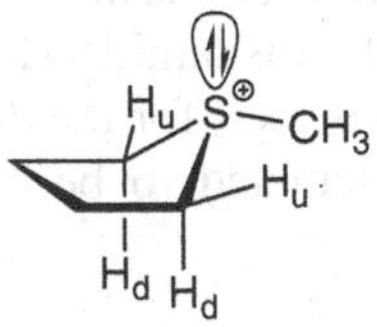

$\overset{+}{S}-C\underline{H}_3$ 1.4×10^{-4} L/mol/sec (35°C)

$\overset{+}{S}-C\underline{H}_d$ 4×10^{-6} L/mol/sec (35°C)
4.3×10^{-5} L/mol/sec (55°C)

$\overset{+}{S}-C\underline{H}_u$ $<1 \times 10^{-7}$ L/mol/sec (55°C)

44

$\overset{+}{S}-C\underline{H}_3 : \overset{+}{S}-C\underline{H}_d = 23:1$

Figure 2.18 Effect of stereochemistry of H–D exchange of sulfonium salt (**44**).

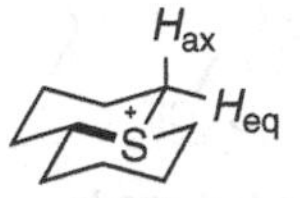

Ratio of kinetic acidity
$H_{eq}:H_{ax} = 35:1$

45

Figure 2.19 Effect of σ^*_{S+-C} to the rate of lithiation of sulfonium salt (**45**).

in a plane for effective interaction to occur. For the carbanion at H_{ax}, there can be no spatial interaction with σ^*_{S+-C}.

A new example of the effect of σ^*_{X-C} came from the isolation and preparation of the positional isomers of **7** (*O-trans*) and **8** (*O-cis*) described in Fig. 2.11. Pseudorotation (BPR) of positional isomers was explained in detail in Section 2.5. Activation energy of BPR of *O-cis*-Ib to *O-trans*-Ib was measured as $\Delta H^{\ddagger} = 21.8$ kcal/mol, $\Delta S^{\ddagger} = -9$ eu [26]. The corresponding values were obtained as $\Delta H^{\ddagger} = 24.4$ kcal/mol, $\Delta S^{\ddagger} = -6.8$ eu for *O-cis*-IIb to *O-trans*-IIb. By comparison of a series of kinetic results on *O*-cis to *O*-trans, it was deduced experimentally and theoretically that the stabilization energy of *O-cis*-IIb against *O-trans*-IIb is 4.4 kcal/mol, which is ascribed to the donation of the unshared electron pair of the nitrogen to σ^*_{O-P} ($n \rightarrow \sigma^*_{O-P}$). This effect can be neglected for the case of *O-trans*-IIb because σ^*_{C-P} has a considerably higher energy than σ^*_{O-P} [36].

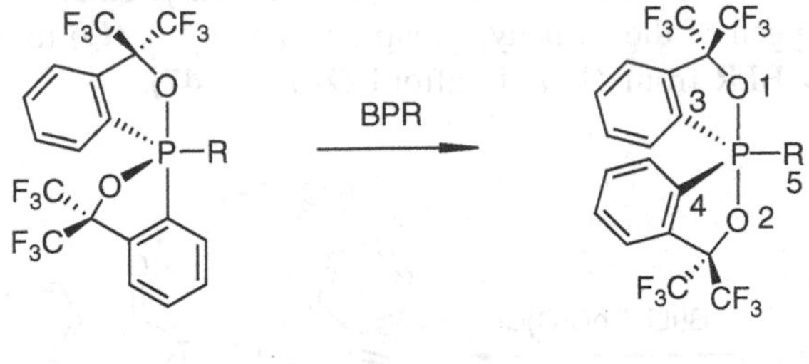

BPR (2.14)

O-cis *O-trans*

I : (a) R = CH_2Ph (b) R = Bu
II : (a) R = NHPh (b) R = NHPr

The next evidence of stabilization of a carbanion adjacent to σ^*_{O-P} is that *O-cis*-Ia is deprotonated with potassium hexamethyldisilamide (KHMDS) and deuterated by D_2O, in contrast to *O-trans*-Ia which is inert for the deprotonation. The stabilization energy of the α-carbanion is calculated to be 14.1 kcal/mol [15b].

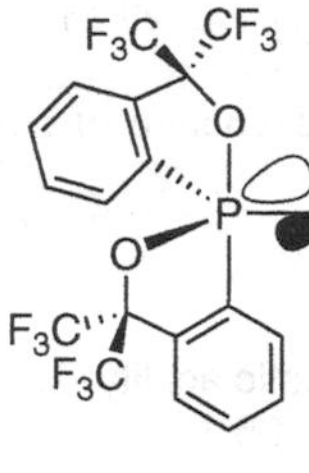

M : *O-cis*-Ia 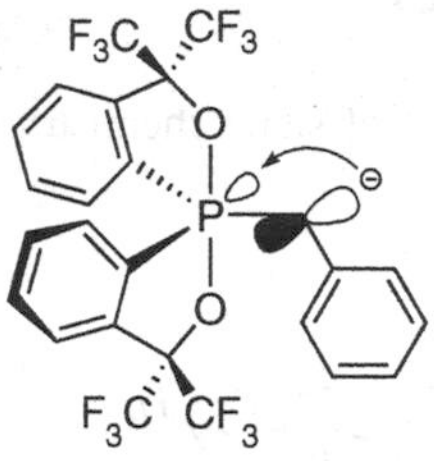 **N** : *O-trans*-Ia

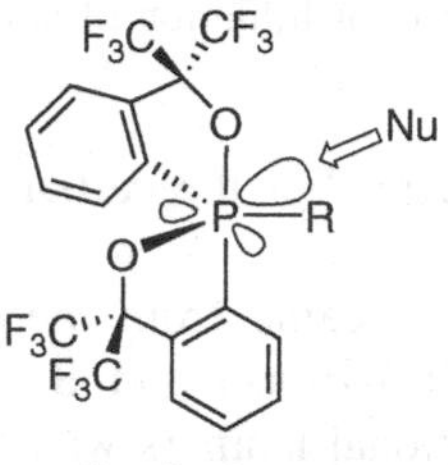

O : *O-cis* 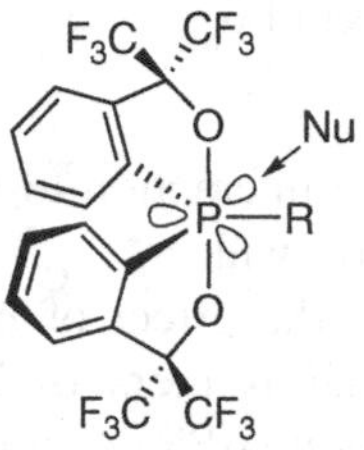 **P** : *O-trans*

Nucleophilic attack by fluoride ion and methyllithium took place from the rear side of σ^*_{O-P} of *O-cis*-I to afford the corresponding phosphoranate (cf. **O**: 12-P-6) but *O-trans*-I is again inert for the nucleophilic attack. The effect of σ^*_{O-P} of *O*-cis was dramatically shown for the benzoylation of *O-cis*-Ia and *O-trans*-Ia. By deprotonation of *O-cis*-Ia and *O-trans*-Ia with BuLi, both anions were generated in situ and benzoyl chloride was added. From the latter, the expected benzoylated product (**46**) was obtained. In contrast, vinyl ether (**47**) was formed in which the oxygen of the carbonyl group attacked σ^*_{O-P} (**Q**) to rearrange to **R** accompanied by BPR from *O-cis* to afford *O-trans* (**47**).

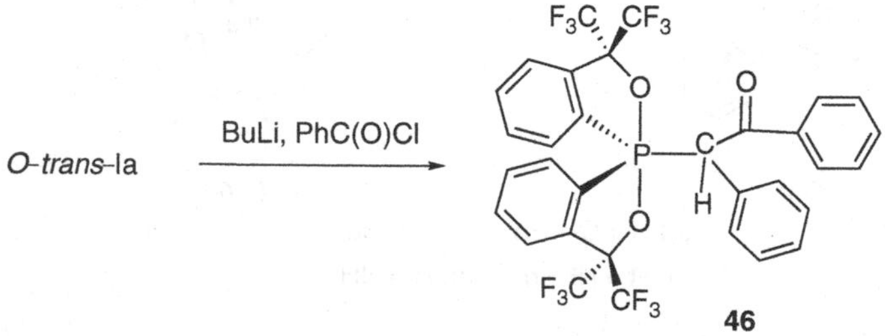

O-trans-Ia $\xrightarrow{\text{BuLi, PhC(O)Cl}}$ **46** (2.15)

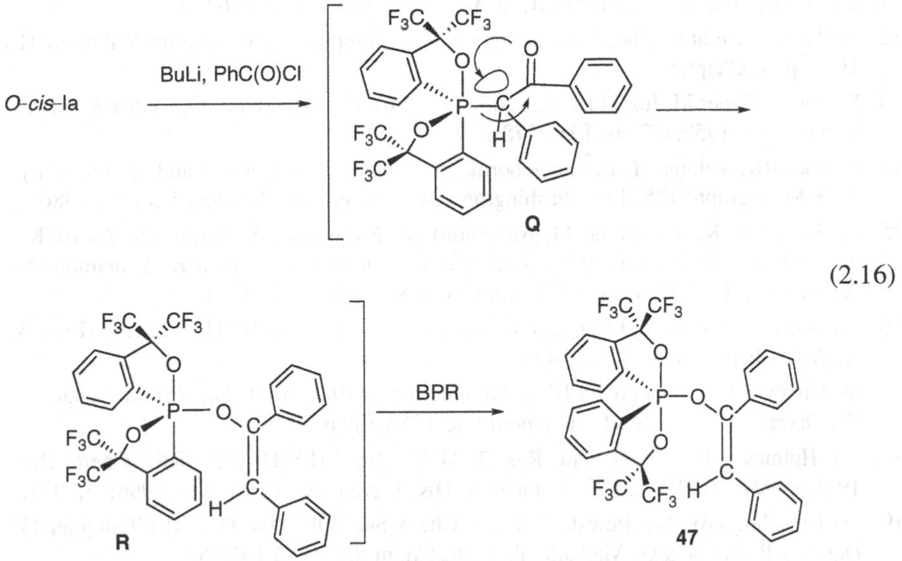

(2.16)

Finally, it can be summarized that stabilizing interaction for carbocation or carbanion takes place effectively when the corresponding σ_{X-C} or σ^*_{X-C} lies in the plane of each other. This is also applicable to combinations of main group elements.

REFERENCES

1. Akiba K.-y. J Synth Org Chem Jpn 1984;42:378, (Japanese).
2. Akiba K.-y., Yamamoto Y. Kagaku Zoukan 115 (ISBN4-7598-0600-8), Organic chemistry of heteroatoms, Chapter 1 and 3, 1988, (Japanese).
3. Kutzelnigg W. Angew Chem Int Ed Engl 1984;23:272.
4. Schmidt MW, Truong PN, Gordon MS. J Am Chem Soc 1987;109:5217.
5. (a) Nagase S. In: Patai S, editor. The chemistry of organic arsenic, antimony, and bismuth compounds. New York: John Wiley & Sons, Inc.; 1994. p. 1, Chapter 1; (b) Nagase S. Kikan Kagaku Sousetu 34 (ISBN 4-7622-1880-4) Akiba K.-y., editor. (Ed. of the series, The Chemical Society of Japan, Publ. By Gakkai Syuppan Center), 1998. p. 113, Chapter 6. (Japanese).
6. Akiba K.-y., editor. Chemistry of hypervalent compounds. Weinheim: Wiley-VCH; 1999.
7. Musher JI. Angew Chem Int Ed Engl 1969;8:54.
8. Perkins CW, Martin JC, Arduengo AJ III, Lau W, Alegria A, Kochi JK. J Am Chem Soc 1980;102:7753.
9. (a) Pimentel GC. J Chem Phys 1951;19:446; (b) Hachland RJ, Rundle RE. J Am Chem Soc 1951;73:4231.
10. Weinhold F, Landis CR. Valency and bonding. Cambridge: Cambridge University Press; 2005, Chapter 3.

11. Cahill PA, Dykstra CE, Martin JC. J Am Chem Soc 1985;107:6359.
12. Akiba K.-y., editor. Chemistry of hypervalent compounds. Weinheim: Wiley-VCH; 1999. p. 9, Chapter 2.
13. Wittig G, Rieber M. Justus Liebigs Ann 1947;562:187; (b) Wittig G, Clauss K. Justus Liebigs Ann 1952;577:26, 1952;578:136.
14. Holmes RR. Volumes I, II, Pentacoordinated phosphorus-structure and spectroscopy, ACS Monograph, 175, 176. Washington (DC): American Chemical Society; 1980.
15. (a) Kajiyama K, Yoshimine M, Nakamoto M, Matsukawa S, Kojima S, Akiba K.-y. Org Lett 2001;3:1873; (b) Matsukawa S, Kojima S, Kajiyama K, Yamamoto Y, Akiba K.-y., Re SY, Nagase S. J Am Chem Soc 2002;124:13154.
16. (a) Bezzi S, Mammi M, Garbuglio C. Nature 1958;182:247; (b) Hansen LK, Hordvik A. Acta Chem Scand 1973;27:411.
17. (a) Culley SA, Arduengo AJ III. J Am Chem Soc 1984;106:1164; (b) Arduengo AJ III, Dixon DA, Roe DC. J Am Chem Soc 1986;108:6821.
18. (a) Holmes RR. Acc Chem Res 2004;37:746; (b) Holmes RR. Chem Rev 1996;96:927; (c) Wong CY, Kennepohl DK, Cavell RG. Chem Rev 1996;96:1971.
19. (a) Kutzelnigg W, Wasilewski J. J Am Chem Soc 1982;104:953; (b) Trinquier G, Daudy J-P, Caruana G, Madaule Y. J Am Chem Soc 1982;104:953.
20. Mutterties EL, Mahler W, Schmutzler R. Inorg Chem 1963;2:613.
21. (a) McDowel RS, Streitwieser A Jr. J Am Chem Soc 1985;107:5849; (b) Wang P, Zhang Y, Glaser R, Reed AE, Schleyer PvR, Streitwieser A Jr. J Am Chem Soc 1991;113:55; (c) Wasada H, Hirao K. J Am Chem Soc 1992;114:16.
22. (a) Trippet S. Pure Appl Chem 1970;40:595; (b) Moreland CG, Doak GO, Littlefield LB, Walker NS, Gilje JW, Braun RW, Cowley AH. J Am Chem Soc 1976;98:2161; (c) Griend LV, Cavell RG. Inorg Chem 1983;22:1817; (d) Nakamoto M, Kojima S, Matsukawa S, Yamamoto Y, Akiba K.-y. J Organomet Chem 2002; 643–644:441.
23. Matsukawa S, Kajiyama K, Kojima S, Furuta S-Y, Yamamoto Y, Akiba K.-y. Angew Chem Int Ed 2002;41:4718.
24. Berry RS. J Chem Phys 1960;32:933.
25. Mislow K. Acc Chem Res 1970;3:321.
26. (a) Kojima S, Kajiyama K, Nakamoto M, Akiba K.-y. J Am Chem Soc 1996;118:12866; (b) Kojima S, Kajiyama K, Nakamoto M, Matsukawa S, Akiba K.-y. Eur J Org Chem 2006;218; (c) Kajiyama K, Yoshimine M, Kojima S, Akiba K.-y. Eur J Org Chem 2006;2739.
27. Kojima S, Nakamoto M, Akiba K.-y. Eur J Org Chem 2008;1725.
28. Jones M Jr. Organic chemistry. 2nd edn. New York: W.W. Norton & Co.; 2000, Chapters 8 and 19.
29. Chuit C, Corriu RJP, Reye C, Young JC. Chem Rev 1993;93:1371.
30. Carey FA, Sundberg RJ. Advanced organic chemistry, Chemical bonding and molecular structure. 5th edn. Springer; 2007, Chapter 1.
31. Peterson DJ, Hays HR. J Org Chem 1965;30:1939.
32. (a) Peterson DJ. J Org Chem 1966;31:950; (b) Parham WE, Motter RF. J Am Chem Soc 1959;81:2146.
33. Doering WvE, Hoffmann AK. J Am Chem Soc 1955;77:521, 514.

34. Barbarella G, Garbesi A, Fava A. Helv Chim Acta 1971;54:341, 2297.
35. (a) Andreeti GD, Bernardi F, Fava F. J Am Chem Soc 1982;104:2176; (b) King JK, Rathore R. J Am Chem Soc 1990;112:2001.
36. Adachi T, Matsukawa S, Nakamoto M, Kajiyama K, Kojima S, Yamamoto Y, Akiba K.-y., Re S, Nagase S. Inorg Chem 2006;45:7269.

NOTES 4

(σ, σ^*) AND (π, π^*): HMO (HUECKEL MOLECULAR ORBITAL) AND ELECTROCYCLIC REACTION

The most fundamental example of the molecular orbital (MO) method is explained here as a reminder by using the interaction of 2p orbitals of atom (**1**) and atom (**2**).

1. When two dumb-bell-shaped 2p orbitals, $\phi 1$ and $\phi 2$ approach along the symmetry axis, two MOs, Φa and Φb, are formed according to their interaction. The lower energy orbital is a bonding orbital (Φb: $\phi 1 + \phi 2$) and the higher energy orbital is an antibonding orbital (Φa: $\phi 1 - \phi 2$). They have an axis of symmetry and thus are σ-type bonds, Φb is σ and Φa is σ^*.
2. When two 2p orbitals, $\phi 1$ and $\phi 2$ approach in parallel fashion, two MOs, $\Phi b = \phi 1 + \phi 2$ and $\Phi a = \phi 1 - \phi 2$, are formed according to their interaction. They are antisymmetric to the plane of symmetry and are π-type bonds. Therefore, Φb is π and Φa is π^*. Stabilization energy (Eσ) of the σ bond is larger than that (Eπ) of the π bond, that is, the σ bond is stronger than the π bond and is more stable. In both cases, two electrons occupy the bonding orbital and are stabilized by 2Eσ or 2Eπ in energy. When one of the two electrons is excited, they occupy separately, one by one, σ and σ^* or π and π^* orbitals. Then, stabilization energy is zero.
3. When the energy of the two 2p orbitals is different by ΔE, the interaction results in an unsymmetrical energy diagram. An example of π bonding is

Organo Main Group Chemistry, First Edition. Kin-ya Akiba.

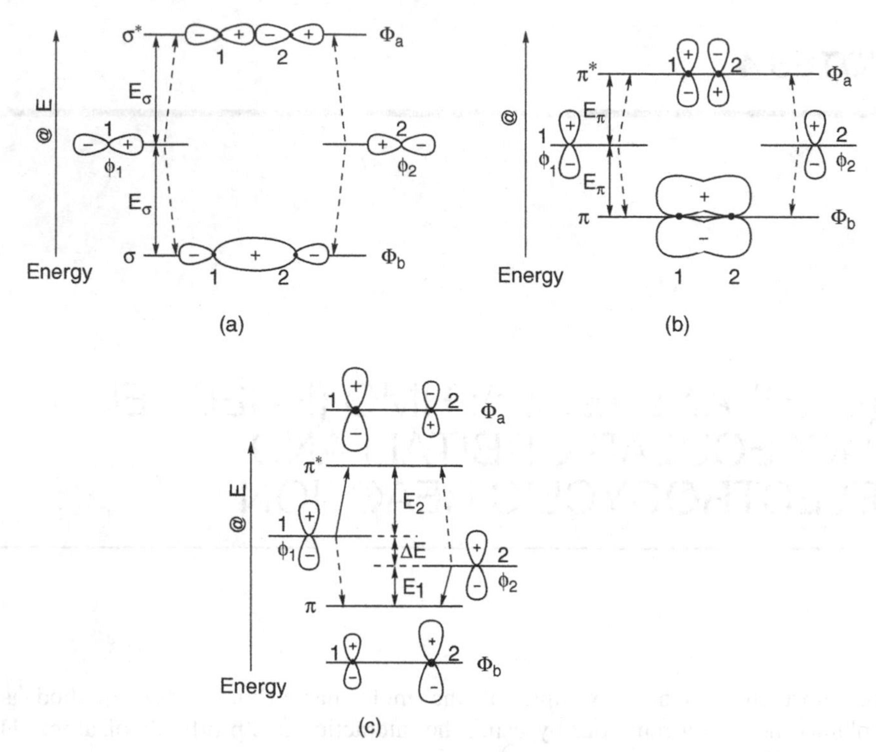

Figure N4.1 Interaction of two 2p orbitals. (a) Interaction of 2p orbitals with axis of symmetry. (b) Interaction of 2p orbitals with plane of symmetry. (c) Interaction of 2p orbitals of different energy with plane of symmetry.

shown as (c) in Fig. N4.1. For bonding orbital (Φb), the contribution of lower energy p orbital ($\phi 2$) is larger, and for antibonding orbital (Φa), that of higher energy p orbital ($\phi 1$) is stronger.

Please do not forget that the antibonding orbital is necessarily formed along with the bonding orbital.

Again, as a reminder and a common knowledge of readers, calculated numerical values of energy (E_i) and atomic coefficient (c_n) of Hueckel Molecular Orbital (HMO) of conjugated systems are listed (Table N4.1) for allyl radical, 1,3-butadiene, 1,3,5-pentadienyl radical, and 1,3, 5-hexatriene. The energy (E_i) is in the β-unit; thus, the energy (E_1) of Φ_1 is the lowest, because the sign of β is minus. The symmetric character of each HMO is apparent from the sign and numerical values of c_i. As an example, relative signs of each 2p atomic orbital are illustrated for 1,3,5-hexatriene in Fig. N4.2. The size of each 2p atomic orbital should be illustrated according to the values (c_i^2) in Table N4.1; however, it is cited equal for simplicity, because the symmetry (relative signs of atomic orbitals) is the most important. In order to clearly show this, the number of nodes and their positions (vertical dotted lines) are illustrated

TABLE N4.1 Energy and Atomic Coefficient of HMO of Conjugate Polyenes

		E_i	c_1	c_2	c_3	c_4	c_5	c_6
Ψ_1	1	1.414	0.500	0.707	0.500			
Ψ_2	2	0.000	0.707	0.000	−0.707			
Ψ_3	3	−1.414	0.500	−0.707	0.500			
Ψ_1	1	1.618	0.372	0.602	0.602	0.372		
Ψ_2	2	0.618	0.602	0.372	−0.372	−0.602		
Ψ_3	3	−0.618	0.602	−0.372	−0.372	0.602		
Ψ_4	4	−1.618	0.372	−0.602	0.602	−0.372		
Ψ_1	1	1.732	0.288	0.500	0.576	0.500	0.288	
Ψ_2	2	1.000	0.500	0.500	0.000	−0.500	−0.500	
Ψ_3	3	0.000	0.576	0.000	−0.576	0.000	0.576	
Ψ_4	4	−1.000	0.500	−0.500	0.000	0.500	−0.500	
Ψ_5	5	−1.732	0.288	−0.500	0.576	−0.500	0.288	
Ψ_1	1	1.802	0.232	0.418	0.521	0.521	0.418	0.232
Ψ_2	2	1.247	0.418	0.521	0.232	−0.232	−0.521	−0.418
Ψ_3	3	0.445	0.521	0.232	−0.418	−0.418	0.232	0.521
Ψ_4	4	−0.445	0.521	−0.232	−0.418	0.418	0.232	−0.521
Ψ_5	5	−1.247	0.418	−0.521	0.232	0.232	−0.521	0.418
Ψ_6	6	−1.802	0.232	−0.418	0.521	−0.521	0.418	−0.232

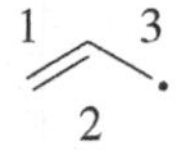

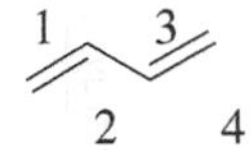

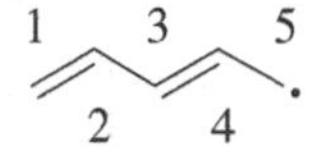

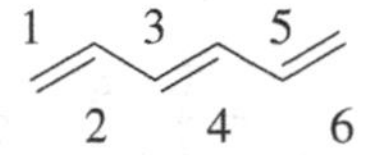

in the Fig. N4.2. Maximum of two electrons can occupy each MO, starting from the lowest energy MO (Φ_1), in which the spin of electrons is opposite.

By the benefit of frontier orbital theory, we need to consider only the HOMO (highest occupied molecular orbital) and LUMO (lowest unoccupied molecular orbital), which control the reaction.

During an electrocyclic reaction, a pericyclic process that involves the cyclization of a conjugated polyene, a bond is formed by the overlap of the two lobes of the same sign at 1, n atoms (head to tail) accompanied by the shift of the π bonds, and there is an equilibrium between the starting compound and the product. Thermal reaction is controlled by HOMO, and photo excited reaction is controlled by LUMO, that is, HOMO of one electron excited state.

For a bond to form, the two orbitals at 1, n (head and tail) positions must rotate so that bonding interaction is achieved by the overlap of the lobes with the same signs. There are two relative directions of rotation, that is, conrotatory and disrotatory, as illustrated in Fig. N4.3. For 1,3,5-hexatriene system, the corresponding lobes at 1,6-positions have the same signs for HOMO and the opposite signs for LUMO. Hence, for bond formation, they should be *disrotatory*

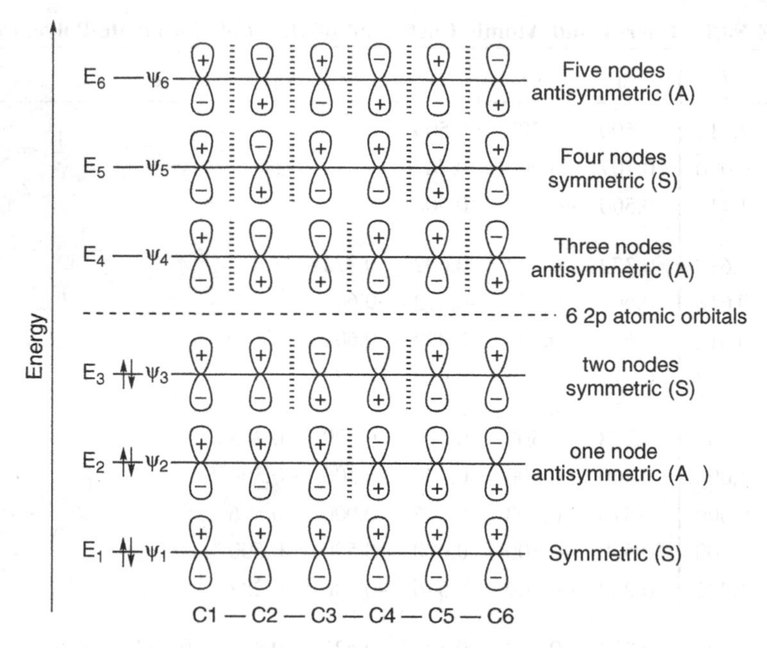

Figure N4.2 HMO of 1,3,5-hexatriene with relative energy, symmetry, and node.

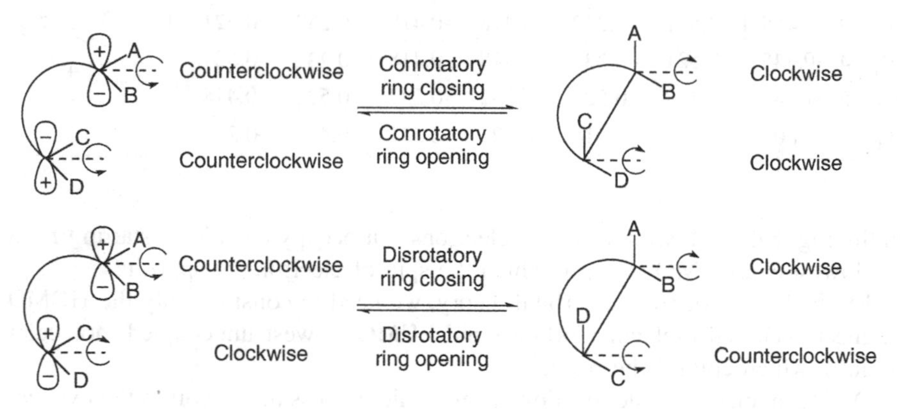

Figure N4.3 Conrotatory and disrotatory ring closure and ring opening of conjugated system.

by thermal reaction and *conrotatory* by photo reaction. For 1,3-butadiene system, HOMO has the opposite signs at 1,4-positions and LUMO has the same signs at 1,4-positions. Thus it is easily understood that thermal reaction is conrotatory and photo reaction is disrotatory.

Experimental results on 1,6-dimethyl-1,3,5-hexatriene system, that is, compounds **1** and **3**, clearly demonstrate that the above rationalization is truly the case for thermal and photo chemical reactions, as are illustrated in Fig. N4.4. We notice here that there are two possibilities for each conrotatory and disrotatory reaction, which cannot be controlled by HOMO or LUMO. If designed to control, this leads to asymmetric reactions.

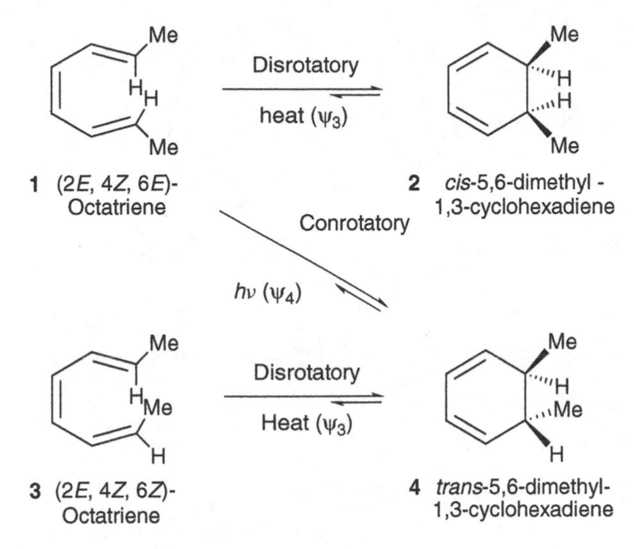

Figure N4.4 Electrocyclic reactions of 1,3,5-hexatriene systems.

CHAPTER 3

LITHIUM, MAGNESIUM, AND COPPER COMPOUNDS

Group / (n) / Period	18 0	1 1A	2 2A	12 2B
1 1s	(1.28)°2372 $^{4}_{2}$**He** Helium [5.50]	(0.37(H-H))1312 (0.32(C-H)) $^{1}_{1}$**H** ← Hydrogen 2.1 435	$(0.66 \times 10^{5+})$ (1.54 −)	
2 [He] 2s2p	2081 $^{20}_{10}$**Ne** Neon [4.84]	1.23 513 (0.78^{+}) $^{7}_{3}$**Li** Lithium 1.0 248	0.89 899 (0.34^{2+}) $^{9}_{4}$**Be** Beryllium 1.5	
3 [Ne] 3s3p	(1.74)°1520 $^{40}_{18}$**Ar** Argon [3.20]	$(1.54)^{\square}$ 496 (0.98^{+}) $^{23}_{11}$**Na** Sodium 0.9	1.36 738 (0.79^{2+}) $^{24}_{12}$**Mg** Magnesium 1.2	
4 $[Ar:3d^{10}]$ 4s4p	1.89 1351 (1.69^{+}) $^{84}_{36}$**Kr** Krypton [2.94]	2.03 419 (1.33^{+}) $^{39}_{19}$**K** Potassium 0.8	1.74 590 (1.06^{2+}) $^{40}_{20}$**Ca** Calcium 1.0	1.25 906 (0.83^{2+}) $^{64}_{30}$**Zn** Zinc 1.6 176
5 $[Kr:4d^{10}]$ 5s5p	2.09 1170 (1.90^{+}) $^{132}_{54}$**Xe** Xenon [2.40]	$(2.48)^{\square}$ 403 (1.49^{+}) $^{85}_{37}$**Rb** Rubidium 0.8	1.92 550 (1.27^{2+}) $^{88}_{38}$**Sr** Strontium 1.0	1.41 867 (1.03^{2+}) $^{114}_{48}$**Cd** Cadmium 1.7 139
6 $[Xe:4f_{14}5d^{10}]$ 6s6p	1040 $^{222}_{86}$**Rn*** Radon [2.06]	2.35 376 (1.65^{+}) $^{133}_{55}$**Cs** Caesium 0.7	1.98 503 (1.43^{2+}) $^{138}_{56}$**Ba** Barium 0.9	1.44 1007 (1.12^{2+}) $^{202}_{80}$**Hg** Mercury 1.9 122

Organo Main Group Chemistry, First Edition. Kin-ya Akiba.

3.1 SYNTHESIS

The fundamental synthetic method for the preparation of alkyllithium compounds involves the reaction of alkyl chloride or alkyl bromide with metallic lithium in diethyl ether or hexane, in which case equimolar amount of lithium halide is formed (Eq. 3.1). The reaction is greatly accelerated by irradiation with ultrasonic wave. Metallic lithium is commercially available as a wire or rod dipped in paraffin. Butyl- and *t*-butyllithium are also commercially available as hexane solutions (ca. 1 M) and are generally used for the synthesis of organolithiums by halogen-lithium exchange with other organohalides (Eq. 3.2). On the other hand, when alkyllithium is prepared from its iodide, the resulting alkyllithium reacts quickly with the remaining iodide to afford the corresponding coupling product (Eq. 3.3).

$$\text{R–X} + 2\text{Li} \longrightarrow \text{R–Li} + \text{LiX} \quad (\text{X} = \text{Cl, Br}) \tag{3.1}$$

$$\text{R–X} + \text{Bu–Li} \longrightarrow \text{R–Li} + \text{BuX} \tag{3.2}$$

$$\text{R–I} + \text{R–Li} \longrightarrow \text{R–R} + \text{LiI} \tag{3.3}$$

Aromatic lithium compounds are synthesized according to Eq. 3.1. Also, they are prepared by halogen-lithium exchange with *t*-butyllithium. The resulting *t*-butyl halide is converted to 2-methylpropene by the second mole of *t*-butyllithium. Butyllithium is also applicable but *t*-butyllithium is better for this.

When an aromatic ring bears a functional group, such as amide, sulfonylamide, or methoxy, butyllithium first coordinates with an unshared electron pair of the functional group and then deprotonates an *ortho*-proton; thus, *ortho*-lithiation takes place cleanly at low temperature. *ortho*-Lithiation also occurs at the adjacent hydrogen of the heteroatom of heteroaromatic compounds. Thus, *ortho*-lithiation is a general method to introduce functional groups to aromatic rings.

$$\text{C}_6\text{H}_5\text{–X} \xrightarrow{2\text{ Li or } t\text{-BuLi}} \text{C}_6\text{H}_5\text{–Li} + \text{LiX (X=Cl, Br, I) or } t\text{-BuX} \tag{3.4}$$

Resulting *t*-BuX is converted to 2-methylpropene by *t*-BuLi

$$\text{C}_6\text{H}_5\text{–X} \xrightarrow[\text{Hexane}]{n\text{-BuLi}} o\text{-Li–C}_6\text{H}_4\text{–X} \quad (\text{X} = \text{CONR}_2, \text{OMe}, \text{SO}_2\text{NR}_2)$$

$$\text{C}_4\text{H}_4\text{X} \xrightarrow[\text{Hexane}]{n\text{-BuLi}} \text{2-Li–C}_4\text{H}_3\text{X} \quad (\text{X} = \text{O, S, NMe}) \tag{3.5}$$

Stereochemistry of double bond is retained by lithiation of alkenyl halide with lithium metal or alkyllithium. This is confirmed by the exchange of metallic substituents such as tributyltin with alkyllithium (transmetallation).

$$\text{(Z)-RCH=CHX},\ \text{(E)-RCH=CHX}\ (X = Cl, Br) \xrightarrow{\text{Li or } t\text{-BuLi}} \text{(Z)-RCH=CHLi},\ \text{(E)-RCH=CHLi} \tag{3.6}$$

$$\text{RCH=CHSnBu}_3 \xrightarrow{n\text{-BuLi}} \text{RCH=CHLi} + \text{Bu}_4\text{Sn} \tag{3.7}$$

Organomagnesium halide has been known as *Grignard reagent* and is generally used in organic synthesis. Grignard reagent is prepared by the reaction of magnesium (commercially available as powder or ribbon) with alkyl halide in diethyl ether or tetrahydrofuran (THF). The reaction is greatly accelerated by the addition of a small amount of iodine or by irradiation with ultrasonic wave. Grignard reagent is also prepared from alkenyl and aryl halide by the same procedure as above. In the case of alkenyl halide, stereochemistry is retained as in lithiation.

$$\text{R–X} + \text{Mg} \longrightarrow \text{RMgX} \quad (X = Cl, Br, I) \tag{3.8}$$

$$\text{RCH=CHBr},\ \text{PhBr} \xrightarrow{\text{Mg}} \text{RCH=CHMgBr},\ \text{PhMgBr} \tag{3.9}$$

Because of higher acidity of alkynyl hydrogen, it is easily metalized by organolithium or Grignard reagent (metal-hydrogen exchange). For this exchange, butyllithium and methylmagnesium bromide are conveniently used.

$$\text{R–C≡C–H} \xrightarrow{n\text{-BuLi}} \text{R–C≡C–Li}; \quad \text{R–C≡C–H} \xrightarrow{\text{MeMgBr}} \text{R–C≡C–MgBr} \tag{3.10}$$

Benzyllithium and benzylmagnesium bromide should be quite useful for the introduction of a benzyl group. However, we cannot prepare them in solution by the conventional method, because they react with the remaining halide quickly to give bibenzyl, that is, coupling product. Instead, benzylpotassium is generated in solution (hexane or benzene) from an equimolar mixture of butyllithium, toluene, and potassium *t*-butoxide. The resulting lithium *t*-butoxide is very low in reactivity (basicity) and it does not affect succeeding synthetic reactions.

$$PhCH_3 + n\text{-}BuLi + t\text{-}BuOK \longrightarrow PhCH_2K + t\text{-}BuOLi + n\text{-}BuH \quad (3.11)$$

Ate complexes of alkyllithiums with copper(I) show complex structures depending on the ratio of the two reagents. They are soft nucleophiles compared to alkyllithiums and are interesting reagents for conjugate addition and substitution of halogens.

$$\begin{aligned} RLi + Cu(I)X &\longrightarrow [RCu]_n + nLiX \\ 3RLi + Cu(I)X &\longrightarrow [R_3CuLi_2] + LiX \\ 2RLi + Cu(I)X &\longrightarrow [R_2CuLi] + LiX \end{aligned} \quad (3.12)$$

3.2 STRUCTURE

Organolithiums are associated usually as 4–6 mers in solution. The number of associations is high in nonpolar solvents (hexane, benzene, etc.) and is low in polar solvents (diethyl ether, THF, etc.). Tetramer of methyllithium was crystallized and the structure was determined as strained cube by X-ray analysis (Fig. 3.1).

Bond distance of carbon and lithium is 2.31 Å and that of lithium-lithium is 2.68 Å. There is no bond between two lithiums, because the coupling of ^7Li cannot be observed by NMR. Dotted wedge lines at the lithium of Fig. 3.1 show that the lithium associates with another cube of methyllithium with a distance of 2.36 Å. The distance is close to that of carbon and lithium in the cube, which means cubes associate with each other as polymer in solid state. In solution, the lithium associates with a solvent in case the cube is retained.

Methyllithium exists as a dimmer associated with solvents in diethyl ether and in TFH and it is further in equilibrium with solvated monomer. Therefore, there

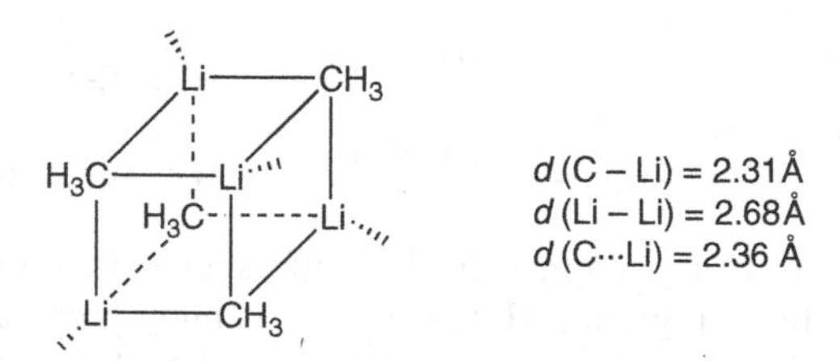

Figure 3.1 Crystal structure of methyllithium.

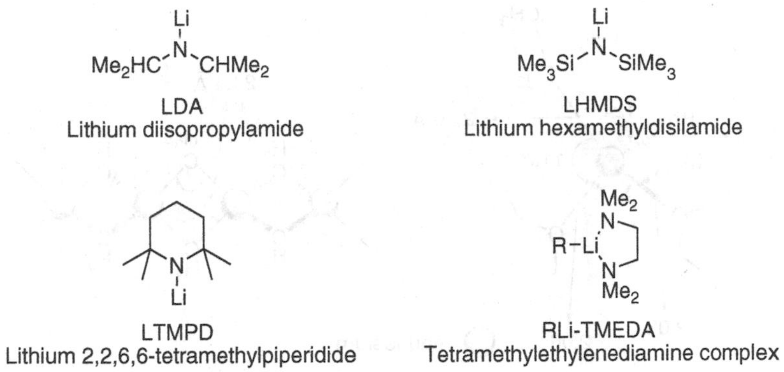

Me₂HC–N(Li)–CHMe₂
LDA
Lithium diisopropylamide

Me₃Si–N(Li)–SiMe₃
LHMDS
Lithium hexamethyldisilamide

LTMPD
Lithium 2,2,6,6-tetramethylpiperidide

RLi-TMEDA
Tetramethylethylenediamine complex

Figure 3.2 Lithium amides and TMEDA complex.

are several kinds of equilibria for alkyllithiums in solution. However, reactions of alkyllithiums with electrophiles are controlled by a solvated monomer, which is the most reactive. A small amount (ca. 0.1 equivalent) of tetramethylethylenediamine (TMEDA) is used to dissociate alkyllithiums in solution, forming a complex (RLi-TMEDA) of monomer as shown in Fig. 3.2. The complex, which is in equilibrium in solution, is used as a reactive reagent for synthesis.

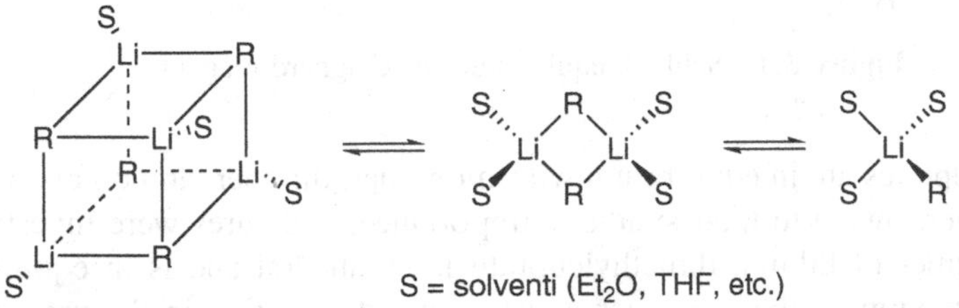

S = solventi (Et_2O, THF, etc.) (3.13)

Strong base without (with very low) nucleophilicity is necessary and useful in organic chemistry. For this purpose, butyllithium is used to deprotonate diisopropylamine, hexamethyldisilylamine, and tetramethylpiperidine, where sterically hindered, nonnucleophilic amides are generated as strong bases such as LDA, LHMDS, and LTMPD in solution. Even these sterically hindered bases mostly exist as dimmers and are in equilibrium with monomer in solution.

Structure of Grignard reagent is well known. The crystal structures of ethylmagnesium bromide and dimethylmagnesium are shown in Figure 3.3. The magnesium is fundamentally in sp^3 hybridization and magnesium-carbon-magnesium bond of the latter is 3-center 2-electron bond which is electron deficient.

Grignard reagent is in equilibrium with several kinds of chemical species in solution and the equilibrium is known as *Schlenk equilibrium* (Fig. 3.4). Dimethylmagnesium is one of the species in equilibrium, but it was isolated and crystallized. Monomeric reagent (RMgX) is the most reactive, and it is the actual species for synthesis and reaction.

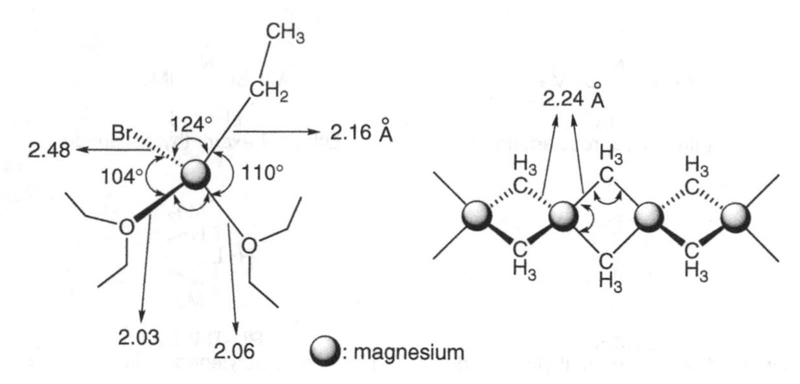

Figure 3.3 Crystal structures of ethylmagnesium bromide and dimethylmagnesium.

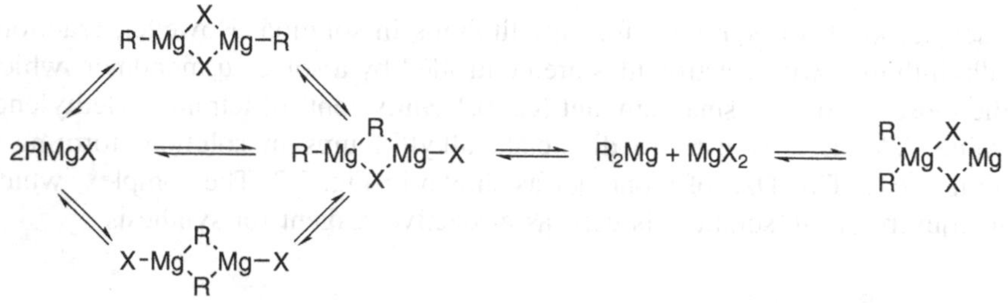

Figure 3.4 Schlenk equilibrium of Grignard reagent.

Organocuprates are in equilibrium with monomer, dimmer, and others in solution. However, owing to their synthetic importance, structures were investigated in detail. Dimer of lithium dimethylcuprate is tetrahedral and is in equilibrium with linear monomer; the monomer is especially dominating in the presence of lithium iodide (Fig. 3.5) [1].

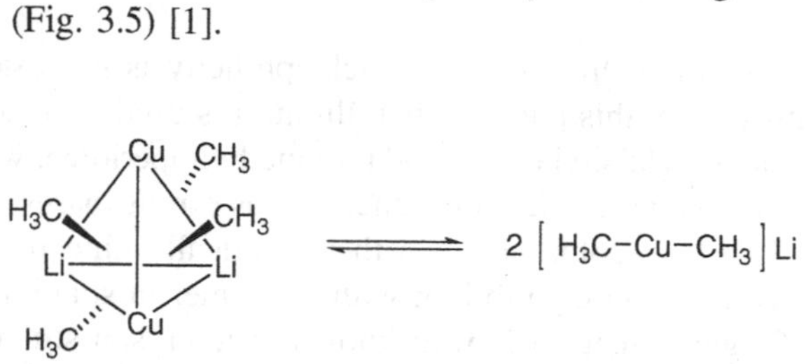

Figure 3.5 Structure of lithium dimethylcuprate: the dimer is tetrahedral and the monomer is linear.

3.3 REACTION

Fundamental reactions of organolithium and Grignard reagent are (i) deprotonation as a base and (ii) reaction as a nucleophile. The ratio of the two types of reactions depends on each reagent and also on the substance to be reacted, but they are usually competitive.

3.3.1 Deprotonation as Base

Organolithium and Grignard reagent are carbo-anion (carbanion) and hence they deprotonate active hydrogens (N–H, O–H, etc). For deprotonation of C–H bond, a variety of reagents shown in Figs. 3.2 and 3.3, and NaH and KH are used depending on the reactivity of C–H bond. There are interesting and useful reactions of enolate generated by deprotonation of α-hydrogen of a carbonyl group, either kinetically or thermodynamically.

Silyl enol ethers (**1** and **2**) of 2-methylcyclohexanone are obtained by trapping the corresponding enolates generated by deprotonation of the ketone with LDA in THF at a low temperature (path a: kinetic method, strong base with large steric hindrance) and by deprotonation with triethylamine in THF at ambient temperature (path b: thermodynamic method, weak base). The ratio of **1** and **2** for method (a) was 99:1 and that for method (b) was 22:78 (Eq. 3.14).

OSiMe$_3$ + OSiMe$_3$

1 **2**

THF; a: LDA, Me$_3$SiCl → 99:1; b: Et$_3$N, Me$_3$SiCl → 22:78 (3.14)

Deprotonation of 2-pentanone gave enolates **3** and **4**. The ratio was 85:15 with LDA (path a: kinetic), and it was 10:90 with lithium alkoxide in t-amyl alcohol (path b: thermodynamic) (Eq. 3.15). Less branched enolate becomes the main product under the kinetic conditions and higher branched enolate is obtained as the major product under the thermodynamic conditions.

OLi + OLi

3 **4**

a: LDA, THF → 85:15; b: t-C$_5$H$_{11}$OLi, t-C$_5$H$_{11}$OH → 10:90 (3.15)

Among reactions of enolates, aldol reaction is the most important. Stereochemistry of C–C bond formation in aldol reaction is controlled by steric repulsion of 1,3-substituents (repulsion between R^2 and H, and not that between R^2 and R^1) of pseudo-cyclohexane ring of transition state (**TS-a** and **TS-b**), which is formed by coordination of lithium (metal) ion with two oxygen atoms (Eq. 3.16). This is experimentally exemplified by the fact that *anti*-adduct is obtained as the major product from *E*-enolate; on the other hand, *syn*-adduct is obtained as the major product from *Z*-enolate. Stereoselectivity is enhanced by the addition of magnesium chloride (bromide) or zinc dichloride. By the addition of metal salt, magnesium or zinc enolate is formed, which is chelated with oxygens much stronger than lithium enolate.

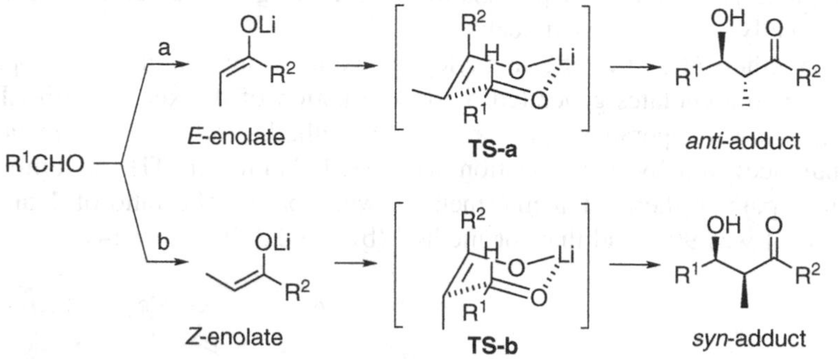

(3.16)

ortho-Lithiation is unique and useful as another type of deprotonation by organolithium. An organolithium coordinates with an unshared electron pair of a functional group and then lithiates an *ortho*-proton of the benzene ring quantitatively. This lithiation supplements halogen–lithium exchange, but it is more useful as a synthetic method because it does not require any halogen in a starting material. The rate of halogen–lithium exchange, however, proceeds much faster at lower temperatures than *ortho*-lithiation.

Butyllithium coordinates with benzyldimethylamine and deprotonates an *ortho*-proton of the benzene ring to yield *ortho*-lithiated amine, and an electrophile is introduced to the ortho position (Eq. 3.17). In contrast, phenylsodium deprotonates a benzyl proton and an electrophile is introduced at the benzyl position.

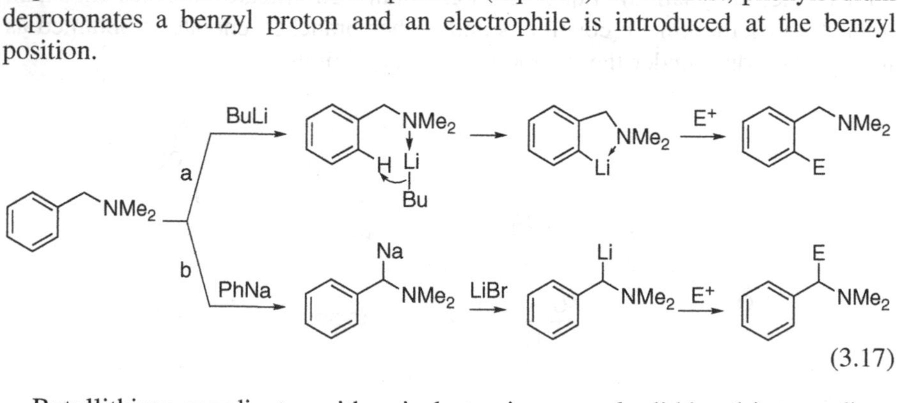

(3.17)

Butyllithium coordinates with anisole to give an *ortho*-lithiated intermediate, to which an electrophile is introduced to yield *ortho*-substituted anisole

(Eq. 3.18). The amide group is the strongest to coordinate with organolithiums and the oxalyl, sulfonyl, amino, and methoxy groups are also effective for coordination [2].

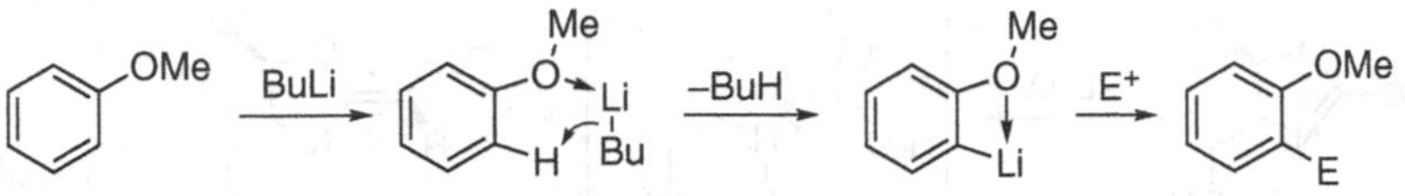

(3.18)

ortho-Lithiation has been used effectively for synthesis. In the presence of both amide and methoxy groups on a benzene ring, *ortho* position to the amide group is lithiated selectively. The formyl and methyl groups thus introduced are converted to yield a five- or six-member ring (Eq. 3.19).

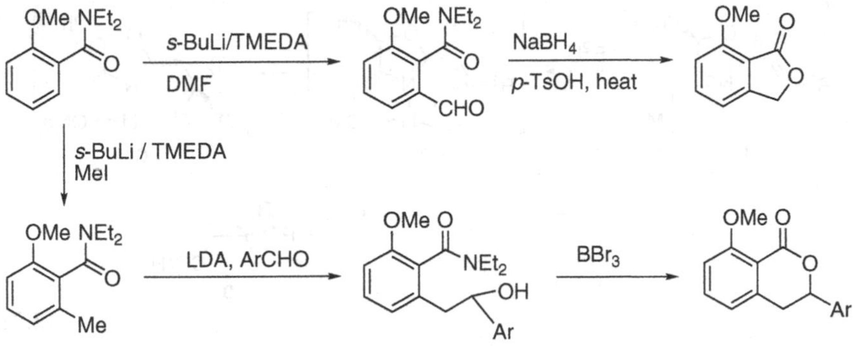

(3.19)

By the reaction of aminoalcohol (commercially available) and propionic acid, oxazoline (**5**) is prepared. Deprotonation of the methylene group of **5** with LDA generates methyleneoxazoline (**A**) as an intermediate, where a pseudo-five-member ring is fused by the coordination of a lithium ion with the methoxy group. Among the two possible stereoisomers, **A** and **B, A** is generated predominantly (**B** is generated only in a small percent) because of smaller steric hindrance between the phenyl and the methyl groups during the deprotonation (Eq. 3.20). Alkylating reagent RX approaches from the upper side of methyleneoxazoline (**A**) to give **6**; thus the reaction is asymmetric. Asymmetric carboxylic acid (**7**) bearing an alkyl group at the α-position is obtained by hydrolysis. The asymmetric yield is as high as 78–86%, depending on the kind of R [3].

On applying the above reaction to a conjugated compound (**8**), conjugated addition of organolithium (R^2Li) takes place through intermediate **C** to give asymmetric carboxylic acid (**9**) having a methylene group between the asymmetric carbon and the carboxylic acid (Eq. 3.21). Surprisingly, the asymmetric yield is as high as 92–99% [4].

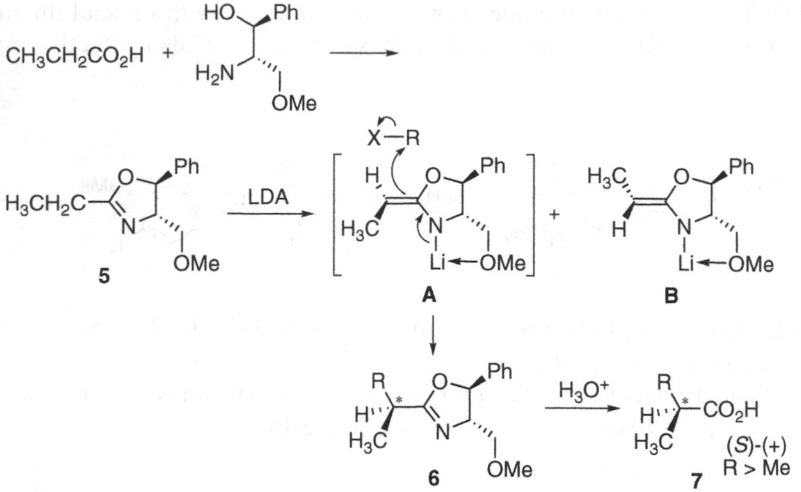

(3.20)

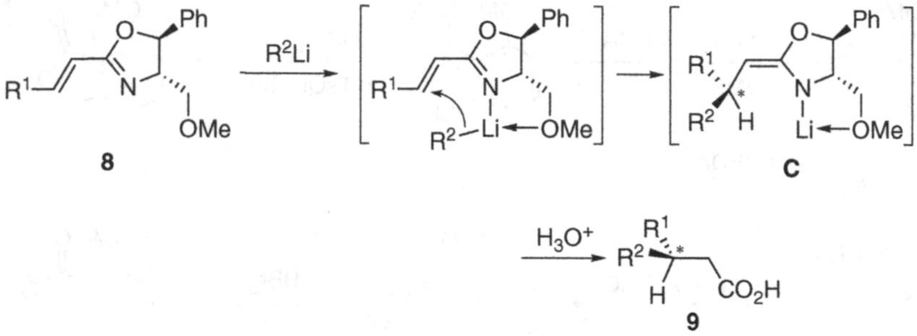

(3.21)

On the basis of the research of Meyers group started in 1975, it was disclosed that asymmetric synthesis can be achieved by the use of easily accessible fused cyclic intermediates, which are generated by the coordination of a lithium ion. Until these fascinating results, asymmetric synthesis had been believed as unique and special phenomenon to bioorganism. These results induced quite active researches on asymmetric synthesis. Other than lithium ion, metallic ions of magnesium, aluminum, and zinc are used to coordinate with unshared electron pair (s) of heteroatoms in order to control stereochemistry of a variety of reactions.

3.3.2 Nucleophilic Reaction

The most fundamental reaction of Grignard reagent and organolithium is nucleophilic addition to carbonyl group. Metal ions of both reagents first coordinate to an unshared electron pair of a carbonyl group and then activate the group by forming a complex. The second reagent attacks the complex generating a six-member transition state (**D**) to yield tertiary alcohol (**10**) (Eq. 3.22). The recovered reagent in this addition process is used in the next reaction. For alkyllithiums, a similar mechanism assuming six-member transition state (**E**) is written to afford **11** (Eq. 3.23). Association of reagents and complexation of solvents to reagents are very complex for each reagent; however, it can be taken for granted that these do not affect the reaction pattern of reagents. Stereochemistry of product alcohols of Eqs. 3.22 and 3.23 is explicitly written as asymmetric for clarity; however it is apparent that these are not at all asymmetric reactions.

RMgX + $R^1C(O)R^2$ ⇌ [RMg(X)···O=CR^1R^2] —RMgX→ [**D**]‡ → RMgX + XMgO–C(R)(R^1)(R^2) **10** (3.22)

$(RLi \cdot S)_2$ + $R^1C(O)R^2$ → → → [**E**]‡ → R–Li·S + S·Li–O–C(R)(R^1)(R^2) **11** (3.23)

S = solvent

As a typical example of Grignard reaction, synthesis of 1,1-diphenylethylene is shown in Eq. 3.24. Addition of phenylmagnesium bromide to ethyl acetate gives acetophenone and the second addition of the Grignard reagent affords diphenylmethylcarbinol almost quantitatively. Dehydration of the carbinol by 20% sulfuric acid gives 1,1-diphenylethylene in 67–70% yield.

However, the fundamental reaction of addition of Grignard reagent to carbonyl group is not so simple and is usually accompanied by reduction. Addition of Grignard reagents to benzophenone is scrutinized and the yields are shown for **12** and **13** in parentheses in Table 3.1. For primary alkyl Grignard reagents, the major product (**13**) is due to reduction, surprisingly. The yield of addition products (**12**) is greatly enhanced in the presence of 10 mol% of $ZnCl_2$.

TABLE 3.1 Alkylation of Benzophenone Catalyzed by Zinc Chloride

R	Yield (%) of **12**[a]	Yield (%) of **13**[a]
Me	94 (91)	0 (0)
Et	84 (25)	15 (72)
pr	71 (14)	29 (86)
i-Pr	75 (62)	10 (14)
Bu	74 (11)	24 (81)
$CH_2{=}CHCH_2$	>99 (>99)	0 (0)

[a]Yields(%) without $ZnCl_2$ are shown in parentheses.

$$PhBr + Mg \xrightarrow[Et_2O]{} PhMgBr$$

$$2\,PhMgBr + CH_3CO_2C_2H_5 \longrightarrow CH_3(Ph)_2COH \xrightarrow{20\%\ H_2SO_4} Ph_2C{=}CH_2 \tag{3.24}$$

Yield 67–70%

$$PhC(O)Ph + \underset{(1.3\ eq.)}{RMgCl} \xrightarrow[THF,\ 0°C]{ZnCl_2\ (10\ mol\%)} \underset{\mathbf{12}}{Ph_2C(OH)R} + \underset{\mathbf{13}}{Ph_2C(OH)H} \tag{3.25}$$

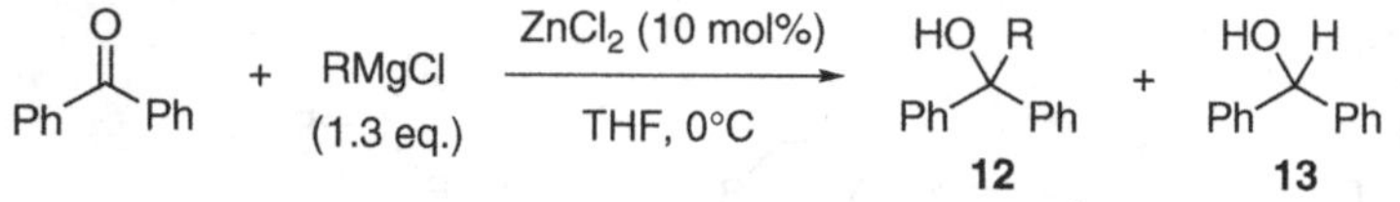

Catalytic cycle of Grignard reaction in the presence of $ZnCl_2$ is shown in Fig. 3.6 [5]. It is conjectured that dialkylzinc (**F**) is generated first to react with Grignard reagent to afford active Zn(II)ate complex (**G**). The complex (**G**) reacts with a carbonyl compound through cyclohexane-type transition state (**H**) to give tertiary alcohol (**I**) and dialkylzinc (**F**), thus forming a catalytic cycle.

Mechanism of Grignard reaction with benzophenone is elucidated to proceed through coupling reaction in cage (**K**), which is initiated by one electron transfer

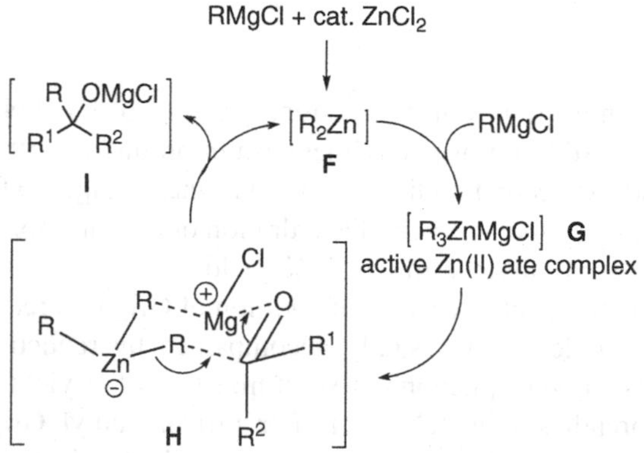

Figure 3.6 Catalytic cycle of Grignard reaction in the presence of $ZnCl_2$.

from Grignard reagent to benzophenone (**J**). Ketyl radical intermediate (**K**) was trapped in a mass spectrometer recently [6].

RMgX + O=C(Ph)Ph —electron transfer→ [**J**] → [**K**]‡ → XMgO–C(R)(Ph)Ph (3.26)

Reaction of Grignard reagent or organolithium with carbonyl compound bearing α-hydrogen proceeds by competition between deprotonation affording enolate and nucleophilic addition affording alcohol. Deprotonation is the major path for highly ionized organometals (R–M; M = K > Na > Li); on the other hand, nucleophilic addition becomes the major path for Grignard reagent. This is kinetic result. Effect of metallic ion is prominent in the reaction of acetophenone; thus phenylpotassium deprotonates (100%) and its Grignard reagent adds nucleophilically (100%).

PhC(O)CH$_3$ + Ph–M → PhC(OM)=CH$_2$ + Ph$_2$C(OM)CH$_3$ (3.27)

M =	
K	100:0
Na	94:6
Li	84:16
MgBr	0:100

Competition between C-alkylation and O-alkylation takes place by the reaction of enolate and electrophile. C-alkylation is preferred by soft electrophile (e.g., R–X; X = I > Br > Cl) and O-alkylation becomes the major path for hard nucleophile (e.g., R–Cl < Me$_3$SiCl < Me$_3$SiOTf).

Enolate of propiophenone (**14**) reacts with pentyl halides in DMSO to give O-alkyl product preferably with chloride and C-alkyl product as the major one with iodide (Eq. 3.28). A similar selectivity is also observed for the reaction of enolate of ethyl acetoacetate (**15**) with butyl halides (Eq. 3.29). C-alkylation is more preferred for **15** than for **14**, because **15** is a softer carbanion than **14**. These are rationalized by soft–hard concept.

14 + n-C$_5$H$_{11}$X —DMSO, NaH→ O-(n-C$_5$H$_{11}$) enol ether + C-(n-C$_5$H$_{11}$) ketone (3.28)

X =	
Cl	1.2:1.0
Br	0.64:1.0
I	0.23:1.0

$$\text{Me-C(OK)=CH-CO}_2\text{Et } (\mathbf{15}) + n\text{-C}_4\text{H}_9\text{X} \xrightarrow[\text{K}_2\text{CO}_3]{\text{DMF}} \text{Me-C(O-}n\text{-C}_4\text{H}_9\text{)=CH-CO}_2\text{Et} + \text{MeCO-CH(}n\text{-C}_4\text{H}_9\text{)-CO}_2\text{Et}$$

X = Cl 0.85:1.0
Br 0.49:1.0
I 0.01:1.0

(3.29)

3.3.3 Conjugate Addition of Lithium Dimethylcuprate

In the presence of Cu(I), organolithiums afford a couple of cuprates, depending on the ratio of the two reagents. Dialkylcuprate is a typical example. Dialkylcuprate reacts with alkyl halide to afford an intermediate (**L**) by oxidative addition and yields a coupling product (R–R′) by reductive elimination. R′–R′ is not obtained at all. Dimethylcuprate can substitute methyl for bromine in good yield (Eq. 3.30).

$$\text{R-X} + \text{R}'_2\text{Cu} \xrightarrow{\text{Oxidative addition}} \left[\text{R-Cu}^{\text{III}}(\text{R}')_2\text{-X}\right] (\mathbf{L}) \xrightarrow{\text{Reductive elimination}} \text{R-R}' + \text{R}'\text{CuX} \quad \text{R}'\text{-R}' \text{ (none)}$$

(3.30)

$$\text{dibromocyclooctatetraene (Br, Br)} + \text{Me}_2\text{CuLi} \longrightarrow \text{dimethylcyclooctatetraene (Me, Me)}$$

(Me)$_2$CuLi:

a: 3-methylcyclohex-2-enone → 3,3-dimethylcyclohexanone

b: PhCH=CHBr → PhCH=CHMe

c: H_2C=C(Me)–CH(R)–OCOMe → Me–CH$_2$–C(Me)=CH–R

d: Me-substituted cyclohexenyl acetate (OCOMe) → dimethylcyclohexene (Me, Me)

(3.31)

Usually, organolithium compounds react at the carbonyl group to give 1,2-addition products in conjugated systems. However, dialkylcuprates commonly give conjugate addition products (Eq. 3.31). Path a shows conjugate addition and path b is substitution of vinyl bromide. Path c is an example of conjugate addition–elimination and path d illustrates that the conjugate addition–elimination is stereospecific, in which the addition of the cuprate occurs from anti-direction to the leaving group in the allylic system. There are ample examples for these reactions mentioned here [7].

REFERENCES

1. (a) Pearson RG, Gregory CD. J Am Chem Soc 1976;98:4098; (b) Lipshutz BH, Kozlowski JA, Breneman CM. J Am Chem Soc 1985;107:3197; (c) Gerold A, Jastrozebski JTBH, Kronenburg CMP, Krause N, van Koten G. Angew Chem Int Ed 1997;36:755.
2. Beak P, Snieckus V. Acc Chem Res 1982;15:306.
3. Meyers AI. Acc Chem Res 1978;11:375.
4. Meyers AI, Whitter CE. J Am Chem Soc 1975;97:6266.
5. Hatano M, Ishihara K. Synthesis 2008:1647.
6. C & EN News, O'Hair RAJ, et al. Angew Chem Int Ed 2008;47:9118.
7. Carey FA, Sundberg RJ. Advanced organic chemistry: Part B reactions and synthesis. 5th ed. Springer, New York; 2007, Chapter 8.1.

CHAPTER 4

BORON AND ALUMINUM COMPOUNDS

Group (*n*) Period	12 2B	13 3B	14 4B
1 1s			
2 [He] 2s2p		0.88 801 $^{11}_{5}$**B** Boron 2.0 372	0.77 1086 $^{12}_{6}$**C** Carbon 2.5 368
3 [Ne] 3s3p		1.25 577 (0.57^{3+}) $^{27}_{13}$**Al** Aluminium 1.5 255	1.17 787 $^{28}_{14}$**Si** Silicon 1.8 301
4 [Ar:3d^{10}] 4s4p	1.25 906 (0.83^{2+}) $^{64}_{30}$**Zn** Zinc 1.6 176	1.25 579 (0.83^{3+}) $^{69}_{31}$**Ga** Gallium 1.6 247	1.22 762 $^{74}_{32}$**Ge** Germanium 1.8 237
5 [Kr:4d^{10}] 5s5p	1.41 867 (1.03^{2+}) $^{114}_{48}$**Cd** Cadmium 1.7 139	1.50 558 (0.92^{3+}) $^{116}_{49}$**In** Indium 1.7 165	1.40 709 $^{120}_{50}$**Sn** Tin 1.8 225
6 [Xe:4f$_{14}$5d^{10}] 6s6p	1.44 1007 (1.12^{2+}) $^{202}_{80}$**Hg** Mercury 1.9 122	1.55 589 (1.05^{3+}) $^{205}_{81}$**Tl** Thallium 1.8 125	1.54 716 $^{208}_{82}$**Pb** Lead 1.8 130

Organo Main Group Chemistry, First Edition. Kin-ya Akiba.

4.1 SYNTHESIS

Organoboranes are easily prepared by the reaction of boron trihalide and organolithium or Grignard reagent (transmetalation) (Eq. 4.1). Each reaction (Eq. 4.1a–c) proceeds quantitatively to give a single product according to the molar ratio of reagents used, in which any mixture of poly substituted organoboranes is not formed. Based on the excellent characteristics of organoboranes, hydroboration and a variety of organoboron reagents are used in organic synthesis quite effectively.

$$\begin{array}{l} BX_3 + 3RM \xrightarrow{a} BR_3 + 3MX \\ (X = \text{Halogen, alkoxy; } M = \text{Li, MgX}) \\ BX_3 + 2RM \xrightarrow{b} R_2BX + 2MX \\ BX_3 + RM \xrightarrow{c} RBX_2 + MX \end{array} \quad (4.1)$$

Borane (BH_3) is generated by the reduction of the boron–halogen bond (B–X bond) with sodium borohydride ($NaBH_4$) in solution and exists as a dimer (diborane, B_2H_6) in solution under ambient conditions (Eq. 4.2). Borane forms stable monomer complexes with strongly coordinating solvents (reagents) such as trimethylamine, diethyl ether, and dimethyl sulfide. Hybridization of boron becomes sp^3 from sp^2 by forming a coordination bond.

$$\begin{array}{c} 3NaBH_4 + 4BF_3{\cdot}Et_2O \longrightarrow 2(BH_3)_2 + 3NaBF_4 + 4Et_2O \\ H_3B \leftarrow NMe_3 \qquad H_3B \leftarrow OEt_2 \qquad H_3B \leftarrow SMe_2 \end{array} \quad (4.2)$$

The most important reaction of borane (B–H bond) is the cis-type addition to carbon–carbon double bond, that is, hydroboration. The reaction proceeds through transition state (**A**), where boron resides on a carbon with less substituent(s), that is, with more hydrogens, thus forming boron–carbon and hydrogen–carbon bonds (Eq. 4.3). This reaction proceeds stepwise according to the number of B–H bonds. Using easily available unsaturated hydrocarbons, useful and characteristic hydroboration reagents have been prepared [1, 2].

In Fig. 4.1, a variety of hydroboration reagents are shown, and it should be easy to write structures of alkenes used for the preparation (most of them are commercially available). By taking a look at the structures of hydroboration reagents, it is apparent that steric effect should be an important factor for stability, synthesis, and reactivity of the reagents. Hence, it is also clearly understood that asymmetric reagents of diisopinocampheylborane and (*R*,*R*)-1-bora-2,5-dimethylcyclopentane afford asymmetric hydroboration products.

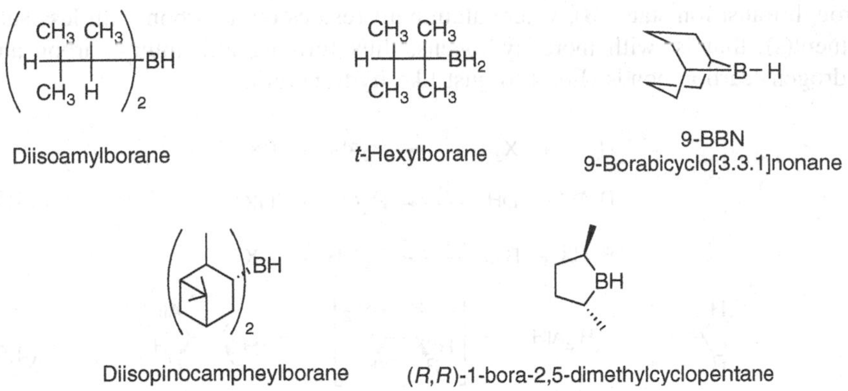

Diisoamylborane

t-Hexylborane

9-BBN
9-Borabicyclo[3.3.1]nonane

Diisopinocampheylborane

(*R*,*R*)-1-bora-2,5-dimethylcyclopentane

Figure 4.1 Organoborane reagents prepared by hydroboration.

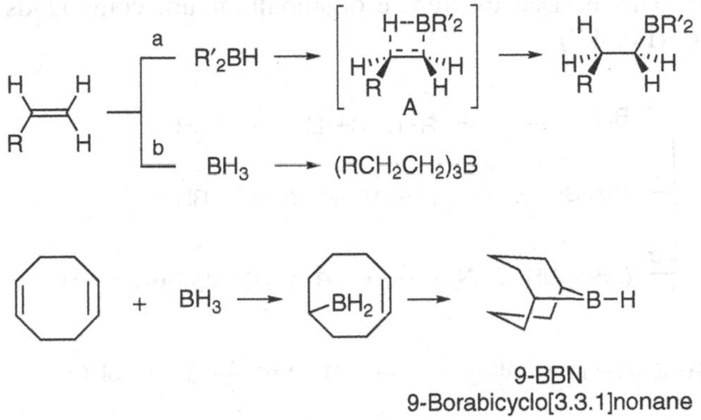

$$RCH=CH_2 \xrightarrow{a:\ R'_2BH} [A] \longrightarrow RCH(H)-CH_2BR'_2 \qquad RCH=CH_2 \xrightarrow{b:\ BH_3} (RCH_2CH_2)_3B \tag{4.3}$$

$$\text{cyclooctadiene} + BH_3 \longrightarrow \text{cyclooctyl-}BH_2 \longrightarrow \text{9-BBN (9-Borabicyclo[3.3.1]nonane)}$$

Organoaluminums are synthesized in large quantities in industry and supplied commercially in hexane and toluene solutions. This is apparently due to their use as catalysts for olefin polymerization by Ziegler–Natta method and its subsequently progressed methods [3].

$$6\ R^1R^2C{=}CH_2 + 2Al + 3H_2 \xrightarrow{\text{High pressure}} 2\,(R^1R^2CHCH_2)_3Al \tag{4.4}$$

$$4R_3Al + 2Al + 3H_2 \longrightarrow 6R_2AlH$$

Aluminum–carbon bond behaves normally as organometallic reagents. That is, the following three reactions proceed quantitatively: (i) scission of Al–R bond by halogen, (ii) reduction of Al–X bond by metal hydrides, (iii) alkylation of Al–X bond by organolithiums (Eq. 4.5)

The most important reaction of the Al–H bond is the cis-type addition to carbon–carbon double bond, that is, hydroalumination. The reaction proceeds through transition state (**B**), where aluminum resides on a carbon with less substituent(s), that is, with more hydrogens, thus forming aluminum–carbon and hydrogen–carbon bonds (Eq. 4.6), just like hydroboration.

$$R_3Al + X_2 \xrightarrow{a} R_2AlX + RX$$
$$R_2AlX + LiH \xrightarrow{b} R_2AlH + LiX \quad (4.5)$$
$$R_2AlX + R'Li \xrightarrow{c} R_2AlR' + LiX$$

$$RCH{=}CH_2 + R'_2AlH \longrightarrow [\mathbf{B}] \longrightarrow RCH_2{-}CH_2AlR'_2 \quad (4.6)$$

B

The carbon–hydrogen bond of acetylene is well known to be acidic enough for exchange by organolithium and Grignard reagent, and it is also activated by triethylamine. This is used to prepare organoaluminum compounds bearing an alkynyl group (Eq. 4.7).

$$R{-}C{\equiv}C{-}H \xrightarrow{a:\ RLi} R{-}C{\equiv}C{-}Li + RH$$
$$R{-}C{\equiv}C{-}H \xrightarrow{b:\ RMgBr} R{-}C{\equiv}C{-}MgBr + RH$$
$$R{-}C{\equiv}C{-}H \xrightarrow{c:\ (i\text{-}Bu)_2AlH\text{ - }Et_3N} R{-}C{\equiv}C{-}Al(i\text{-}Bu)_2 + H_2 + Et_3N \quad (4.7)$$

$$3\ R{-}C{\equiv}C{-}Li + AlCl_3 \longrightarrow (R{-}C{\equiv}C)_3Al + 3LiCl$$

4.2 STRUCTURE

The primary characteristics of the structure of boron and aluminum compounds are that borane and trimethylaluminum exist as dimers in solution. This is enabled by the formation of an electron-deficient bridging bond between two central atoms.

The formation of the bond is rationalized by the fact that boron and aluminum are sp^3 hybridized and electrons are donated from the B–H or Al–C bond to one vacant orbital of the central atom to form a σ-type 3-center 2-electron bond (3c–2e) in the plane of the paper (Fig. 4.2). This is also understood in another way that a hydride ion or a methyl carbanion donates electrons to two vacant orbitals (sp^3) of boron or aluminum to form a 3-center 2-electron bond. In Fig. 4.2, a solid line denotes a bond in plane and a wedge line, a bond in perpendicular plane, where a solid line directs to the front and a dotted line, to the back. This is the same way as to depict the stereochemistry of sp^3 orbitals of carbon.

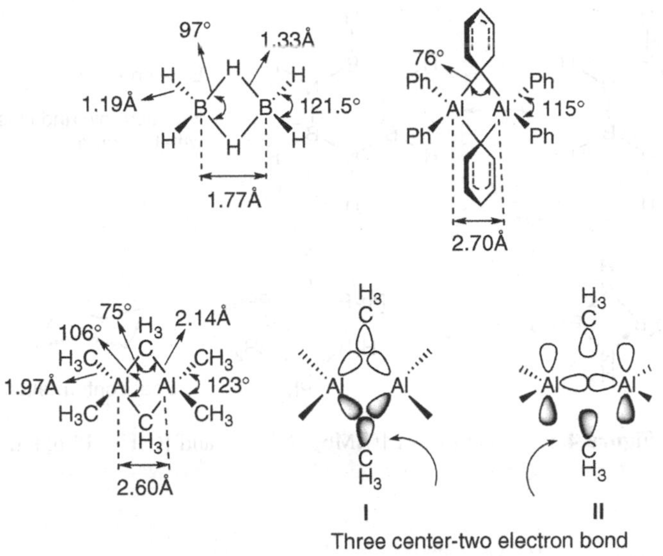

Figure 4.2 Structures of $(BH_3)_2$ and $[Al(CH_3)_3]_2$.

A bridging bond (solid line in plane) is longer than a single bond (wedge line), and bond order of the former is fundamentally 0.5. Actually, the B–H bond length of diborane is 1.19 Å for single bond and 1.33 Å for bridging bond. Al–C bond length of the dimer of trimethylaluminum is 1.97 Å for single bond and 2.14 Å for bridging bond. The bridging bond is longer by 10% than the single bond; thus, the former is weaker than the latter. In the dimer of trimethylaluminum, Al–Al bond length is 2.60 Å and is almost equal (slightly shorter) to the sum of a single bond radius of aluminum [4]. Thus, it can be understood in another way that aluminum is sp^2 hybridized, forming Al–Al bond, and two sets of 3-center 2-electron bonds are formed by using vacant 3p π-orbitals of the two aluminum and the unshared electron pair of a methyl carbanion above and below the molecular plane (Fig. 4.2**II**). Triphenylaluminum is also a dimer, and the bridging phenyl group is perpendicular to the plane of the four-member ring.

Boron compounds have a fertile chemistry. Only a couple of examples are introduced here. Borazine $(HB{=}NH)_3$ is prepared from ammonium chloride and sodium borohydride (Eq. 4.8). Borazine is quite stable thermally and is planar, isoelectronic, and isostructural with benzene. By coordination of a lone pair of nitrogen or phosphorus to boron, stable four-member and six-member rings are formed (Fig. 4.3).

$$NH_4Cl + NaBH_4 \xrightarrow{-NaCl,\ -H_2} H_3N \cdot BH_3 \xrightarrow{\Delta} \underset{\text{Borazine}}{(HB = NH)_3}$$

$$BCl_3 + RNH_3Cl \xrightarrow{145^\circ C} (ClBNR)_3 \xrightarrow{NaBH_4} (HB = NR)_3 \quad (4.8)$$

Borazine

Isoelectronic and isostructual with benzene

Boat conformation

Figure 4.3 Structures of $[(BMe_2{-}NH_2)]_2$ and $[(BH_2{-}PPh_2)]_3$.

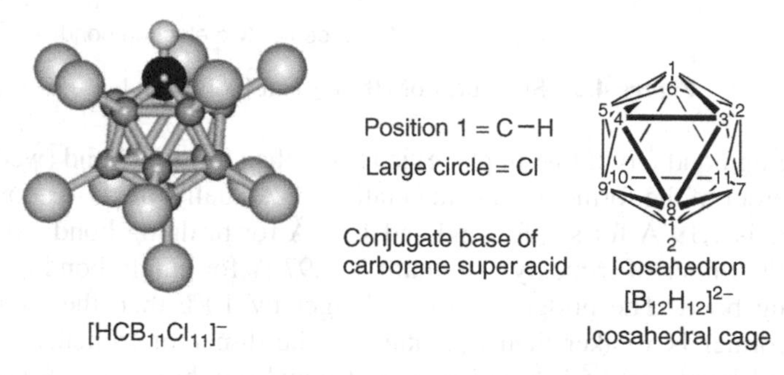

Figure 4.4 Structures of conjugate base of carborane super acid and icosahedral cage.

There are numerous cage compounds of boron, and the chemistry is quite complex. Among them, a dianion of icosahedral cage, $[B_{12}H_{12}]^{2-}$ $2K^+$, is typical and well investigated. The icosahedral cage stabilizes lone pair(s) inside the cage, and halogens can substitute for hydrogens, while the skeleton is intact [5]. This chemistry led to the preparation of carborane super acid, H^+ $[HCB_{11}Cl_{11}]^-$. In the super acid, H–C group sits at position 1 of the icosahedron and 11 B–Cl groups reside at the other vertices. The super acid is the strongest acid ever known as a single molecule. The acidity of the super acid is stronger than 100% sulfuric acid and is much stronger than triflic and fluorosulfonic acid. A combination of MeI and Ag^+ $[HCB_{11}Cl_{11}]^-$ can be used as a strong methylating reagent (Fig. 4.4) [6].

4.3 REACTION

The reaction of borane and alkene proceeds through a four-member transition state as was explained by Eq. (4.3). The stereochemistry of the reaction is explicitly realized by the addition of borane to triple bond, affording a stereochemically rigid alkene (Eq. 4.8).

$$\mathrm{R{-}C{\equiv}C{-}H} + \mathrm{R'_2BH} \longrightarrow \text{(R)(H)C=C(H)(BR'_2)} \tag{4.9}$$

The reaction of trialkylborane with alkaline hydrogen peroxide affords anti-Markovnikov type (less-branched) alcohol quantitatively (Eq. 4.10). The selectivity of formation of isomeric alcohols is contrary to the hydration of alkenes with mercuric acetate or protic acid (Markovnikov type, Eq. 4.11). At the intermediate **C**, the carbon-bearing substituent reacts with water preferably, because the carbocation is more stabilized at the substituted carbon than at the other, and the corresponding C–Hg bond is slightly longer and weaker than the other.

$$3\,\mathrm{RCH{=}CH_2} + \mathrm{BH_3} \longrightarrow \mathrm{(RCH_2CH_2)_3B} \xrightarrow{\mathrm{H_2O_2/NaOH\ aq.}} 3\mathrm{RCH_2CH_2OH} + \mathrm{B(OH)_3} \tag{4.10}$$

$$\mathrm{RCH{=}CH_2} + \mathrm{Hg(OAc)_2} \longrightarrow \left[\text{RHC}\overset{+}{\underset{\mathrm{Hg(OAc)}}{-}}\text{CH}_2\right]_{\mathbf{C}} \xrightarrow{\mathrm{H_2O}} \mathrm{RCH(OH){-}CH_2HgOAc} \xrightarrow{\mathrm{NaBH_4/NaOH\ aq.}} \mathrm{RCH(OH){-}CH_3} \tag{4.11}$$

The mechanism of Eq. 4.10 is totally different from that of Eq. 4.11. Alkaline hydrogen peroxide (pK_a = 11.4) attacks the boron to give borate anion (**D**). Then, one of the alkyl groups of **D** rearranges to the neighboring oxygen as carbanion to afford **E**, keeping the stereochemistry of the carbanion. This process is repeated to give boric ester, and the ester is hydrolyzed to afford three moles of alcohol (Eq. 4.12).

$$\mathrm{R_3B} \xrightarrow{^{-}\mathrm{OOH}} \left[\mathrm{R_3B^{\ominus}{-}O{-}OH}\right]_{\mathbf{D}} \longrightarrow \left[\mathrm{R_2B{-}OR}\right]_{\mathbf{E}} \longrightarrow \mathrm{B(OR)_3} \xrightarrow{\mathrm{H_2O}} 3\mathrm{R{-}OH} + \mathrm{B(OH)_3} \tag{4.12}$$

Borate anion (**F**) can be generated by the reaction of carbanion and borane. One of the alkyl groups at the borate rearranges to alkylate the neighboring carbon, keeping the stereochemistry of the rearranging carbanion (Eq. 4.13). An electron pair of borate shifts in plane antiparallel to the leaving group (Br^-) to alkylate the α-carbon.

$$BrCH_2CO_2Et \xrightarrow{t\text{-BuOK}} Br\overset{-}{C}HCO_2Et \xrightarrow{R_3B} [\mathbf{F}] \xrightarrow{-Br^-} EtO_2C(H)(R)C{-}BR_2 \xrightarrow{H_2O} RCH_2CO_2Et \quad (4.13)$$

Suzuki–Miyaura coupling is notable here, although it is catalyzed by palladium (0). A variety of halides (R^1–X) are coupled with organoboronic acid (R^2–$B(OH)_2$) to yield coupling products (R^1–R^2) in the presence of palladium catalyst (Eq. 4.14) [7].

Suzuki–Miyaura coupling

$$R^1\text{-}X + R^2\text{-}BY_2 \xrightarrow{Pd^0 \text{ (cat.), ligand, base}} R^1\text{-}R^2 \quad (4.14)$$

R^1, R^2 = Aryl, alkenyl, alkyl

Even secondary alkyl halides, which are quite difficult for use in coupling reactions, could be coupled with boronic acid under very carefully selected conditions (Eq. 4.15). Aromatic chlorides are also employed successfully for coupling under almost neutral conditions (Eq. 4.16)

$$R^1R^2CHX + (HO)_2BR^3 \xrightarrow[sec\text{-BuOH, 60 °C}]{4\%\ Ni(cod)_2,\ 8\%\ \text{Bathophenanthroline},\ 1.6\ \text{eq. KO}t\text{-Bu}} R^1R^2CHR^3 \quad (4.15)$$

X = Br, I

Bathophenanthroline

$$\text{2,6-dimethylchlorobenzene} + (HO)_2B\text{-}(2\text{-fluorophenyl}) \xrightarrow[\text{Toluene, 90 °C}]{Pd(OAc)_2,\ \text{S-Phos},\ K_3PO_4} \text{2-fluoro-2',6'-dimethylbiphenyl} \quad (4.16)$$

The catalytic cycle is illustrated in Fig. 4.5. Oxidative addition of R^1–X to palladium (0) generates the first Pd(II), on which ligand exchange of the halide to

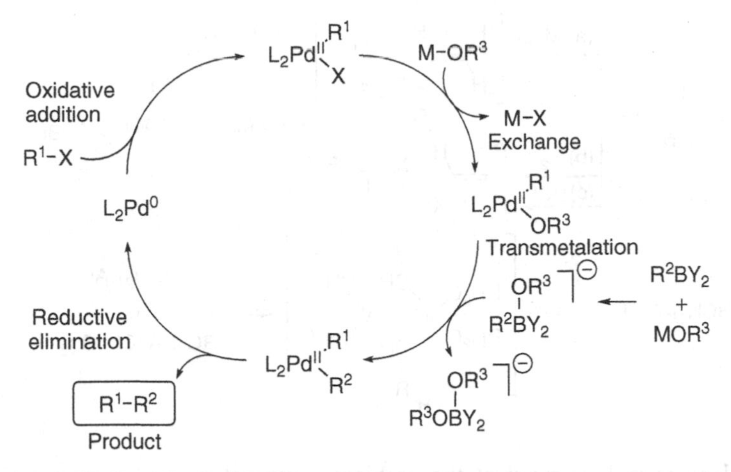

Figure 4.5 Catalytic cycle of Suzuki–Miyaura coupling.

alkoxide takes place to give the second activated Pd(II). Transmetallation of the second Pd(II) with a borate (activated borane with an alkoxide) occurs to give the third Pd(II). Ligand coupling takes place on the third Pd(II) to afford a coupled product ($R^1{-}R^2$) by reductive elimination, and palladium (0) is recovered for the next reaction.

The addition of alane (Al–H bond) to triple bond gives a *cis*-adduct (Eq. 4.17); however, the orientation of the adduct is controlled by the electronic effect of substituent (Z). The orientation depends on whether substituent (Z) is electron-donating (i.e., efficient resonance effect is possible with the aluminum through the double bond) or electron-withdrawing (no resonance effect with Z). This is probably because the Al–C bond is longer than the B–C bond, as the orientation of hydroboration is determined solely by steric effect.

$$R{-}C{\equiv}C{-}Z + (i\text{-}C_4H_9)_2AlH \ (\text{DIBAL}) \longrightarrow \begin{cases} \text{a: } (i\text{-Bu})_2Al(R)C{=}CH(Z) & Z = NR'_2,\ OR' \\ \text{b: } R(H)C{=}C(Z)Al(i\text{-Bu})_2 & Z = AlR'_2,\ SiR'_3,\ SO_2R' \end{cases} \tag{4.17}$$

Reactivity of ate complex of alane (**G**), generated by the addition of methyllithium to vinylalane, is enhanced to react with electrophiles, with retention of stereochemistry (Eq. 4.18). Halogen and hydrogen halide can react with alane without activation by forming the ate complex. In the case of vinylalane, it was shown that the stereochemistry of the double bond is retained. Borane is similarly activated by forming a borate as is shown in Fig. 4.5.

Trialkylaluminum bearing different substituent(s) exchanges ligand (substituent) in solution through dimer (**H**) as an intermediate. This type of exchange does not take place with borane at all.

(4.18)

(4.19)

Ethylene is polymerized at the ambient temperature and pressure (20°C, 1 bar) by using the Ziegler–Natta catalyst, which consists of titan chloride(s) and trialkylaluminum. The successful industrial uses developed organoaluminum chemistry greatly. Recently, metallocene [8] and phenoxy-imine [9] catalysts have been developed to show several hundred times of catalytic activity compared to the Ziegler–Natta catalyst. For this process, methylaluminoxane (MAO) is used as a cocatalyst. MAO is obtained by condensation of hydrolysis products of trimethylaluminum under suitable conditions and has a molecular weight of 1000–1500. Zirconium metallocene (Cp_2ZrCl_2) and MAO system exchanges the methyl group of MAO and the chlorine of metallocene in toluene solution to afford the 14-electron Zirconium species (**I**). Subsequent addition and insertion of ethylene takes place on the active Zirconium complex (**I**) for polymerization (Eq. 4.20). The process has been industrialized for production of synthetic rubber and polyethylene.

$$Me_3Al + H_2O \xrightarrow{-MeH} \text{MAO (Polymethylaluminoxane)}$$

$$Cp_2ZrCl_2 \xrightarrow{MAO} Cp_2Zr(CH_3)_2 \rightleftharpoons [Cp_2Zr(CH_3)]^{\oplus} [(CH_3)Al_{MAO}]^{\ominus} \quad \mathbf{I} \qquad (4.20)$$

Cp = Cyclopentadienyl anion

REFERENCES

1. Brown HC. Organic synthesis via boranes. New York: John Wiley; 1975. pp. 29–32.
2. Sonderquist JA, Brown HC. J Org Chem 1981;46:4599.
3. Ziegler K, Gellert H-G, Martin H, Nagel K, Schneider J. Ann Chem 1954;589:91.

4. Hendrickson CH, Eyman DP. Inorg Chem 1967;6:1461.
5. Housecroft CE, Sharpe AG. Inorganic chemistry. 3rd edn. London, Pearson-Prentice Hall; 2008, Chapters 6. 9 and 7. 13.
6. Juhasz M, Hoffmann S, Stoyanov E, Kim K-C, Reed CA. Angew Chem Int Ed 2004;43:5352.
7. (a) Miyaura N, Suzuki A. Chem Rev 1995;95:2457; (b) Chemler SR, Trauner D, Danishefsky SJ. Angew Chem Int Ed 2001;40:4544; (c) Yamamoto Y, Takizawa M, Yu X-Q, Miyaura N. Angew Chem Int Ed 2008;47:928.
8. Sinn H, Kaminsky W, Vollmer HJ, Woldt R. Angew Chem Int Ed 1980;19:390.
9. Matsui S, Mitani M, Saito J, Tohi Y, Makio H, Matsukuwa N, Takagi Y, Tsuru K, Nitabaru M, Nakano T, Tanaka H, Kashiwa N, Fujita T. J Am Chem Soc 2001;123:6847.

CHAPTER 5

SILICON, TIN, AND LEAD COMPOUNDS

Group (n) / Period	13 3B	14 4B	15 5B
1 1s			
2 [He] 2s2p	0.88 801 $^{11}_{5}$**B** boron 2.0 372	0.77 1086 $^{12}_{6}$**C** carbon 2.5 368	0.70 1402 $^{14}_{7}$**N** nitrogen 3.0 292
3 [Ne] 3s3p	1.25 577 (0.57^{3+}) $^{27}_{13}$**Al** aluminium 1.5 255	1.17 787 $^{28}_{14}$**Si** silicon 1.8 301	1.10 1012 $^{31}_{15}$**P** phosphorus 2.1 264
4 [Ar:3d^{10}] 4s4p	1.25 579 (0.83^{3+}) $^{69}_{31}$**Ga** gallium 1.6 247	1.22 762 $^{74}_{32}$**Ge** germanium 1.8 237	1.21 947 $^{75}_{33}$**As** arsenic 2.0 200
5 [Kr:4d^{10}] 5s5p	1.50 558 (0.92^{3+}) $^{116}_{49}$**In** indium 1.7 165	1.40 709 $^{120}_{50}$**Sn** tin 1.8 225	1.41 834 $^{121}_{51}$**Sb** antimony 1.9 215
6 [Xe:4f^{14}5d^{10}] 6s6p	1.55 589 (1.05^{3+}) $^{205}_{81}$**Tl** thallium 1.6 125	1.54 716 $^{208}_{82}$**Pb** lead 1.8 130	1.52 703 $^{209}_{83}$**Bi** bismuth 1.9 143

5.1 SYNTHESIS

The major component of rocks on the earth is silicate; hence natural resources of silicon are abundant. Actually, while the Clarke number of oxygen is the largest (46.6%), that of silicon is the second largest (27.7%).

Silicon–carbon bond (single bond energy = 301 kJ/mol) is weaker than carbon–carbon bond (368 kJ/mol); however, it is quite stable under ambient conditions. Silicon–oxygen bond (458 kJ/mol) is quite strong but silicate can be converted to silicon and halosilicons and are used industrially in large scale.

Organosilicon compounds are produced commercially by heating alkyl chlorides (or chlorobenzene) with metallic silicon containing a small amount of copper at high temperatures (ca. 300°C). For the chloromethane case, the resulting mixture of methylchlorosilanes (Me_3SiCl, Me_2SiCl_2, and $MeSiCl_3$) is fractionally distilled to give pure products, using efficient distillation tower (Eq. 5.1). This process is called *Rochow method* and fundamental organosilicons are supplied by this method. Polysilanes with Si–Si bond are obtained as a distillation residue. More than 1000 organosilicons are available commercially.

$$\text{R–Cl} + \text{Si (alloy with Cu)} \xrightarrow{\Delta} R_nSiCl_{4-n} \quad (\text{R = alkyl, phenyl}) \tag{5.1}$$

A great variety of organosilicons are prepared by reactions of chlorosilanes and organolithiums or Grignard reagents (Eq. 5.2). Hydrosilanes (**1**) with Si–H bond(s) are obtained by the reduction of silicon–chlorine bond(s) with lithium aluminum hydride (Eq. 5.3) [1].

$$\text{RM (M = Li, MgX)}: \begin{cases} \text{(a)}\ SiCl_4 \rightarrow R_4Si \\ \text{(b)}\ R'_2SiCl_2 \rightarrow R'_2Si(R)Cl \rightarrow R'_2SiR_2 \\ \text{(c)}\ R'_3SiCl \rightarrow R'_3SiR \end{cases} \tag{5.2}$$

$$R_nSiCl_{4-n} + LiAlH_4 \longrightarrow \underset{\mathbf{1}}{R_nSiH_{4-n}} \tag{5.3}$$

Silicon–chlorine bond is converted to silyl triflate (**2**) by heating with trifluoromethane sulfonic acid, where hydrogen chloride is evolved. Silyl triflate (**2**) is a stronger silylating reagent than silyl chloride. Silyl enol ether (**3**) is prepared by the reaction of carbonyl compound bearing α-hydrogen and silylating reagent.

$$RMe_2SiCl + TfOH \xrightarrow{} \underset{\mathbf{2}}{RMe_2SiOTf} + HCl \quad (Tf = CF_3SO_2)$$

$$\text{(a) } RMe_2SiCl \text{ or (b) } RMe_2SiOTf \ (\text{R = Me, } t\text{-Bu, Ph}) \xrightarrow{R^1COCH_2R^2} \xrightarrow{B:} \underset{\mathbf{3}}{R^1C(OSiMe_2R){=}CHR^2} \tag{5.4}$$

Silicon–chlorine bond is rapidly hydrolyzed by water to afford silanol (**4**), and two molecules of silanol (**4**) easily condense to give siloxane (**5**); hence, **4** cannot be isolated. This contrasts the stability of carbon–chlorine bond to water, and alcohols are quite stable to be isolated by distillation.

$$R_3SiCl \xrightarrow{H_2O} \underset{\mathbf{4}}{R_3SiOH} \xrightarrow{R_3SiOH} \underset{\mathbf{5}}{R_3SiOSiR_3} + H_2O \qquad (5.5)$$

Dichlorosilane, when heated with sodium in toluene, affords polysilanes having interesting character and structure. On the other hand, trichlorosilane carrying a bulky silyl group gives silacubane under similar conditions to those described above (Eq. 5.6).

$$R_2SiCl_2 \xrightarrow{Na} \left(\begin{array}{c}R\\|\\-Si-\\|\\R\end{array}\right)_n \text{ Polysilane}$$

$$RSiCl_3 \xrightarrow{Na} \text{Silacubane } (R_8Si_8) \qquad R = t\text{-}BuMe_2Si \qquad (5.6)$$

Silyl anion (sodium or lithium salt) cannot be generated in solution by the reaction of chlorosilane and a metal, because of rapid coupling of the anion and the remaining chlorosilane. Accordingly, silyl anion is generated in situ by the reaction of disilane (path a) or silylstannane (paths b and c) with methyllithium (Eq. 5.7).

$$\begin{array}{l}Me_3Si{-}SiMe_3 \ \text{(a)}\\ Me_3Si{-}SnMe_3 \ \text{(b)}\end{array} \xrightarrow{MeLi} Me_3SiLi + (Me_4Si;\ Me_4Sn) \xrightarrow{CuI} (Me_3Si)_2CuLi$$

$$Y{-}SiPh_2{-}SnMe_3 \xrightarrow{\text{(c) } MeLi} Y{-}SiPh_2{-}Li + Me_4Sn \qquad (Y = Et_2N,\ t\text{-}BuO) \qquad (5.7)$$

By method (c), strongly nucleophilic silyl anion carrying amino or methoxy group can be generated. Silyl anions form complexes with copper(I) to become stabilized and handy reagents for synthetic work [2].

Hydrosilane (Si–H bond) is more stable and less reactive than borane or alane and cannot add directly to unsaturated bond(s). Therefore, a platinum or rhodium complex is used to activate hydrosilane in order to realize the occurrence of catalytic hydrosilylation (Eq. 5.8). Stereochemistry of hydrosilylation is shown to be cis-addition. This has been revealed by the stereochemistry of adducts of silanes with acetylenic compounds (Eq. 5.9).

$$R^1HC{=}CH_2 \xrightarrow[\text{cat. } H_2PtCl_6]{R_3SiH} R^1CH_2CH_2SiR_3 \tag{5.8}$$

$$R{-}C{\equiv}C{-}H \xrightarrow[\text{cat. } H_2PtCl_6]{\text{(a) } HSiCl_3} (R)(H)C{=}C(H)(SiCl_3); \quad \xrightarrow[\text{cat. } RhCl(PPh_3)_3]{\text{(b) } R_3SiH} (R)(H)C{=}C(H)(SiR_3) \tag{5.9}$$

Reactive silyl anion substitutes for the halogen of allyl halide at the α-position (path a). Alternatively, silyl anion stabilized as copper complex adds selectively at the γ-position followed by departure of the halide anion (path b). It is interesting and useful as reagents for synthesis that positional isomers of allylsilanes can be prepared by controlling the reactivity of silyl anions.

$$R{-}CH{=}CH{-}CH_2{-}X \;(\gamma, \alpha) \xrightarrow{\text{(a) } Me_3SiLi} R{-}CH{=}CH{-}CH_2{-}SiMe_3; \quad \xrightarrow{\text{(b) } (Me_3Si)_2CuLi} R{-}CH(SiMe_3){-}CH{=}CH_2 \tag{5.10}$$

5.2 REACTION

The fundamental structure of a saturated organosilicon is tetrahedron, and silicon is sp^3 hybridized, just like carbon. Therefore, there is no special structural characteristic for organosilane. On the other hand, the energy level of σ_{Si-C} becomes higher and that of σ^*_{Si-C} becomes lower for Si–C single bond compared to C–C single bond. Stereoelectronic effect based on this fact appears on the reactivity of the Si–C bond. The idea was explained in Chapter 2 and is important to realize the following reactions.

Peterson reaction is one of most important reactions of organosilicon for synthesis. This is olefin synthesis, which is similar to Wittig reaction (Eq. 5.11). Instead of phosphorus ylide, an α-proton of trimethylsilyl group is deprotonated to react with a carbonyl compound affording oxide anion (**A**) (path a); then, an alkene is produced followed by the elimination of silanol. The oxide anion (**A**) is also generated through paths b and c. By following path b, Grignard reagent is generated in situ from α-chlorotrimethylsilane to react with a carbonyl compound.

In path c, organolithium or Grignard reagent is added to the β-carbonyl group of trimethylsilyl compound in order to afford the oxide anion (**A**). From the intermediate (**A**) generated through either path, a mixture of alkenes is obtained as expected, followed by the elimination of silanol.

$$\mathrm{Me_3SiCH_2R} \xrightarrow[\mathrm{TMEDA}]{\mathrm{BuLi}} [\mathrm{Me_3Si\bar{C}HRLi^+}] \xrightarrow{(a)\ R^1R^2C{=}O} \mathbf{A}$$

$$\mathrm{Me_3SiCH_2Cl} \xrightarrow{\mathrm{Mg}} [\mathrm{Me_3SiCH_2MgCl}] \xrightarrow{(b)} \mathbf{A}$$

$$\mathrm{Me_3SiCH_2COR^1 + R^2Li} \xrightarrow{(c)} \mathbf{A}\ \left[\mathrm{Me_3Si{-}CHR{-}C(R^1)(R^2){-}O^-X^+}\right] \longrightarrow \mathrm{(R)(H)C{=}C(R^1)(R^2)} + \mathrm{(R)(H)C{=}C(R^1)(R^2)} + \mathrm{Me_3SiOX\ (X = Li, MgCl)} \quad (b,c\ ;\ R = H) \tag{5.11}$$

Aldol reaction can take place with silyl enol ether when it is activated by Lewis acids such as $TiCl_4$, $SnCl_4$, $ZnCl_2$, Me_3SiOTf, and Ph_3CClO_4 (Eq. 5.12). The transition state of this reaction is considered to be noncyclic, which is different from cyclohexane-type cyclic transition state where a metal ion coordinates with two oxygens (cf. Eqs. 3.14–3.16). According to the noncyclic transition state, it is difficult to control the ratio of syn:anti isomers of the hydroxyl group to the vicinal substituent, which is greatly affected by the combination of silyl enol ether, aldehyde, and Lewis acid [3].

$$\text{1-(trimethylsilyloxy)-2-methylcyclohexene} + \mathrm{PhCHO} \xrightarrow{\mathrm{TiCl_4}} \text{2-methyl-2-(hydroxy(phenyl)methyl)cyclohexanone}$$

$$\text{1-(trimethylsilyloxy)-6-methylcyclohexene} + \mathrm{PhCHO} \xrightarrow{\mathrm{TiCl_4}} \text{2-(hydroxy(phenyl)methyl)-6-methylcyclohexanone} \tag{5.12}$$

An enolate anion (**B**) is generated as an intermediate, when trimethylsilyl enol ether is activated by tetrabutylammonium fluoride (Eq. 5.13). Then, the enolate (**B**) reacts with an aldehyde through noncyclic transition state (**C**) of the least steric hindrance, where anti-periplanar orientation of the double bond of enolate anion (**B**) and the carbonyl group of aldehyde is preferred. On the basis of this preference, *syn*-adduct is selectively produced (the ratio: syn:anti = 95:5), irrespective of stereochemistry (E and Z) of silyl enol ether used as a starting material.

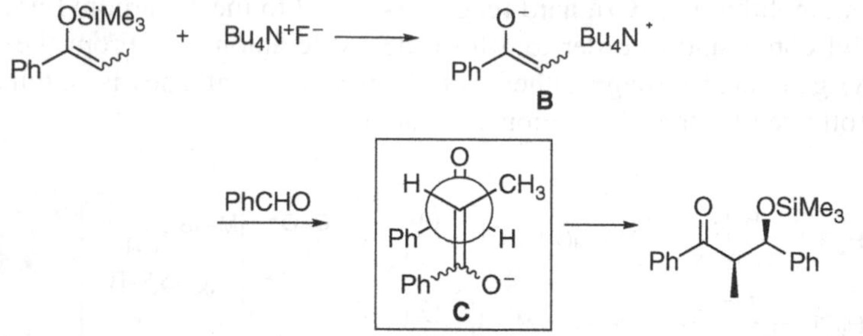

(5.13)

Stereochemistry of the reaction of vinylsilane and allylsilane with electrophiles is characteristic. The electrophile reacts with vinylsilane at the α-position to substitute the silyl group, retaining stereochemistry of the double bond. It is rationalized that a carbocation emerges at the β-position and the vacant p-orbital of the cation is stabilized by HOMO (highest occupied molecular orbital) of C–Si bond which should be in plane with the p-orbital (**D**). Three examples are shown, that is, (i) halogenation; (ii) acylation; and (iii) protonation (Eq. 5.14).

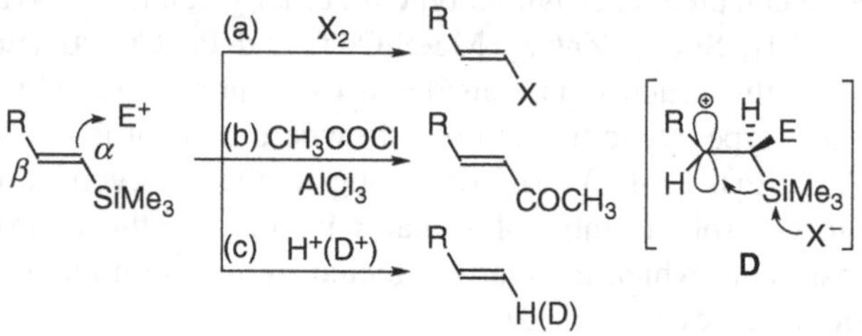

(5.14)

The silyl group of silylbenzene is eliminated quite rapidly by treating with an acid. This is due to the stabilization of the conjugated cation (**E**) which is effected by donation of electrons from the HOMO of the C–Si bond (Eq. 5.15).

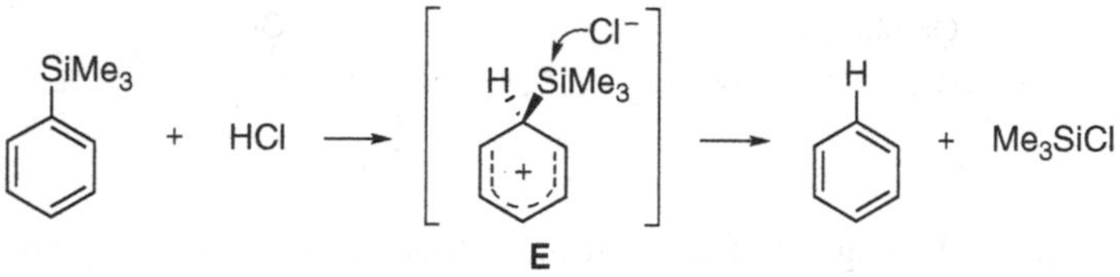

(5.15)

It is also established that the electrophile attacks allylsilane at the γ-position to give an alkene, accompanied by the elimination of the silyl group (Eq. 5.16). It is also rationalized that a carbocation emerges at the β-position and the vacant p-orbital is stabilized by donation of electrons from the HOMO of the C–Si bond, which is in plane with the p-orbital (**F**). There are three examples cited in Eq. (5.16), that is, (i) acylation; (ii) alkylation; and (iii) protonation.

E^+, β, γ, α, R, $SiMe_3$ → (a) CH_3COCl / $AlCl_3$; (b) R′X / cat. $TiCl_4$; (c) $H^+(D^+)$ → R, $COCH_3$; R, R′; R, H; [R, R, E, H, H, $SiMe_3$, X'^-, H] **F** (5.16)

The aldehyde activated with titan tetrachloride attacks allylsilane also at the γ-position to afford δ-hydroxyolefin, followed by the elimination of trimethylsilyl chloride (Eq. 5.17). At the transition state (**G**), trimethylsilyl group and the carbonyl group complexed with titan tetrachloride reside trans to each other to avoid steric hindrance; thus, the resulting hydroxyl group and the substituent (R^2) of the product are syn to each other [4, 5].

R^2, R^1, $SiMe_3$, γ, α, H, H + RCHO → $TiCl_4$ → [H, R^2, R^1, $SiMe_3$, $Cl_4Ti\cdots O$, R, H, H] **G** (5.17)

→ OH, R, δ, α, R^1, R^2

The allylsilane bearing a strong electron-withdrawing group such as fluoro or alkoxy at the silyl group reacts with an aldehyde to give δ-hydroxyolefin in the presence of fluoride ion (Eq. 5.18). The fluoride ion coordinates with the silyl group to generate a pentacoordinate silyl anion; then, carbonyl oxygen coordinates with the silyl anion to form a six-member transition state (**H**). Relative stereochemistry of the hydroxyl group and R^2 is anti for this δ-hydroxyolefin. The formation of highly coordinate silicon species is characteristic of hypervalent species, as was described in Chapter 2.

R^2, R^1, SiF_3, γ, α, H, H → CsF/RCHO → [H, R^1, F, F, Si, F, F, R^2, O, R]$^\ominus$ $Cs^\oplus$ **H** (5.18)

→ OH, R, R^1, R^2

When hydroxyketone is used, the hydroxyl group coordinates with the silyl group in collaboration with the carbonyl group to form the bicyclic transition state (**I**) (Eq. 5.19). Hence, the C–C bond formation becomes an asymmetric reaction.

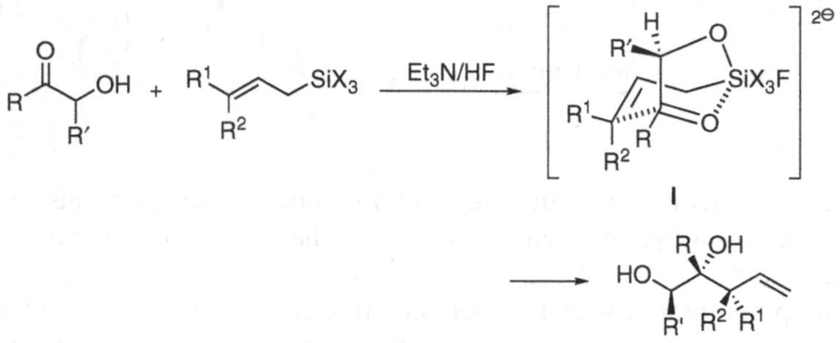

(5.19)

The formation and characteristics of highly coordinate main group compounds are explained in Chapter 2. Here, some more examples of synthetically useful reactions utilizing highly coordinate organosilicon compounds (intermediates) are described.

When a silyl group bears strong electron-withdrawing group(s) such as fluorine and chlorine (Eqs. 5.18 and 5.19), penta- or hexacoordinate silicate is generated and the silicate can further coordinate with lone pair electrons. Furthermore, these silicates are still Lewis acidic. One of the fundamental reactions using the Lewis acidic character of silicate is to convert sp^3 carbon bonded to silicon (**J**) to alcohol with alkaline hydrogen peroxide (Eq. 5.20). Stereochemistry of the migrating R group is retained, because the R group rearranges as carbanion. This is certainly similar to the conversion of borane [$(RCH_2CH_2)_3B$] to alcohol with alkaline hydrogen peroxide (**K**). Reaction of alkyltrifluorosilane with amine oxide (**L**) can be understood similarly (Eq. 5.21).

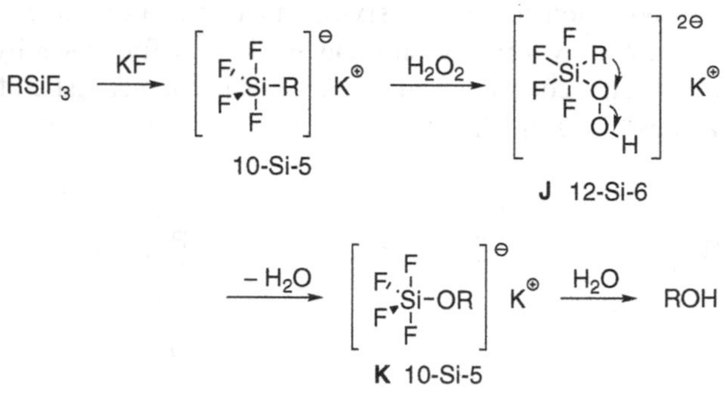

(5.20)

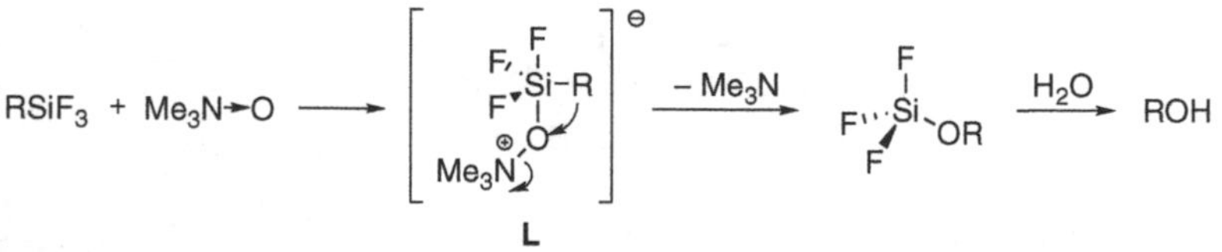

(5.21)

An extremely strong fluoride ion (**6**: countercation is extremely stable) can afford hexacoordinate difluoroallylsilicate (**M**: 12-Si-6), where the starting allylsilane does not have any electron-withdrawing group at the silicon (Eq. 5.22). As **M** is a dianion, the allyl group of **M** is reactive enough to directly substitute the iodine of alkyl iodide (Eq. 5.22).

$$Me_3Si\text{-}CH_2CH{=}CH_2 + 2\,[(Me_2N)_3P{=}\overset{\oplus}{N}{=}P(NMe_2)_3]\,F^{\ominus} \longrightarrow$$

6

$$\left[Me_2SiF_2(CH_2CH{=}CH_2) \right]^{2\ominus} 2M^{\oplus} \xrightarrow{n\text{-}C_{11}H_{23}I} n\text{-}C_{11}H_{23}CH_2CH{=}CH_2 \quad (5.22)$$

M 12-Si-6

$$M^{\oplus} = (Me_2N)_3P{=}\overset{\oplus}{N}{=}P(NMe_2)_3$$

Alkylpentafluorosilicate (**7**) reacts with bromine to give alkyl bromide by S_N2-type mechanism and the carbon is inverted (Eq. 5.23). On the other hand, vinylpentafluorosilicate (**8**) is brominated with bromine by retention of stereochemistry (Eq. 5.24).

$$\left[RSiF_5 \right]^{2\ominus} \xrightarrow{Br_2} R{-}Br + [BrSiF_5]^{2\ominus} \quad (5.23)$$

7

$$\left[F_5Si\text{-}CH{=}CH\text{-}R \right]^{2\ominus} \xrightarrow{Br_2} R\text{-}CH{=}CH\text{-}Br + [BrSiF_5]^{2\ominus} \quad (5.24)$$

8

When ketene silyl acetal (**9**) is activated with a fluoride anion (e.g., $(Me_3SiF_2)^-$, HF_2^-) in the presence of methyl methacrylate, the latter is polymerized (Eq. 5.25). Firstly, a fluoride ion coordinates with the trimethylsilyl group of **9** to form 10-Si-5-type silicate, keeping the ketene acetal structure; then, the silicate attacks to add to methyl methacrylate in conjugate fashion to afford a hexacoordinate silicate (**N**: 12-Si-6), in which the trimethylsilyl group is transferred to the carbonyl group of the ester. Thus, 10-Si-5-type silicate is regenerated within **N**. Again, methyl methacrylate coordinates with the 10-Si-5-type silicate with the ester carbonyl to generate 12-Si-6 silicate and the same type of reaction is successively repeated to produce a polymer. The polymerization is called *group transfer polymerization* because trimethylsilyl group is transferred to the monomer (methyl methacrylate) in each step.

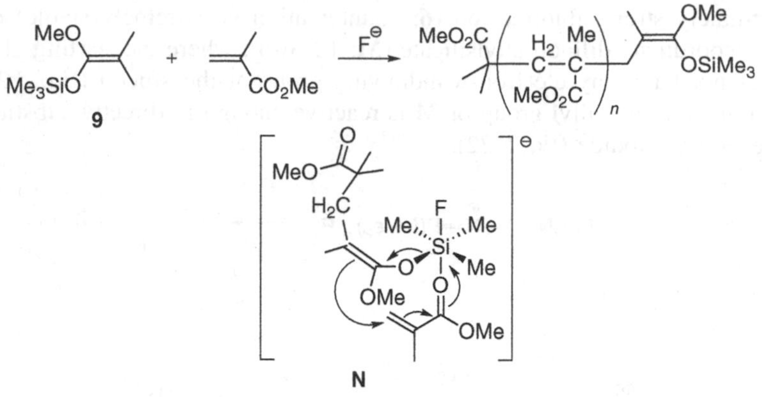

(5.25)

5.3 ORGANOTIN AND LEAD COMPOUNDS

Organotin compounds are produced industrially by direct reaction of tin powder and alkyl chloride at high temperatures (Eq. 5.26). Tetraethyllead was the only organolead compound prepared and sold industrially, and it was also produced by the direct reaction of ethyl chloride with sodium–lead alloy (Eq. 5.27). The compound was used as an antiknock material for gasoline, but it is not used at present. These are similar types of reactions as Rochow method for the preparation of organosilicons.

$$R-Cl + Sn\ (powder) \xrightarrow{\Delta} R_2SnCl_2 + R_3SnCl \quad (5.26)$$

$$4C_2H_5Cl + 4Na-Pb \xrightarrow{\Delta} (C_2H_5)_4Pb + 3Pb + 4NaCl \quad (5.27)$$

Organotin compounds with four organic groups are prepared by the reaction of tin tetrachloride and Grignard reagent or alkyllithium (Eq. 5.28) [6].

$$SnCl_4 + 4RMgX \longrightarrow R_4Sn + 4MgClX \quad (5.28)$$

R = alkyl (Me, Et, Bu, *i*-Pr, $CH_2{=}CH$, $CH_2{=}CHCH_2$)
aryl (Ph)

On the other hand, organolead compounds with four organic groups (**10**: R_4Pb) are produced by a one-pot reaction of lead dichloride ($PbCl_2$) with three equivalents of organometallic reagent (RM) and one equivalent of organic halide (RX) (Eq. 5.29a).

Firstly, divalent organolead (R_2Pb) is produced in situ by the reaction of $PbCl_2$ with two equivalents of RM, and the resulting R_2Pb reacts with RX to give R_3PbX by oxidative addition. Then, R_3PbX and the remaining one equivalent of RM are converted to afford R_4Pb (**10**) and MX (Eq. 5.29a). By the reaction of two equivalents of $PbCl_2$ and four equivalents of RM, R_2Pb is generated in the first step and it disproportionates with the remaining $PbCl_2$ to afford R_2PbCl_2 and Pb. Then, R_2PbCl_2 and two equivalents of RM afford the final product R_4Pb (**10**) in excellent yield (Eq. 5.29b).

$$PbCl_2 + RX + 3\,RM \xrightarrow{(a)} \underset{\mathbf{10}}{R_4Pb} + 2\,MCl + 2\,MX$$

RX - RM: EtI–EtMgBr; BuBr–BuLi; PhI - PhLi, etc. (5.29)

$$2\,PbCl_2 + 4\,RM \xrightarrow{(b)} \underset{\mathbf{10}}{R_4Pb} + Pb + 4MCl$$

Among organotin compounds, Bu_2SnX_2 (stabilizer for polyvinyl chloride), and Bu_3SnX and $Bu_3SnOSnBu_3$ (disinfectant and germicide) are produced industrially. R_4Sn is stable, easily prepared, and used as a starting material for disproportionation with $SnCl_4$ as follows (Eq. 5.30).

$$3\,R_4Sn + SnCl_4 \xrightarrow{(a)} 4\,R_3SnCl$$

$$R_4Sn + SnCl_4 \xrightarrow{(b)} (R_3SnCl + RSnCl_3) \longrightarrow 2\,R_2SnCl_2 \qquad (5.30)$$

$$R_4Sn + 3\,SnCl_4 \xrightarrow{(c)} 4\,RSnCl_3$$

note: R = Bu is commonly used; R = Me and Ph react the same way as Bu

As Bu_3SnCl (**11**) is most generally used, reactions of organotin compounds are explained using the Bu_3Sn group as an example. Bu_3SnCl reacts with a variety of nucleophiles (MeONa, NaOH, Me_2NLi, and RM) through formally nucleophilic displacement (Eq. 5.31). Reduction of **11** with metal hydrides gives tin hydride (**12**, Bu_3SnH) and that with sodium affords bis-stannane (**14**). Organometals give tetraorganotins (**13**, Bu_3SnR) on reaction with **11**.

Bu_3SnCl (**11**) →
- (a) NaOMe → $Bu_3SnOMe + NaCl$
- (b) 2NaOH → $(2\,Bu_3SnOH) \longrightarrow Bu_3SnOSnBu_3 + H_2O$
- (c) Me_2NLi → Bu_3SnNMe_2
- (d) $LiAlH_4$ ($NaBH_4$) → Bu_3SnH **12**
- (e) RM (M = MgBr, Li) → Bu_3SnR **13**
- (f) Na → $Bu_3SnSnBu_3$ **14**

(5.31)

Reduction of alkyl halide (R–X) with tin hydride gives R–H in high yield—a reaction in which AIBN (azobisisobutyronitrile) is used to initiate free radical chain reactions (Eq. 5.32). In the propagation step, tin radical ($Bu_3Sn\cdot$)

specifically abstracts halogen to generate alkyl radical (R·), and subsequently R· abstracts hydrogen from tin hydride (Bu_3SnH), affording R–H. Moreover, tin radical ($Bu_3Sn·$) can add to alkene and alkyne to initiate free radical chain reactions; thus, the reaction is useful as a synthetic method for a variety of organotin compounds (Eq. 5.33). Equation (5.33) looks formally similar to hydroboration and hydrosilylation; however, the products are a mixture of stereoisomers because the reaction proceeds via free radical chain reactions

$$\mathrm{Bu_3SnH} + \underset{\text{AIBN}}{\mathrm{Me_2C(CN)-N{=}N-C(CN)Me_2}} \xrightarrow{(a)} \mathrm{Bu_3Sn\bullet} \quad \mathrm{H-CMe_2(CN)} + \mathrm{N_2}\uparrow$$

$$\mathrm{Bu_3Sn\bullet} + \mathrm{R-X} \xrightarrow{(b)} \mathrm{Bu_3SnX} + \mathrm{R\bullet} \qquad (5.32)$$

$$\mathrm{R\bullet} + \mathrm{Bu_3SnH} \xrightarrow{(c)} \mathrm{R-H} + \mathrm{Bu_3Sn\bullet}$$

note : X = Cl, Br, I ; reduction of R–X by radical chain reactions

$$\mathrm{Bu_3SnH} + \mathrm{AIBN} \longrightarrow [\mathrm{Bu_3Sn\bullet}] \begin{cases} \xrightarrow{(a)\ \mathrm{RCH{=}CH_2}} \mathrm{RCH_2CH_2SnBu_3} \\ \xrightarrow{(b)\ \mathrm{R-C{\equiv}C-H}} \mathrm{R(H)C{=}C(H)SnBu_3} \end{cases} \qquad (5.33)$$

Lithium enolate reacts with tributyltin chloride (**11**) to yield tin enolate (**15**) (Eq. 5.34). Lithium chloride is inevitably present under these conditions, and the characteristics of **15** are influenced by the salt. Instead, pure tin enolate (**15**) is obtained by the reaction of vinyl acetate and tributylmethoxytin, where the resulting methyl acetate can be removed under reduced pressure. Tin enolate (**15**) is more reactive than silyl enolate (**3**), but it is known that **15** is in equilibrium with its keto form. The reactivity of tin enolate can be modified by the addition of anionic species, because **15** easily coordinates with anionic species to form hypervalent species (Section 2.3 of Chapter 2).

$$\mathrm{R^1R^2C{=}C(OLi)R^3} + \mathrm{Bu_3SnCl} \xrightarrow{(a)\ -\mathrm{LiCl}} \underset{\mathbf{15}}{\mathrm{R^1R^2C{=}C(OSnBu_3)R^3}} \xleftarrow{(b)\ -\mathrm{MeC(O)OMe}} \mathrm{R^1R^2C{=}C(O-C(O)-Me)R^3} + \mathrm{Bu_3SnOMe} \qquad (5.34)$$

Cross-coupling of organotin compound (**13**) and organic halide is catalyzed by palladium (0) to form R–R′. This is of use for carbon–carbon bond formation between vinyl and/or aryl groups (Eq. 5.35).

$$Bu_3SnR + R'{-}X \xrightarrow{Pd(0)} R{-}R' + Bu_3SnX \tag{5.35}$$

note: reactivity of R; ethenyl> vinyl> aryl> benzyl>alkyl

An ate complex (**O** or **P**) is quite easily generated in situ by the reaction of organotin with organolithium at low temperature, and one of the substituents is released in the solution as another kind of organolithium. Thus, this is useful to generate organolithium that is not easy to prepare (Eq. 5.36)

$$Ph_3SnCH{=}CH_2 + PhLi \xrightarrow{(a)} [Ph_4SnCH{=}CH_2]^{\ominus} \ (\mathbf{O}) \longrightarrow LiCH{=}CH_2 + Ph_4Sn\downarrow$$

$$(CH_2{=}CHCH_2)_4Sn + 4BuLi \xrightarrow{(b)} [(CH_2{=}CHCH_2)_4SnBu]^{\ominus} Li^{\oplus} \ (\mathbf{P}) \longrightarrow 4\ CH_2{=}CHCH_2Li + Bu_4Sn \tag{5.36}$$

The lead–carbon bond of an organolead compound (**16**) is cleaved by hydrogen chloride or halogen to give lead halide (**17a** and **b**) and R–H or R–X (Eq. 5.37). The halides (**17a** and **b**) can be converted to the tetravalent organolead compound (**18**) bearing different substituents (Eq. 5.38) [7]. Lead tetraacetate has been used as an oxidizing reagent for a long time; however, lead tetra (trifluoroacetate) (**19**) is a more powerful electrophile as expected. Electrophilic substitution of aromatic hydrocarbon with **19** can take place to afford aryllead tris(trifluoroacetate) (**20**) (Eq. 5.39).

$$\underset{\mathbf{16}}{R_4Pb} + HCl \xrightarrow{(a)} \underset{\mathbf{17a}}{R_3PbCl} + RH$$

$$R_4Pb + Br_2 \xrightarrow{(b)} \underset{\mathbf{17b}}{R_3PbBr} + RBr \tag{5.37}$$

R = alkyl, phenyl

$$\underset{\mathbf{17b}}{R_3PbBr} + R'M \longrightarrow \underset{\mathbf{18}}{R_3PbR'} + MBr \tag{5.38}$$

R, R′ = Et, Bu, Ph ; M = MgBr, Li

$$ArH + \underset{\mathbf{19}}{Pb(OCOCF_3)_4} \longrightarrow \underset{\mathbf{20}}{ArPb(OCOCF_3)_3} + CF_3COOH \tag{5.39}$$

REFERENCES

1. (a) Colvin EW. Silicon in organic synthesis. Butterworths; 1981; (b) Colvin EW. Silicon reagents for organic synthesis. Academic Press; 1988; (c) Lennon PJ, Mack DP, Thompson QE. Organometallics 1989;8:1121; (d) Sommer LH, Whitmore FC. J Am Chem Soc 1946;68:481.
2. (a) Kawachi A, Tamao K. J Synth Org Chem Jpn 2001;59:892, (Japanese); (b) Kawachi A, Tamao K. Bull Chem Soc Jpn 1997;70:945.
3. Fleming I, Paterson I. Synthesis 1979: 737.
4. (a) Sakurai H. Synlett 1999; 1; (b) Fleming I, Dunogues J, Smithers R. Org React 1989;37:57.
5. Kira M, Zhang LC. In: Akiba K.-y., editor. Chemistry of hypervalent compounds. Weinheim, Wiley-VCH; 1999, Chapter 5.
6. Pereyre M, Quintard J-P, Rahm A. Tin in organic synthesis. Butterworth; 1987.
7. Leeper RW, Summers L, Gilman H. Chem Rev 1954;54:101.

NOTES 5

STABLE CARBENE AND ITS COMPLEX

Carbene is a neutral species featuring a divalent carbon atom with only six electrons in its valence shell. Carbene is the only species which has two stable electronic configurations in the ground state, that is, triplet and singlet. Fundamentally, the triplet carbene is linear and sp hybridized and the singlet carbene is bent (120°) and sp^2 hybridized (Fig. N5.1). Two singlet carbenes can dimerize to afford an alkene by nonleast motion (Fig. N5.2).

In order to stabilize simple reactive carbene, electronic effects (inductive and mesomeric) and steric effect have been utilized. A persistent triplet carbene (**2**) with a half life of 40 min, which is the most stable until now, was reported recently [1].

CF3 N2 Br / Ph / Br CHF3 / Ph — **1** —($h\nu$, $-N_2$)→ CF3 Br / Ph / Br CF3 / Ph — **2** Triplet persistent carbene (N5.1)

On the other hand, major interest in carbene chemistry has been focused in stabilizing the bent singlet carbene, which has unshared electron pair and a vacant 2p orbital. Carbene, therefore, should be nucleophilic and can also be electrophilic. In early 1960s, Wanzlick obtained an electron-rich olefin (**4**) by the

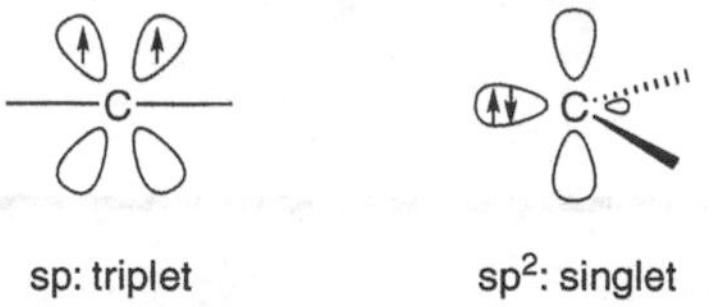

sp: triplet   sp^2: singlet

Figure N5.1 Carbene: electronic configurations of the ground state.

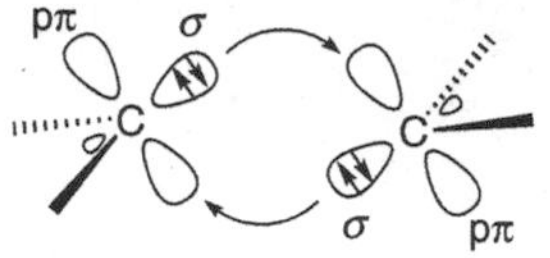

pπ σ σ pπ

Figure N5.2 Dimerization of singlet carbenes by nonleast motion.

thermolysis of **3** accompanied by the elimination of chloroform. He conjectured the presence of diaminocarbene (**A**) as an intermediate [2].

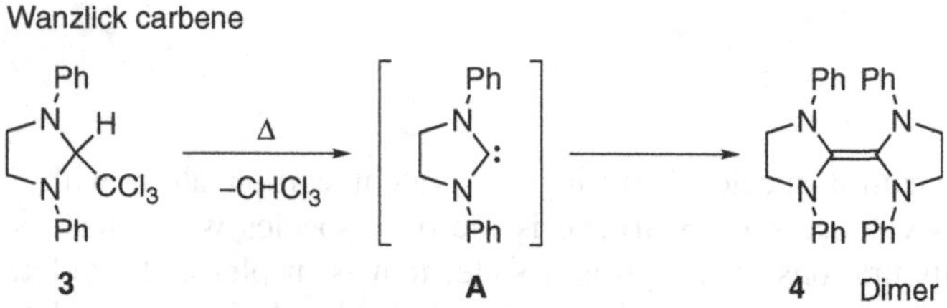

Wanzlick carbene

3 —Δ, $-CHCl_3$→ [**A**] → **4** Dimer (N5.2)

Wanzlick also obtained a dimer (**6**) by deprotonation of the corresponding imidazolium salt (**5**) but could not detect a carbene, that is, imidazol-2-ylidene (**B**) [3]. Almost two decades later, however, Arduengo could prepare a stable carbene **B** by the use of a bulky adamantyl group as R [4]. On the basis of this success, he reported the preparation of stable carbenes of **C** and **D**. This showed that the nitrogen is quite effective in stabilizing carbenes, and the effect is ascribed to the electron-withdrawing (inductive) and electron-donating (mesomeric) effects of nitrogen and the latter effect is more influential than the former (Arduengo carbene).

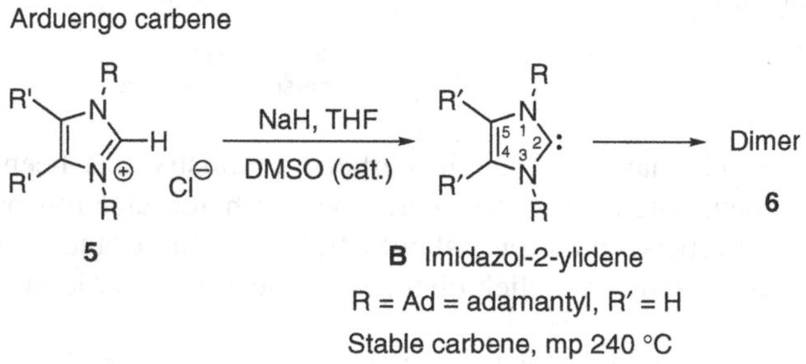

Arduengo carbene

5 —NaH, THF; DMSO (cat.)→ **B** → Dimer **6** (N5.3)

B Imidazol-2-ylidene

R = Ad = adamantyl, R′ = H

Stable carbene, mp 240 °C

Mes
N
H
N⊕ Cl⊖
Mes
7 Mes = mesityl
KH, THF
Mes
N
N
Mes
R
C Imidazolin-2-ylidene
stable
Dimer
8
(N5.4)

Me S H Me N⊕ Cl⊖ Dipp
9
Dipp = 2,6-diisopropylphenyl
KH, THF
Me S Me N Dipp
D
Thiazol-2-ylidene
stable
[Lewis acid]
Dipp Me S N Me Me N S Me Dipp
10
(N5.5)

The general resonance structures of amino carbenes are shown in Eq. (N5.6) and the idea was extended to prepare amino-aryl, amino-thio, amino-oxy carbenes (Eq. 5.7).

(N5.6)

Aminocarbene : resonance structures

Amino-aryl   Amino-thio   Amino-oxy
(N5.7)

Bertrand employed phosphorus and silicon substituents to attach to a carbene carbon and obtained a variety of stable carbenes (several weeks at 25°C) by photolysis of the corresponding diazo compounds. These types of carbenes are bent, and four contributing structures are shown in Fig. N5.3 [5].

The lone pair electrons of a singlet carbene coordinate to afford stable, heavier carbene analogs of group 14 and 15 elements. Moreover, complexes with transition metals have been prepared and used as catalysts for organic syntheses. This branch of chemistry has been energetically and fruitfully developed by Grubbs and Schrock, and recent advances are compiled well in *Chem. Rev*. 2009 (cf. Fig. 6.2 in Chapter 6.) [6a,b]. Boryl anion (**15**) is isoelectronic similar to a carbene and can have similar chemical scopes Fig. N5.4 [7].

Bertrand carbene

$(R_2N)_2\ddot{P}-\ddot{C}-SiMe_3$ ⟷ $(R_2N)_2\overset{\oplus}{P}=\overset{\ominus}{\ddot{C}}-SiMe_3$ ⟷ $(R_2N)_2P\equiv C-SiMe_3$

λ^3-Phosphacarbene λ^4-Phosphavinylidene λ^5-Phosphaacetylene

11 R = cyclohexyl, *i*-Pr ⟷ $(R_2N)_2\overset{\oplus}{P}=C=\overset{\ominus}{Si}Me_3$

λ^4-Phosphavinylidene

Figure N5.3 Possible resonance structures of Bertrand carbene.

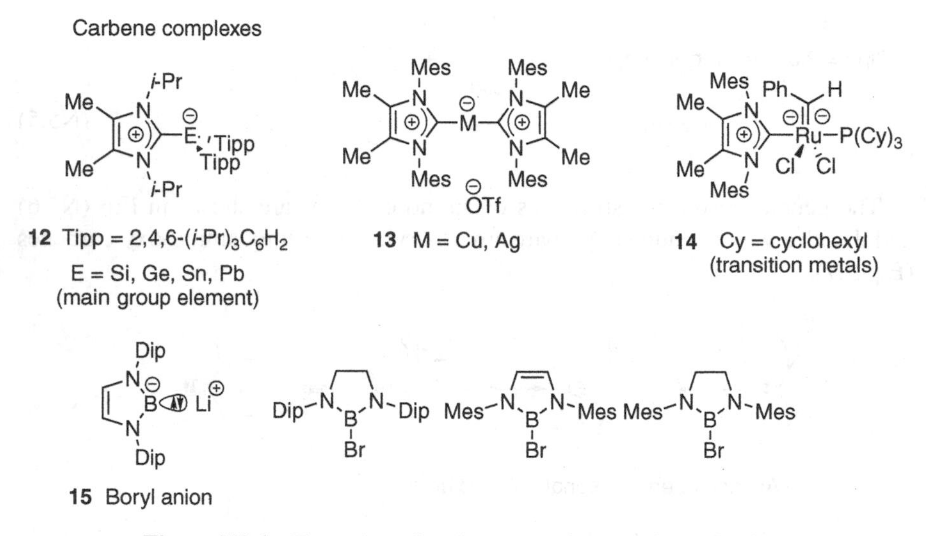

Figure N5.4 Examples of carbene complexes and boryl anion.

REFERENCES

1. Itoh T, Nakata Y, Hirai K, Tomioka H. J Am Chem Soc 2006; 128: 957.
2. (a) Wanzlick HW, Schokola E. Angew Chem 1960; 72: 494; (b) Wanzlick HW, Kleiner H. Angew Chem 1961; 73: 493.
3. Schoenherr HJ, Wanzlick HW. Chem Ber 1970; 103: 1037.
4. (a) Arduengo AJ III, Harlow RL, Kline M. J Am Chem Soc 1991; 113: 361; (b) Arduengo AJ III. Acc Chem Res 1999; 32: 913.
5. (a) Igau A, Grutzmacher H, Baceiredo A, Bertrand G. J Am Chem Soc 1988; 110: 6463; (b) Bertrand G, Reed R. Coord Chem Rev 1994; 137: 323; (c) Bourissou D, Gurret O, Gabbai FP, Bertrand G. Chem Rev 2000; 100: 39.

6. (a) Schrock RR. Chem Rev 2009; 109: 3211; (b) Mizuhata Y, Sasamori T, Tokitoh N. Chem Rev 2009; 109: 3479.
7. (a) Segawa Y, Yamashita M, Nozaki K. Science 2006; 314: 113; (b) Yamashita M, Nozaki K. Synth Org Chem Jpn 2010; 68: 359 (Japanese).

CHAPTER 6

PHOSPHORUS, ANTIMONY, AND BISMUTH COMPOUNDS

Group / (n) / Period	14 4B	15 5B	16 6B
1 1s			
2 [He] 2s2p	0.77 1086 $^{12}_{6}$**C** Carbon 2.5 368	0.70 1402 $^{14}_{7}$**N** Nitrogen 3.0 292	0.66 1314 (1.40^{2-}) $^{16}_{8}$**O** Oxygen 3.5 351
3 [Ne] 3s3p	1.17 787 $^{28}_{14}$**Si** Silicon 1.8 301	1.10 1012 $^{31}_{15}$**P** Phosphorus 2.1 264	1.04 1000 (1.74^{2-}) $^{32}_{16}$**S** Sulfur 2.5 272
4 [Ar:3d[10]] 4s4p	1.22 762 $^{74}_{32}$**Ge** Germanium 1.8 237	1.21 947 $^{75}_{33}$**As** Arsenic 2.0 200	1.17 941 (1.91^{2}) $^{80}_{34}$**Se** Selenium 2.4 245
5 [Kr:4d[10]] 5s5p	1.40 709 $^{120}_{50}$**Sn** Tin 1.8 225	1.41 834 $^{121}_{51}$**Sb** Antimony 1.9 215	1.37 869 (2.24^{2-}) $^{130}_{52}$**Te** Tellurium 2.1
6 [Xe:4f$_{14}$5d[10]] 6s6p	1.54 716 $^{208}_{82}$**Pb** Lead 1.8 130	1.52 703 $^{209}_{83}$**Bi** Bismuth 1.9 143	1.53 812 $^{209}_{84}$**Po*** Polonium 2.0

Organo Main Group Chemistry, First Edition. Kin-ya Akiba.

6.1 PHOSPHORUS COMPOUNDS

Phosphoric acid and its derivatives are quite important and essential materials for the life of animals and humans; hence there has been a vast amount and a great variety of research related to them. For example, there are phosphate fertilizers, nucleic acids including DNA and RNA, reactions related to adenosine triphosphate–diphosphate (ATP–ADP) which is the source of biological energy, and so on.

Phosphoric acid [$(HO)_3P{=}O$] is a pentavalent acid bearing the phosphoryl group ($P{=}O$). Phosphorus acid [$(HO)_3P$] is a reduced type of phosphoric acid without the phosphoryl group. Both have their esters (the HO group is successively substituted for the RO group) and reduced derivatives (HO is successively substituted for H).

Phosphine (PH_3) and phosphorane (PH_5) are prototypes of organophosphorus compounds bearing a phosphorus–carbon bond. In this chapter, their reactions of interest are mainly described.

Organophosphorus compounds can have several valences such as 2, 3, 4, 5, and 6, and they are classified as follows:

phosphide: $R_2P^-M^+$
phosphine: R_3P
phosphite: $(RO)_3P$
phosphonium salt: $R_4P^+\ Y^-$
phosphonium ylide: $R_3P^+\text{-}C^-R^1R^2$
(alkylidene phosphorane): ($R_3P{=}CR^1R^2$)
phosphine oxide and sulfide: $R_3P{=}O$ and $R_3P{=}S$
phosphate and thiophosphate: $(RO)_3P{=}O$, $(RS)_3P{=}O$, and $(RO)_3P{=}S$
phosphorane: R_5P; [$(RO)_5P$]
phosphoranate: $R_6P^-\ M^+$; $R_nP^-(OR)_{6-n}\ M^+$

Each class of compounds has its own character and reactivity. It is taken for granted that organophosphorus compounds founded the basis for organosilicon and organosulfur compounds.

Fundamental and characteristic reactions of organophosphorus compounds are as follows and they are explained in this chapter.

1. Tertiary phosphine and its nucleophilic reaction
2. Arbuzov reaction
3. Perkow reaction
4. Synthesis of optically active phosphorus compounds
5. Phosphonium salt and Wittig reaction
6. Structure and reactivity of phosphorane
7. Stability of phosphoranate

8. Formation of transition-metal complexes of phosphine and their synthetic application

6.2 SYNTHESIS OF ORGANOPHOSPHORUS COMPOUNDS

Most organophosphorus compounds for general-purpose use are commercially available. Tertiary phosphines such as tributyl- and triphenylphosphine, triethyl phosphite, and triphenyl phosphate are examples. However, their fundamental synthetic methods are explained briefly.

Primary and secondary phosphine are effectively prepared by the reduction of mono- and dihalophospine (which should be obtained commercially) with lithium aluminum hydride (LAH) or sodium metal (Eq. 6.1 and 6.2).

$$R_nPX_{3-n} + LiAlH_4 \longrightarrow R_nPH_{3-n} + LiAlX_4 \qquad (6.1)$$

$$R_nPX_{3-n} + Na \longrightarrow R_nPH_{3-n} + NaX \qquad (6.2)$$

Phosphonite $[RP(OR')_2]$, phosphonate $[RP(=O)(OR')_2]$, phosphinite $[R_2P(OR')]$, phosphinate $[R_2P(=O)(OR')]$ can also be reduced with LAH to afford the corresponding primary and secondary phosphine (Eq. 6.3 and 6.4).

$$R^1{}_nPCl_{3-n} + (R^2OH)_{3-n} \xrightarrow{Et_3N} R^1{}_nP(OR^2)_{3-n}$$

$$R^1{}_nP(=O)(Cl)_{3-n} + (R^2OH)_{3-n} \xrightarrow{Et_3N} R^1{}_nP(=O)(OR^2)_{3-n} \qquad (6.3)$$

$$R^1{}_nP(OR^2)_{3-n} + LiAlH_4 \longrightarrow R^1{}_nPH_{3-n}$$

$$R^1{}_nP(=O)(OR^2)_{3-n} + LiAlH_4 \longrightarrow R^1{}_nPH_{3-n} \qquad (6.4)$$

Tertiary phosphine is prepared most generally by the Grignard method; that is, it is obtained by dropwise addition of halophosphine to the Grignard reagent in ether. When unsymmetrical tertiary phosphine is required, phosphinite chloride (R^1R^2PCl: dialkylchlorophosphine) or phosphonite dichloride (R^1PCl_2: alkyldichlorophosphine) is used for the Grignard method (Eq. 6.5 and 6.6).

$$PX_3 + RM \longrightarrow R_3P + 3\,MX\ (M = MgX \text{ or } Li) \qquad (6.5)$$

$$RR^1PCl + R^2M \longrightarrow RR^1PR^2 + MX\ (M = MgX \text{ or } Li) \qquad (6.6)$$

Esters of phosphoric acid $[(P(=O)(OR)_3]$ and also derivatives of similar structures mentioned above can be applied for the Grignard method (R^1MgX) to substitute the ester group(s) to R^1 (Eq. 6.7). This is explained in more detail for the synthesis of asymmetric phosphine.

$$R_2P(OR) + R^1MgX \longrightarrow R_2PR^1$$
$$R_2P(=O)(OR) + R^1MgX \longrightarrow R_2P(=O)R^1 \tag{6.7}$$

Alkalimetal phosphide (R^1R^2PM) reacts with an alkylating reagent to give tertiary phosphine and also with an unsaturated halide to afford unsaturated tertiary phosphine with retention of the configuration of the unsaturated part of the halide (Eq. 6.8).

$$R^1{}_2PM + R^2X \xrightarrow{(a)} R^1{}_2PR_2 + MX \quad (R^2 = \text{alkyl, aryl, acyl})$$
$$R^1{}_2PM + H_2C\overset{X}{\triangle}CH_2 \xrightarrow{(b)} R^1{}_2PCH_2CH_2XH \quad (X = NR^2, O, S) \tag{6.8}$$
$$R^1{}_2PM + \ce{>C=C<}^{H}_{X} \xrightarrow{(c)} \ce{>C=C<}^{H}_{PR^1_2}$$

Tertiary phosphine oxide and sulfide are reduced with LAH, Si_2Cl_6, and some others to the corresponding phosphine (Eq. 6.9). Friedel–Crafts reaction of PCl_3 with aromatic hydrocarbon is also a useful synthetic method for Ar_3P.

$$R_3P\,(=X) \xrightarrow{\text{LAH, etc.}} R_3P \tag{6.9}$$

Fundamental synthetic methods for tertiary phosphine oxide are Grignard method and oxidation of the obtained tertiary phosphine with air, hydrogen peroxide, and hydroperoxide (Eqs. 6.10 and 6.11). Tertiary phosphine sulfide is prepared by reaction of tertiary phosphine with sulfur and also by the Grignard method (Eq. 6.12). By the Grignard method, alkyl Grignard reagent affords diphosphane disulfide instead of monosulfide which is the case for the aryl Grignard reagent.

$$R_nP(O)X_{3-n} + (3-n)R^1M \longrightarrow R_nP(O)R_{3-n} + (3-n)MX \ (X = MgX \text{ or } Li) \tag{6.10}$$

$$R_3P \xrightarrow{[O]} R_3P{=}O \tag{6.11}$$

$$RP(S)Cl_2 + 2\,R^1MgX \longrightarrow RP(S)R^1{}_2 \text{ or } RR^1P(S)P(S)RR^1 + 2\,MgCl \tag{6.12}$$

6.3 TERTIARY PHOSPHINE AND ITS NUCLEOPHILIC REACTION

Phosphine has some similarity with amine, because phosphine bears an unshared electron pair. Basicity of phosphine is, however, weaker than that of amine; on the other hand; its nucleophilicity is stronger than that of amine. Accordingly, phosphine reacts with alkyl halide to form phosphonium salt easily (Eq. 6.13), and with ammonium salt it affords phosphonium salt and amine (Eq. 6.14).

$$R_3\ddot{P} + R'X \longrightarrow R'R_3P^+X^- \qquad (6.13)$$

$$R_4N^+X^- + R'_3\ddot{P} \longrightarrow RR'_3P^+X^- + R_3\ddot{N} \qquad (6.14)$$

Nucleophilic reaction of phosphine with saturated alkyl halide proceeds via the S_N2 reaction and the configuration of the carbon is inverted. This is explicitly shown by stereochemistry of deoxygenation of *cis*-2-butene epoxide (**1**) by tributylphosphine to afford *trans*-2-butene (**2**) (Eq. 6.15). Carbon–carbon bond rotation takes place at zwitter ion (**A**) to yield oxaphosphetane (**B**), then phosphine oxide is eliminated to give the olefin **2**. The structure of oxaphosphetane (**B**) is the same as that of the intermediate of Wittig reaction.

$$\mathbf{1} + Bu_3P \longrightarrow \mathbf{A}\ (Bu_3P^{\oplus}\text{–}CH(Me)\text{–}CH(Me)\text{–}O^{\ominus}) \longrightarrow \mathbf{B}\ (Bu_3P\text{–}O) \longrightarrow \mathbf{2}\ (\text{trans-}MeCH{=}CHMe) + Bu_3P{=}O \qquad (6.15)$$

Recently, tertiary phosphines with bulky alkyl groups were prepared and used to coordinate with transition metals. The complexes show remarkable catalytic reactivity and are quite efficient reagents for organic synthesis. Trialkylphosphine is more electron-donating than triphenylphosphine and hence it coordinates with transition metal more strongly than the latter. Moreover, steric demands can be complied with alkyl groups easily. As transition-metal complexes, in comparison with those of triphenylphosphine, they (i) accelerate oxidative addition of organic halides and (ii) also accelerate reductive elimination of substrates by steric repulsion. For instance, oxidative addition of aryl halide takes place to palladium(0) complex, and amination of benzene ring occurs because of the reductive elimination effected after halogen–amine exchange in the presence of the base. Catalytic amination and alkoxylation of benzene ring have been developed actively by employing a variety of bulky trialkylphosphines (Fig. 6.1). [1]

Tricyclohexylphosphine is used as ruthenium complexes, which are extremely useful for the olefin metathesis reaction (Fig. 6.2). The reaction cannot take

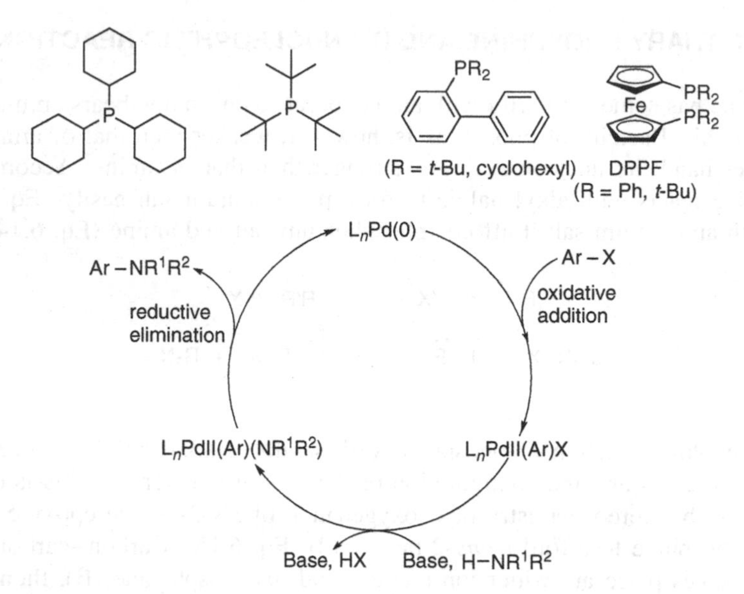

Figure 6.1 Tertiary phosphine with bulky substituent and its catalytic reaction.

place with triphenylphosphine. These ruthenium catalysts strongly coordinate with the carbon–carbon double bond but not with usual functional groups such as alcohol, amide, carboxylic acid, and so on. Therefore, they can be used with olefins bearing a variety of functional groups. Synthesis of cinnamic acid from acrylic acid and styrene is shown as an example. The Nobel Prize for chemistry in 2005 was awarded for olefin metathesis and the reaction has been focused intensively and the catalytic cycle is also illustrated (Fig. 6.2) [2].

PCy3, Cl, Ru, H, Ph, PCy3

Cy = cyclohexyl

Ruthenium complex

$$R^1CH{=}CH_2 + H_2C{=}CHR^2 \rightleftharpoons R^1CH{=}CHR^2 + H_2C{=}CH_2$$

$$PhHC{=}CH_2 + H_2C{=}CHCO_2H \longrightarrow PhHC{=}CHCO_2H + H_2C{=}CH_2$$

Figure 6.2 Ruthenium complexes of Grubbs and olefin methathesis and the addition mechanism of olefin methathesis.

Grubbs olefin methathesis

Figure 6.2 (*Continued*)

6.4 ARBUZOV REACTION

Trialkyl phosphite (**3**) reacts with alkyl halide to afford dialkyl phosphonate (**4**) (Eq. 6.16). This is called *Arbuzov reaction* basically and is more generally used for conversion of trivalent P–(OR) type compound (**5**) to pentavalent phosphoryl (R′–P=O) type compound (**6**) (Eq. 6.17). This reaction proceeds through two steps. Firstly, the unshared electron pair of phosphorus attacks the carbon of alkyl halide (R^4X) to form phosphonium salt (**C**) via S_N2 reaction; then the resulting halide ion (X^-) reacts with the carbon of R^3 by S_N2 to eliminate R^3X. Accordingly, the configuration of the carbon of R^3 and R^4 is inverted. When R^3X is more reactive than R^4X, the product becomes a mixture of $R^4R^1R^2P{=}O$ and $R^3R^1R^2P{=}O$.

$$\underset{\mathbf{3}}{P(OR)_3} + R'X \longrightarrow \underset{\mathbf{4}}{R'{-}\overset{O}{\overset{\|}{P}}(OR)_2} + RX \qquad (6.16)$$

$$\underset{\mathbf{5}}{R^1R^2P(OR^3)} + R^4X \longrightarrow \underset{\mathbf{C}}{R^4R^1R^2\overset{\oplus}{P}(OR^3)\overset{\ominus}{X}} \longrightarrow \underset{\mathbf{6}}{R^4R^1R^2P{=}O} + R^3X \qquad (6.17)$$

One typical example of Arbuzov reaction involves heating triethyl phosphite with a small amount of ethyl iodide to give diethyl ethylphosphonate in high yield (Eq. 6.18). The following three facts are experimentally established: (i) the concentration of ethyl iodide is maintained constant throughout the reaction; (ii) the reaction rate is proportional to the concentration of ethyl iodide; (iii) the reaction is faster in solvents with larger dipole moment than in nonpolar solvent. These results imply that the rate-determining step of Arbuzov reaction is the first step to generate the ionic intermediate (**C**).

$$P(OEt)_3 + EtI \longrightarrow Et\overset{+}{P}(OEt)_3\,I^- \longrightarrow Et-\overset{O}{\overset{\|}{P}}(OEt)_2 + EtI \qquad (6.18)$$

The reactivity order of tertiary phosphine to ethyl iodide is as follows: Et_3P > Ph_3P > $Ph_2P(OEt)$ > $PhP(OEt)_2$ > $P(OEt)_3$. For the order, the inductive effect of the substituent exerts the primary effect and the electron-donating effect of oxygen(s) (resonance effect) does not affect the order significantly.

A variety of phosphonic esters (**4**) have been prepared by Arbuzov reaction and they are used as reagents for carbonyl olefination (Wadsworth–Emmons reaction; one of revised methods of Wittig reaction, see Fig. 6.7).

6.5 PERKOW REACTION

Trialkyl phosphite (**3**) reacts with α-halocarbonyl compound to give enol phosphate (**7**) and alkyl halide (Eq. 6.19). This unique reaction is called *Perkow reaction*.

$$P(OR^1)3 + R^2\overset{O}{\overset{\|}{C}}CR^3R^4X \longrightarrow \underset{\mathbf{D}}{(R^1O)_3\overset{\oplus}{P}-\overset{O^{\ominus}}{\overset{|}{\underset{R^2}{\underset{|}{C}}}}-CR^3R^4X}$$

$$\longrightarrow \underset{\mathbf{E}}{(R^1O)_3\overset{\oplus}{P}-O-\underset{R^2}{C}=CR^3R^4\ X^{\ominus}} \longrightarrow \underset{\mathbf{7}}{(R^1O)_2\overset{O}{\overset{\|}{P}}-O-\underset{R^2}{C}=CR^3R^4} + R^1X \qquad (6.19)$$

First, an unshared electron pair of phosphite attacks the carbon of the carbonyl group to yield the zwitter ion (**D**); then the resulting oxide ion attacks the phosphonium group intramolecularly to rearrange to the second phosphonium salt (**E**), which is similar to the phosphonium salt (**C**) of Eq. (6.17). The halide ion generated in situ reacts with alkoxyl group (R^1O) in an S_N2 manner to give the final product, that is, enol phosphate (**7**) and alkyl halide (R^1X). Attention should be paid to the movement of the electron within the zwitter ion (**D**) to effect three-center rearrangement.

Trimethyl phosphite reacts with 2-chloroacetoacetic ester **8** at the carbon of the acetyl group from the least hindered direction to form the zwitter ion (**F**). By a three-center rearrangement, **F** gives the phosphonium salt (**G**), which is the same type of phosphonium salt as **E**. Enol phosphate of the trans type (**9**) is obtained finally, accompanied with methyl chloride yielded by Arbuzov reaction.

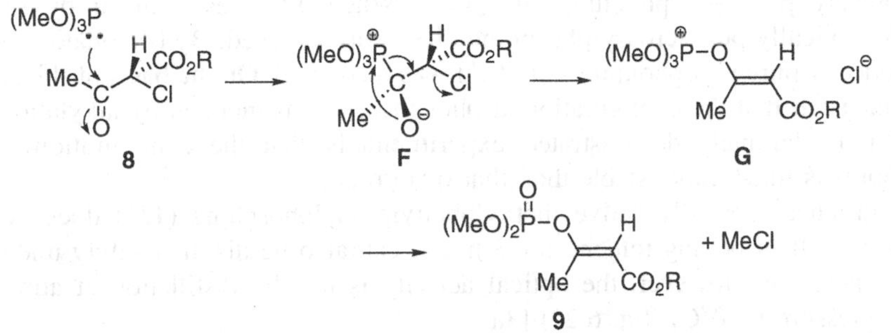

(6.20)

Trimethyl phosphite and α-diketone (**10**) form the five-member cyclic phosphorane (**11**), which is called the *Ramirez compound* (Eq. 6.21). During the reaction, the phosphite attacks the carbon of a carbonyl group to generate the zwitter ion (**H**) followed by the three-center rearrangement to afford the zwitter ion, that is, a vinyl-type phosphonium salt (**I**). The latter gives the five-member cyclic phosphorane by intramolecular reaction. It has long been conjectured as the direct attack of phosphite at the oxygen of carbonyl group to generate the phosphonium salt (**E, G, I**) directly. But at present, on the basis of the substituent effect and kinetic investigations of these reactions, it is realized that phosphite attacks the carbon of the carbonyl group to generate zwitter ion (**D, F, H**) first, and the zwitter ion affords phosphonium salt (**E, G, I**) by three-center rearrangement.

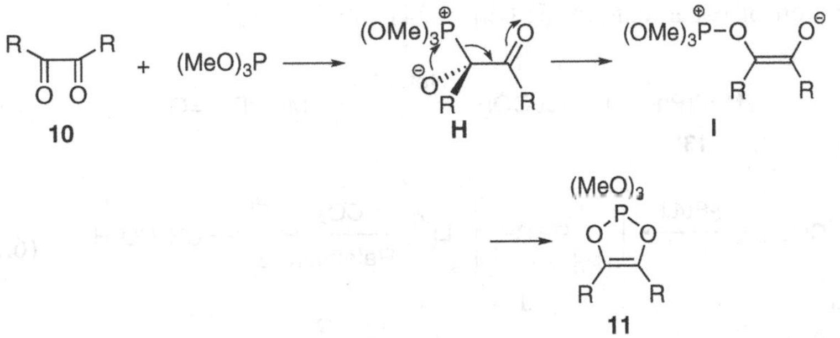

(6.21)

6.6 SYNTHESIS OF OPTICALLY ACTIVE PHOSPHINES

The first synthesis of an optically active tertiary phosphine was accomplished by electrochemical reduction of the benzylphosphonium salt. That is, phosphonium salt bearing a benzyl group (**12**) was treated with D-dibenzoyltartaric acid and the resulting complex was purified by recrystallization, thereby obtaining the optically pure phosphonium salt (**12***). When **12*** was reduced by the cathode, optically pure tertiary phosphine (**13***) was obtained. By benzylation of **13***, optically pure phosphonium salt (**12***) was recovered. On the basis of this, it was concluded that the configuration at phosphorus is maintained by alkylation. Hence it is definitely demonstrated experimentally that the configuration of phosphorus is much more stable than that of nitrogen.

For instance, optically active methylphenylpropylphosphine (**13***) does not racemize at all in boiling toluene for 3 h and can also be distilled safely under reduced pressure. However, the optical activity is lost by distillation at atmospheric pressure (230°C) (Eq. 6.22) [3a].

$$[\mathrm{MePrPhPCH_2Ph}]^{\oplus}\mathrm{Br}^{\ominus}\ (\mathbf{12}) \xrightarrow[\text{D-Dibenzoyl tartaric acid}]{\text{Optical resolution}} (+)\text{-}[\mathrm{MePrPh\overset{*}{P}CH_2Ph}]^{\oplus}\mathrm{Br}^{\ominus}\ (\mathbf{12^*})$$

* Denotes optically active compound

$$\xrightarrow[-\,\mathrm{PhCH_2\bullet}]{\text{Electrode reduction}} (+)\text{-}\ \mathrm{MePrPh\overset{*}{P}}\ (\mathbf{13^*}) \xrightarrow{\mathrm{PhCh_2Br}} (+)\text{-}[\mathrm{MePrPh\overset{*}{P}CH_2Ph}]^{\oplus}\mathrm{Br}^{\ominus}\ (\mathbf{12^*}) \qquad (6.22)$$

Using the optically active tertiary phosphine, stereochemistry of a couple of reactions at phosphorus was investigated. Stereochemistry of both reactions at phosphorus proceeds by retention: (i) oxidation with hydroperoxide (Eq. 6.23); (ii) generation of α-carboanion (**J**) (Eq. 6.24) [3b].

$$\mathrm{MePrPh\overset{*}{P}}\ (\mathbf{13^*}) + t\text{-BuOOH} \xrightarrow{\text{Retention}} \mathrm{MePrPh\overset{*}{P}{=}O} \qquad (6.23)$$

$$\mathrm{(Pr)(Ph)\overset{*}{P}{-}CH_3}\ (\mathbf{13^*}) \xrightarrow{t\text{-BuLi}} \left[\mathrm{(Pr)(Ph)\overset{*}{P}{-}CH_2}\right]^{\ominus}\mathrm{Li}^{\oplus}\ (\mathbf{J}) \xrightarrow[\text{Retention}]{\mathrm{CO_2}} \mathrm{(Pr)(Ph)\overset{*}{P}{-}CH_2CO_2H} \qquad (6.24)$$

Phenyldichlorophosphine (phenylphosphonous dichloride) and two moles of menthol gave the diester **14** (dimenthyl phenylphosphonite) in the presence of pyridine, and **14** afforded menthyl methylphenylphosphinate (**15**) by Arbuzov reaction with methyl iodide (Eq. 6.25). Compound **15** is a mixture of diastereomers and one of the diastereomer is obtained by recystallization; hence optically active **15*** is prepared. The Grignard reagent substituted the menthyloxy group and optically active tertiary phosphine oxide (**16***) was obtained. Stereochemistry at the phosphorus was completely inverted by this substitution [4].

$$\mathrm{PhPCl_2} + 2\,\mathrm{MenOH} \xrightarrow{2\ \text{pyridine}} \underset{\mathbf{14}}{\mathrm{PhP(OMen)_2}} \xrightarrow{\mathrm{MeI}}$$

$$\underset{\mathbf{15}}{\mathrm{MePhP(O)(OMen)}} \xrightarrow{\text{Resolution}} \underset{\mathbf{15^*}}{\mathrm{MePh\overset{*}{P}(O)(OMen)}} \xrightarrow[\text{Inversion}]{\mathrm{RMgX}} \underset{\mathbf{16^*}}{\mathrm{RMePh\overset{*}{P}{=}O}} \quad (6.25)$$

MenOH = Menthol

Stereochemistry of phosphorus was retained during the reaction of P–H of phosphinate ester (**17***) with sulfur in the presence of a base, giving optically active **18*** (Eq. 6.26). Substitution of the methylthio group (MeS) by the Grignard reagent proceeded with retention of configuration at phosphorus; however, substitution of the alkoxyl group (menthyl-*O*) took place to give **20** by inversion. These reactions are stereospecific (100%) and hence they are quite useful synthetically.

$$\underset{\mathbf{17^*}}{\mathrm{O{=}\overset{*}{P}(Ph)(H)(OMen)}} \xrightarrow{\mathrm{S_8},\ (\mathrm{C_6H_{11}})_2\mathrm{NH}} \xrightarrow[\text{Retention}]{\mathrm{MeI}} \underset{\mathbf{18^*}}{\mathrm{O{=}\overset{*}{P}(Ph)(SMe)(OMen)}} \xrightarrow[\text{Retention}]{\mathrm{MeMgBr}}$$

$$\underset{\mathbf{19^*}}{\mathrm{O{=}\overset{*}{P}(Ph)(Me)(OMen)}} \xrightarrow[\text{Inversion}]{\mathrm{PrMgBr}} \underset{\mathbf{20^*}}{\mathrm{(Ph)(Me)(Pr)\overset{*}{P}{=}O}} \quad (6.26)$$

The double bond of phosphine oxide (P=O) is quite strong and it is generally difficult to reduce the oxide to tertiary phosphine. However, optically active tertiary phosphine is important and a variety of methods have been tried to reduce the optically active oxide to optically active phosphine.

Optically active tertiary phosphine oxide (**21***) reacts with trichlorosilane to generate the zwitter ion (**K**) (Fig. 6.3). When heated, the zwitter ion (**K**) affords phosphonium ion (**L**) by intramolecular hydride transfer to give tertiary phosphine (**22***), retaining the stereochemistry. Optical purity of the product is 75–90%. When the same reaction is carried out in the presence of triethylamine (a strong base), a complex (zwitter ion) with trichlorosilane is first formed and the complex attacks the zwitter ion (**K**) from the rear side donating hydride ion to afford **M**, where the P–O bond is cleaved. The stereochemistry on the phosphorus is inverted to give **23***. Optical purity of **23*** is 100% (inversion) with

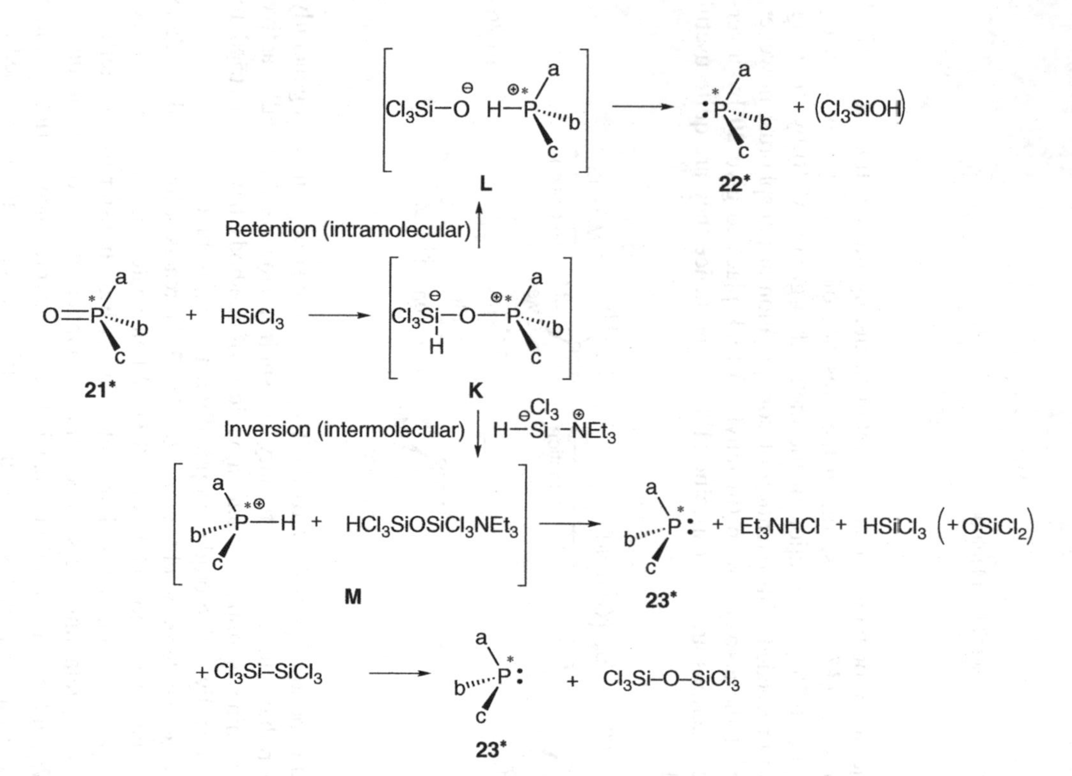

Figure 6.3 Stereochemistry of reduction of phosphine oxide.

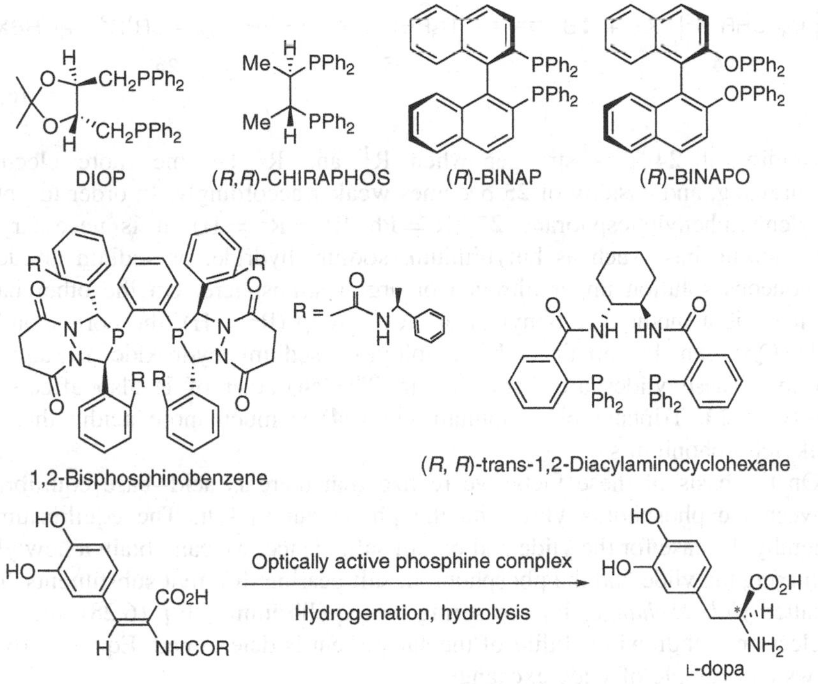

Figure 6.4 Optically active phosphine ligand and asymmetric reduction.

triethylamine, but it is lowered with weaker bases such as *N*,*N*-dimethylaniline, and others [5].

On the other hand, the reduction of **21*** to give **23*** takes place quantitatively (100%) by inversion with the use of hexachlorodisilane ($Cl_3Si{-}SiCl_3$).

In Section 6.3, the characteristic reactions on the phosphorus atom have been described, in which due attention was paid to the stereochemistry. On the other hand, recent attention is focused on tertiary phosphines connected to optically active carbon skeletons. These do not contain any optical activity on phosphorus but they form optically active transition-metal complexes, being coordinated with rhodium, ruthenium, and palladium, and so on. They are utilized as catalysts for asymmetric synthesis quite effectively. Several examples are shown in Fig. 6.4. Professor Noyori was awarded the Nobel Prize (2001) for successful contributions to catalytic asymmetric reduction using 2,2′-bis (diphenylphosphino)-1,1′-binaphthyl (BINAP) and others [6].

6.7 YLIDE AND WITTIG REACTION AND RELATED ONES

As described in Chapter 2, phosphonium salt (**24**) exchanges proton much more rapidly than ammonium salt and phosphorus ylide (**25**) is more stable than nitrogen ylide (Eq. 6.27). Phosphine-alkylene or alkylidenephosphorane (**25′**) is its resonance form.

$$[R_3PCHR^1R^2]^{\oplus} X^{\ominus} + :B \rightleftharpoons R_3\overset{\oplus}{P}-\overset{\ominus}{C}R^1R^2 \longleftrightarrow R_3P{=}CR^1R^2 + HBX \qquad (6.27)$$

24 → **25** ↔ **25′**

Acidity of **24** gets stronger when R^1 and R^2 become more electron-withdrawing, and basicity of **25** becomes weaker accordingly. In order to obtain metylenetriphenylphosphorane **25** ($R = Ph$, $R^1 = R^2 = H$), it is necessary to use a strong base such as butyllithium, sodium hydride, or sodium amide in nonaqueous solution under nitrogen or argon atmosphere. On the other hand, when R^1 is a methoxycarbonyl or an acyl group ($R^2 = H$), the corresponding ylide (**25**) can be obtained by employing sodium hydroxide in aqueous solution. These ylides are stable in air. The character of R also affects the acidity of **24**. Triphenylphosphonium salt (**24**) is much more acidic than the trialkylphosphonium salt.

On the basis of these facts, we realize that there is acid–base equilibrium between the phosphorus ylide and the phosphonium salt. The equilibrium is generally the case for the ylide and onium salt. Hence, we can obtain a new ylide by mixing the ylide and the phosphonium salt bearing different substituents. This is called *ylide exchange*. By measuring the equilibrium of Eq. (6.28), the order of electron-withdrawing ability of the substituent is determined. Equation (6.29) shows an example of ylide exchange.

$$R_3P{=}CHR^1 + [R_3PCH_2R^2]X \rightleftharpoons R_3P{=}CHR^2 + [R_3PCH_2R^1]X \qquad (6.28)$$

$$2\,Ph_3P{=}CH_2 + PhCOCl \rightleftharpoons Ph_3P{=}CHCOPh + [Ph_3PCH_3]Cl \qquad (6.29)$$

Order of electron-withdrawing ability: $PhCO > CO_2Me > Ph > alkyl$

Phosphorus ylide is generally obtained by deprotonation of phosphonium salt with base and ylide exchange (Eqs. 6.27 and 6.28). Other than these, phosphorus ylide can be formed by the reaction of tertiary phosphine as follows: (i) with carbene (Eq. 6.30); (ii) with diazo compound (Eq. 6.31); (iii) with conjugated olefin (Eq. 6.32); and (iv) with tetrahalogenomethane (Eq. 6.33).

$$Ph_3P + :CXY \longrightarrow Ph_3P{=}CXY \qquad (6.30)$$

(X, Y = H, Cl, Br)

$$Ph_3P + R_2CN_2 \longrightarrow Ph_3P{=}N{-}N{=}CR_2 \xrightarrow{\Delta} Ph_3P{=}CR_2 + N_2 \qquad (6.31)$$

$$Ph_3P + H_2C{=}CHR \longrightarrow Ph_3\overset{\oplus}{P}CH_2\overset{\ominus}{C}HR \longrightarrow Ph_3P{=}CHCH_2R \qquad (6.32)$$

($R = CN, CO_2R^1, CONH_2$)

$$2\,Ph_3P + CX_4 \longrightarrow Ph_3P{=}CX_2 + Ph_3PX_2 \quad (X = Cl, Br) \tag{6.33}$$

Wittig reaction is to convert the carbonyl group to the carbon–carbon double bond by the reaction of alkylidenetriphenylphosphorane (**26**) with a variety of carbonyl compounds (e.g., aldehyde, ketone, and ester of aliphatic and aromatic as well as cyclic and acyclic acids) (Eq. 6.34). In the case of conjugated carbonyl compounds, the carbonyl group is converted to the carbon–carbon double bond cleanly. The resulting alkene is a mixture of trans and cis isomers and the ratio depends on reaction conditions. The ratio of *trans*- and *cis*-stilbenes was 75:25 in ether using PhLi as base and was 47:53 in ethanol employing EtONa as base (Eq. 6.35).

$$\underset{\mathbf{26}}{Ph_3P{=}CR^1R^2} + R^3R^4C{=}O \longrightarrow \underset{\mathbf{27}}{R^1R^2C{=}CR^3R^4} + Ph_3P{=}O \tag{6.34}$$

$$[Ph_3PCH_2Ph]X \xrightarrow[\text{EtONa/EtOH}]{\text{PhLi/Et}_2\text{O}} \xrightarrow{\text{PhCHO}} \text{PhCH=CHPh} \quad \textit{trans:cis} \; 75{:}25 \; (\text{PhLi/Et}_2\text{O}),\; 47{:}53 \; (\text{EtONa/EtOH}) \tag{6.35}$$

Figure 6.5 illustrates reaction mechanism of the Wittig reaction using ethoxycarbonylmethylenephosphorane (**28**) and benzaldehyde [7]. First, benzaldehyde and **28** generates a zwitter ion (**N**), and **N** cyclizes to form an oxaphosphetane (**O**); then **O** decomposes to the product alkene **29** and phosphine oxide (path a).

Reaction of an epoxide (**30**) and tertiary phosphine (R_3P) gives the same alkene **29** and this confirms the initial formation of a zwitter ion (**N**) (path b). When *m*-chlorobenzaldehyde is added to the above reaction (path b) from the beginning, the product alkene becomes a mixture of **31** containing *m*-chlorophenyl group and **29** (path c). On the basis of this, we can conclude that the zwitter ion (**N**) partially dissociates to the starting materials (**28** and benzaldehyde) in equilibrium and the resulting **28** follows path c. The cis–trans ratio of the product alkene depends on reaction conditions and this is rationalized by the presence of dissociation equilibrium of the zwitter ion (**N, P**).

An intermediate oxaphosphetane (**R**) is observed at low temperature by NMR (Eq. 6.36). At −70°C, a sterically more hindered *cis*-oxaphosphetane (**R**) compared to the trans form is generated and **R** decomposes to a *cis*-alkene at around 0°C. During the process, **R** generates a zwitter ion (**S**) partially, and hence dissociation equilibrium (d) can take place. This was certified by the formation of a mixture of *cis*- and *trans*-alkene at −5°C. The dissociation equilibrium of oxaphosphetane via zwitter ion takes place only when substituent(s) of an ylide is an alkyl group. The equilibrium between oxaphosphetane (**R**) and the zwitter ion (**S**) does not take place and the decomposition of the oxaphosphetane to an

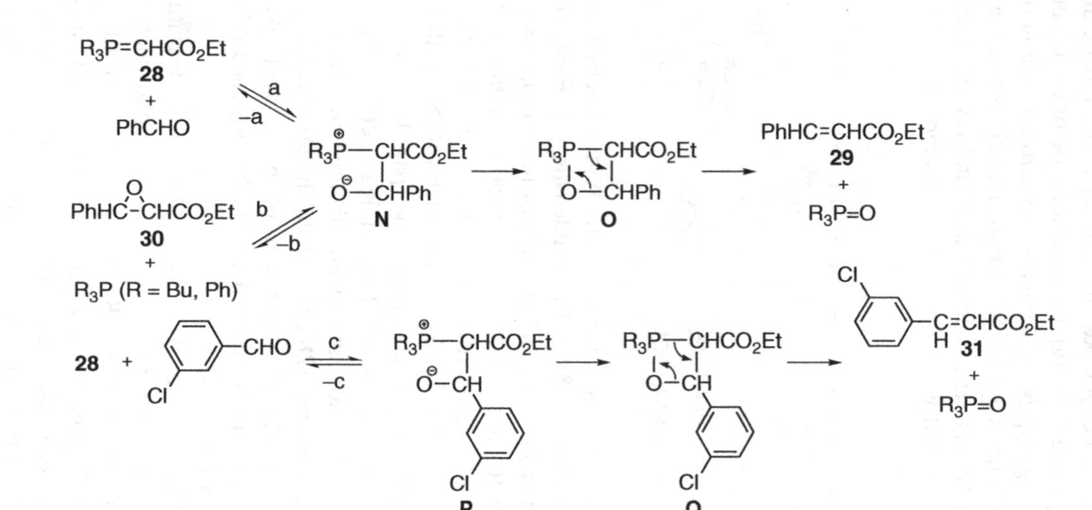

Figure 6.5 Mechanism of Wittig reaction: dissociation of the zwitter ion.

alkene undergoes quite rapidly when substituent bears an electron-withdrawing group (**O, Q** etc.; see Fig. 6.5).

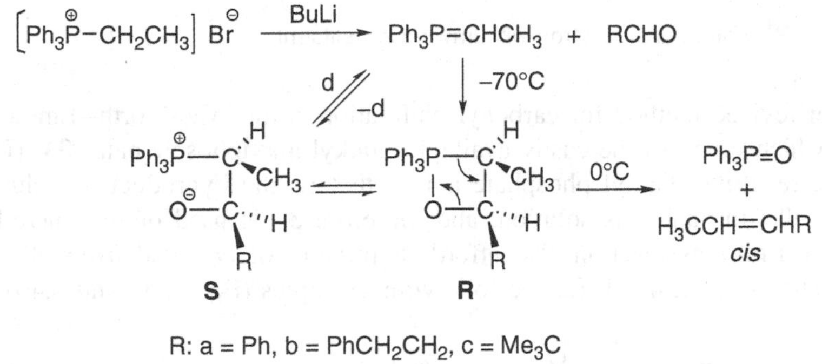

$$[Ph_3\overset{\oplus}{P}-CH_2CH_3]\ \overset{\ominus}{Br} \xrightarrow{BuLi} Ph_3P{=}CHCH_3 + RCHO \xrightarrow{-70^\circ C} \mathbf{S} \rightleftharpoons \mathbf{R} \xrightarrow{0^\circ C} Ph_3P{=}O + H_3CCH{=}CHR\ (cis) \quad (6.36)$$

R: a = Ph, b = $PhCH_2CH_2$, c = Me_3C

How about stereochemistry at phosphorus? Retention at phosphorus is confirmed for alkylation and oxidation using an optically active methylphenylpropylphosphine ((+)-**13***). It is also the case for Wittig reaction because a phosphonium salt (+)-**12*** gives a phosphine oxide (+)-**20*** quantitatively. But inversion takes place by alkaline hydrolysis of phosphonium salt because the phosphonium salt (+)-**12*** affords the oxide (-)-**20*** quantitatively (Fig. 6.6).

α-Carbanion is stabilized by the phosphonium group in ylide and it is also stabilized by the phosphoryl group (P=O). This means that alkyldiphenylphosphine oxide (**32**) can also be used for carbonyl olefination (Eq. 6.37). In this case, the stabilization of α-carbanion is weaker than ylide, and hence the reactivity of α-carbanion (**T**) is higher than ylide and reacts with carbonyl compounds under milder conditions. Moreover, diphenylphosphinate anion (**33**), that is the byproduct, is water soluble and therefore it is much easier to be separated from the product alkene than triphenylphosphine oxide. Thus, **32** is also used for carbonyl olefination. This method is called the *Horner reaction*.

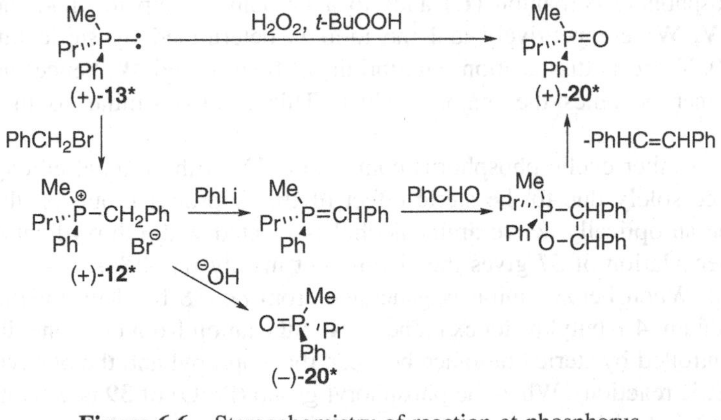

Me(Pr)(Ph)P: (+)-**13*** —H_2O_2, *t*-BuOOH→ Me(Pr)(Ph)P=O (+)-**20***

(+)-**13*** —$PhCH_2Br$→ $[Me(Pr)(Ph)\overset{\oplus}{P}-CH_2Ph]\ \overset{\ominus}{Br}$ (+)-**12*** —PhLi→ Me(Pr)(Ph)P=CHPh —PhCHO→ Me(Pr)(Ph)P(–CHPh–O–CHPh) —(-PhHC=CHPh)→ (+)-**20***

(+)-**12*** —$^{\ominus}OH$→ O=P(Me)(Pr)(Ph) (−)-**20***

Figure 6.6 Stereochemistry of reaction at phosphorus.

$$Ph_2P(=O)\text{-}CH_2R^1 \ (\mathbf{32}) \xrightarrow{\text{Base}} Ph_2P(=O)\text{-}\overset{\ominus}{C}HR^1 \ (\mathbf{T}) \xrightarrow{R^2R^3C=O} R^2R^3C=CHR^1 + Ph_2P(=O)O^{\ominus} \ (\mathbf{33}) \quad (6.37)$$

R^1 = alkyl, aryl, electron-withdrawing substituent

Another revised method for carbonyl olefination is the Wadsworth–Emmons reaction, which employs the easily available dialkyl alkylphosphonate (**34**) (Eq. 6.38). The resulting dialkyl phosphate anion, that is, the byproduct, is soluble in weakly alkaline aqueous solution, and therefore **34** is used quite generally. Wadsworth–Emmons reaction also affords a mixture of *cis*- and *trans*-alkene [8]. The ratio is about 40:60 for the following examples (Eqs. 6.39 and 6.40).

$$(RO)_2P(=O)\text{-}CH_2R^1 \ (\mathbf{34}) \xrightarrow{\text{Base}} (RO)_2P(=O)\text{-}\overset{\ominus}{C}HR^1 \ (\mathbf{U}) \xrightarrow{R^2R^3C=O} R^2R^3C=CHR^1 + (RO)_2P(=O)O^{\ominus}$$

R^1 = alkyl, aryl, electron-withdrawing substituent

(6.38)

$$(MeO)_2P(=O)\text{-}\overset{\ominus}{C}H CO_2Me + EtCOMe \longrightarrow \text{Et(Me)C=CH(CO}_2\text{Me)}$$

cis-**35** 43% (Et, CO_2Me / Me, H) + *trans*-**35** 57% (Et, H / Me, CO_2Me) (6.39)

$$(EtO)_2P(=O)\text{-}\overset{\ominus}{C}(Me)CO_2Me + Me_2CHCHO \longrightarrow Me_2HC\text{-}CH=C(Me)CO_2Me$$

cis-**36** 40% (Me_2HC, CO_2Me / H, Me) + *trans*-**36** 60% (Me_2HC, Me / H, CO_2Me)

(6.40)

The mechanism of the reaction is similar to that of the aldol reaction. Alkylphosphonate carbanion (**U**) adds to a carbonyl group to yield the oxide anions (**V, W**) competitively and the ratio is determined by steric hindrance (Fig. 6.7). There is dissociation equilibrium in both **V** and **W**; hence the stable trans isomer becomes the major product. This is also similar to the Wittig reaction.

A six-member cyclic phosphorus compound (**37**) with an axial ethoxy group is obtained solely due to the steric effect of the isopropyl group on the nitrogen when an optically active aminoalcohol is reacted with ethoxydichlorophosphine. Benzylation of **37** gives the cis product mainly (*cis*-**38**:*trans*-**38** = **15:1**) (Fig. 6.8). When benzyl anion is generated from *cis*-**38** by deprotonation with *t*-butyllithium, 4-*t*-butylcyclohexanone adds to the anion from only one direction, being controlled by steric hindrance between the isopropyl and the phenyl groups (asymmetric reaction). When the phosphoryl group (P=O) of **39** is activated with trityl (triphenylmethyl) cation and treated with lutidine successively, an optically

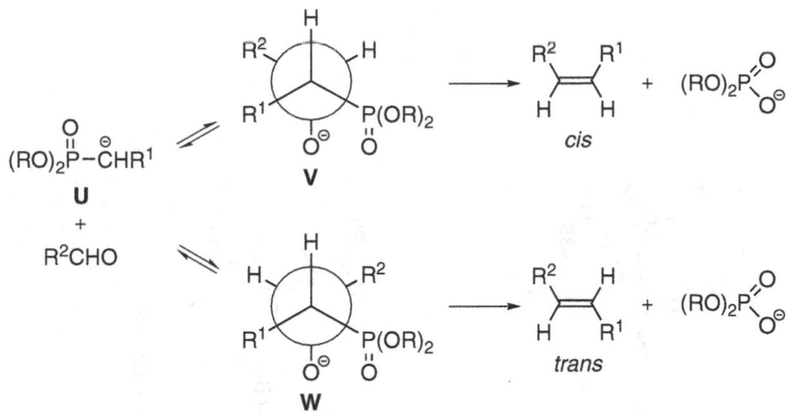

Figure 6.7 Mechanism of Wordsworth–Emmons reaction.

active olefin (**40**) is obtained in high yield (60–80%), where **X** should be the intermediate. The optical purity of **40** is 99% [9].

Using optically active 1,2-dimethylaminocyclohexane, a bicyclic allylphosphoryl compound (**41**) is easily obtained. α-Phosphoryl anion generated by deprotonation of the allyl proton of **41** with t-butyllithium adds to cyclopentenone (**42**) in a conjugative manner to afford **43**, which is obtained after alkylation with R^4X. In **43**, consecutive three carbons are optically active (i.e., R^1, R^2, R^3). By ozone oxidation and reduction, **43** gives aldehyde (**44**), in which the configuration of the consecutive three carbons are kept optically active [10]. The aldehyde is a useful precursor for natural product synthesis. Wittig reaction employing optically active phosphoryl compounds has been developed as a new tool for organic synthesis.

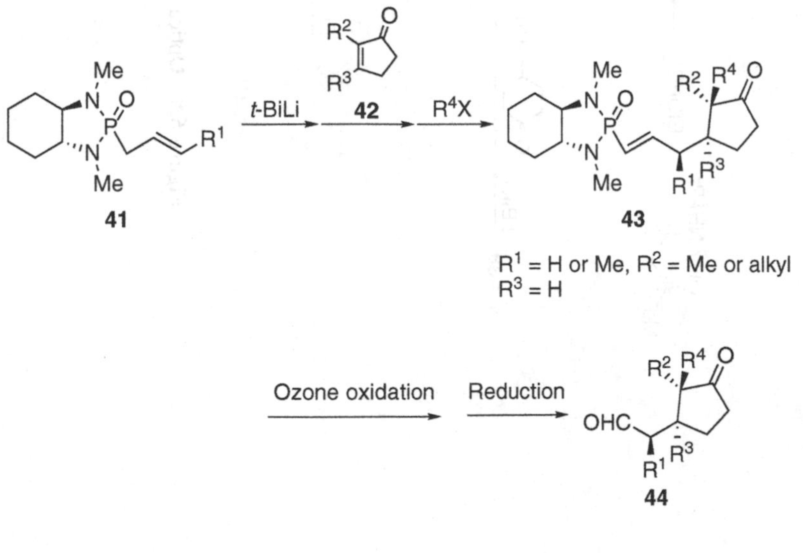

R^1 = H or Me, R^2 = Me or alkyl (6.41)
R^3 = H

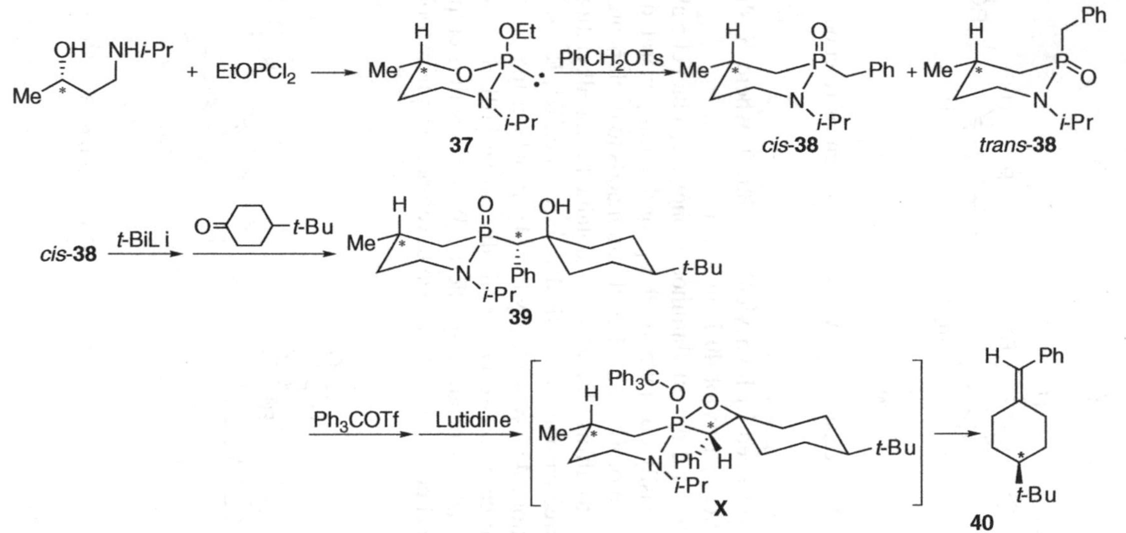

Figure 6.8 Optically active olefin synthesis using optically active amino alcohol.

6.8 REACTIONS OF PHOSPHONIUM SALTS AND FORMATION OF PHOSPHORANES

Wittig prepared pentaphenylphosphorane (**45**, M = P) as a stable solid by the reaction of tetraphenylphosphonium salt with phenyllithium (Eq. 6.42). This is pentavalent and is the first example of organic hypervalent compound (cf., Notes 6). Pentaphenyl compounds of arsenic, antimony, and bismuth were also prepared by Wittig [11]. Wittig conjectured and called the first compound as homopolar pentavalent phosphorane because he did not know at that time that the structure is trigonal bipyramid and the character of axial and equatorial bond is different. He could prepare pentamethylantimony (**46**) but could not obtain pentamethyl compounds of nitrogen, phosphorus, and arsenic probably because of steric hindrance among methyl groups; they are still not prepared yet.

$$Ph_4MI + PhLi \longrightarrow Ph_5M + LiI \quad M = P\ (\mathbf{45}),\ As,\ Sb,\ Bi \tag{6.42}$$

$$Me_4AsI + MeLi \longrightarrow Me_3As{=}CH_2 + CH_4 + LiI \tag{6.43}$$

$$Me_3SbBr_2 + 2\ MeLi \longrightarrow \underset{\mathbf{46}}{Me_5Sb} + 2\ LiBr \tag{6.44}$$

Pentavalent phosphorane is formed when nucleophile adds to the phosphoryl group (P=O) of a $R_3P{=}O$ type compound. When there are different kinds of substituents on phosphorus, the one with the highest leaving ability leaves to regenerate the phosphoryl group. Hydrolysis of phosphate ester is a good example (Eq. 6.45). The rate of acidic and alkaline hydrolysis of the five-member cyclic phosphate (**47**) is enormously rapid, that is by 10^8 times, compared to the corresponding acyclic phosphate (**48**). The rate of hydrolysis of **49**, that is, methyl ester of **47**, under the same conditions as above is faster than that of acyclic phosphate ester (**50**) by 10^6 times.

$$\mathbf{49} + H_2O \xrightarrow{H^+} \mathbf{47} + \mathbf{51} \qquad 30 : 70 \tag{6.45}$$

49, **47**, **51**; **52** ($n = 3, 4$); $(MeO)_2P(O)OH$ **48**; $(MeO)_2P(O)OMe$ **50**

These facts are explained as in Eq. (6.46). Nucleophile H_2O attacks the phosphoryl group of tetrahedral phosphate (**49**) from the rear side of the group (P=O) and resides as an apical group to form **Y**. **Y** undergoes Berry pseudorotation

(BPR) to become **Z** and an apical P–O bond is cleaved to give **51**. **Z** again undergoes BPR to form **A′** from which a methoxy group is eliminated from an apical position to give **47** [12]. With tetrahedral compounds such as phosphonium salt and phosphoryl compounds, the nucleophile attacks the central phosphorus atom from the rear side to form trigonal bipyramidal phosphorane (**Y, Z, A′** in bracket). From the trigonal bipyramidal intermediates, a leaving group (LG) is eliminated from an apical position. Intermediate phosphoranes are not in a transition state, but they are stable enough to undergo BPR, hence hydrolysis of phosphate derivatives can take place under mild conditions (at ambient temperature).

49 + H_2O $\xrightarrow{H^+}$ [MeO–P(OH)(OH)(–O–CH₂CH₂–O–) **Y** $\xrightarrow{BPR}$ MeO–P(OH)(OH)(–O–CH₂CH₂–O–) **Z** $\xrightarrow{BPR}$ HO–P(OMe)(OH)(–O–CH₂CH₂–O–) **A′**]

Z → **51** (HOCH₂CH₂O–P(=O)(OMe)OH)

A′ → **47** + MeOH

(6.46)

Apicophilicity and pseudorotation are new concepts to understand the reactivity of main group element compounds and they were explained in detail as a part of main group element effect in Chapter 2. These are apparently effective for understanding the hydrolysis of phosphate esters. A couple of examples to make the concept clearer are illustrated below.

Synthetic method of phosphinic acid ($R^1R^2P(O)OH$) and thiophosphinic acid ($R^1R^2P(S)OH$) and their esters are well established. Therefore, phosphonium salt (**53**) and its optically active compound (**53***) can be prepared as starting materials for mechanistic study. The reaction of **53** with sodium hydroxide in water–dioxane afforded a mixture of **54** and **55**, from which thiol (R^2SH) and alcohol (R^1OH) were eliminated from **53** competitively (Eq. 6.47).

$[Me–P(SR^2)(Ph)–OR^1]^+ SbCl_6^-$ (**53**) $\xrightarrow[\text{rt}]{H_2O\text{-dioxane (1:1)}, \text{NaOH (0.01 M)}}$ $Me–P(=O)(Ph)–OR^1$ (**54**) + $Me–P(=O)(Ph)–SR^2$ (**55**)

note: $Me–P(=S)(Ph)–OH \xrightarrow{R^2X} Me–P(SR^2)(Ph)=O \xrightarrow{R^1{}_3O^+SbCl_6^-}$ **53**

$Me–P(=S)(Ph)–Cl \xrightarrow{R^1OH} Me–P(=S)(Ph)–OR^1 \xrightarrow{R^2{}_3O^+SbCl_6^-}$ **53**

(6.47)

The experimental results are summarized as follows [13]:

1. When R^1 is Me, the ratio of **54:55** is kept constant as 2:1 even if R^2 is changed (i.e., R^2 = Me, Et, *i*-Pr). When R^1 is Et, the ratio is constant at 4:1, and when R^1 is *i*-Pr, the ratio is constant at 9:1. From these, we can conclude that the ratio **54:55** remains constant irrespective of the change of R^2, but it depends on the kind of R^1.
2. When optically active (*S*)-**53*** is hydrolyzed under the same alkaline conditions, (*R*)-**54*** is obtained by the elimination of thiol (R^2SH), in which the stereochemistry on the phosphorus is retained. On the other hand, (*R*)-**55*** is obtained by the elimination of alcohol (R^1OH), where the stereochemistry on the phosphorus is inverted (Eq. 6.48). On the basis of the stereochemical result, reaction mechanism is illustrated as follows:

[Me(Ph)P*(OMe)(SMe)]⁺ $SbCl_6^-$ (*S*)-**53*** —NaOH→ Me(Ph)P*(OMe)=O (*R*)-**54*** Retention + Me(Ph)P*(SMe)=O (*R*)-**55*** Inversion (6.48)

Me(Ph)P*(OR^1)(SR^2) (*S*)-**53*** —⁻OH→ **B′** [Me,Ph-P(OR^1)(SR^2)(OH)] —BPR (Rate determining)→ **C′** [Ph,R^1O-P(SR^2)(Me)(OH)]

B′ —Rate determining, − R^1OH→ Me(Ph)P*(SR^2)=O (*R*)-**55*** Inversion

C′ —Rapid, − R^2SH→ Me(Ph)P*(OR^1)=O (*R*)-**54*** Retention (6.49)

3. First, phosphorane (**B′**) is generated by the reaction of **53*** with hydroxide ion.
4. Alcohol (R^1OH) leaves **B′** accompanied by the inversion of phosphorus. The elimination of alcohol becomes more reluctant according to the increase of basicity of alcohol. This means that the step is rate-determining. This is in accordance with the experimental result (1).
5. Phosphorane (**B′**) yields **C′** by pseudorotation in competition with the formation of (*R*)-**55***.

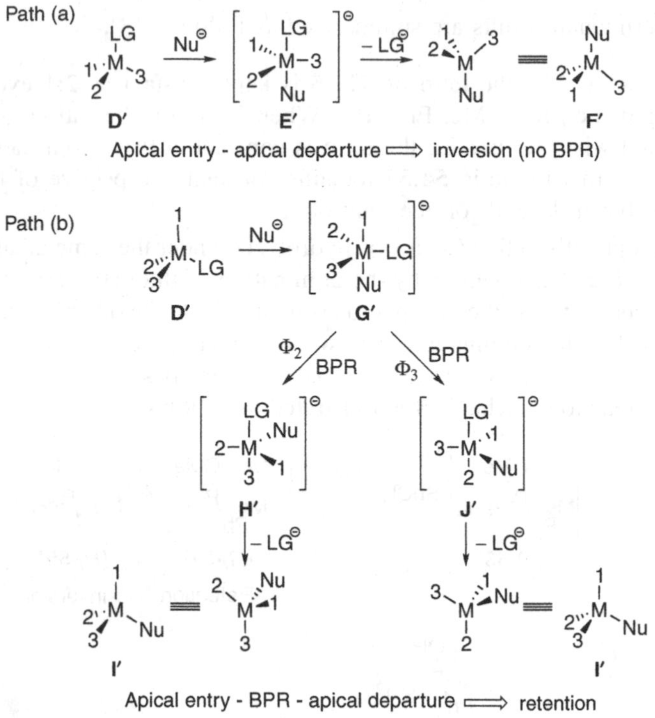

Figure 6.9 Stereochemistry of substitution via hypervalent intermediate (10-M-5).

6. Thiol (R^2SH) leaves **C**′ to give (*R*)-**54***, where the stereochemistry on phosphorus is retained. The rate of elimination of thiol does not depend on R^2, therefore BPR is the rate-determining step. This is because the acidity of thiol is large enough to be eliminated easily and this rationalization is also in accordance with the experimental result (1).
7. Apical entry–apical departure (i.e., the nucleophile (Nu^-) attacks a tetrahedral phosphorus to form a trigonal bipyramidal phsophorane and a LG departs from the phosphorane from an apical position) is the basis of the stereochemistry on phosphorus. This results in inversion when both LG and Nu are on apical positions. On the other hand, it results in retention when LG leaves the phosphorane-bearing LG and Nu on apical–equatorial positions. These relations are generally illustrated in Fig. 6.9.

There are fundamentally two paths, that is, path (a) and path (b), for the substitution of tetrahedral starting material **D**′.

Path (a): Nu^- attacks **D**′ from the opposite face of LG (through plane 1, 2, 3) to form a hypervalent intermediate (**E**′: trigonal bipyramid, 10–M–5). In **E**′, both LG and Nu are on apical positions and **F**′ results from departure of LG^-. Accordingly, inversion takes place at the central atom (M) by the apical entry–apical departure.

Path (b): Nu^- attacks **D′** from the opposite face of substituent 1 (through plane 2, 3, LG) to form a hypervalent intermediate **G′**. BPR (ϕ2) of **G′** with substituent 2 as pivot gives **H′**, and then **I′** is formed by departure of LG^- from the apical position. Hence, retention at the central atom (M) results from apical entry–BPR–apical departure. Then it is clear that BPR (ϕ3) of **G′** with substituent 3 as pivot gives **J′** and the same result is obtained as ϕ2. It is also understood, however, that inversion results from BPR (ϕLG) of **G′** with LG as pivot (see Notes 7). The starting material (**D′**) of Fig. 6.9 can be a neutral silicon compound (8-Si-4), positively charged phosphonium compound (8-P-4), or neutral hypervalent sulfurane (10-S-4). Examples of silicon compounds are mentioned below.

Substitution at silicon (8-Si-4) has been believed to proceed by inversion, just like at carbon. However, it has recently been shown that the stereochemistry of substitution at silicon is more complex, depending on kinds of nucleophile, substituent, and solvent. Optically active silicon (8-Si-4) contained in the tetralin skeleton reacts with organolithiums (MeLi, $PhCH_2Li$, PhLi) via almost complete retention (90–100%), irrespective of a kind of LGs (F, Cl, OMe) (Eq. 6.50, Table 6.1). There is only one exception proceeding through inversion (85%), in which $PhCH_2Li$ and Cl are used. On the other hand, inversion (80–95%) is commonly observed when alkyl (R = Me, $PHCH_2$) Grignard reagent is employed, with one exception of retention (94%) in the case of MeMgBr (nucleophile) and a methoxy group (LG) (Table 6.2). In the case of PhMgBr, retention (95%) was generally observed but selectivity (56%) was poor in the case of Cl as a LG.

It is quite difficult to foresee these apparently complex experimental results. We can, however, rationalize the result by invoking a hypervalent intermediate (10-Si-5) according to the mechanism of Fig. 6.9 (see Notes 7).

*Si(1-Naph)(X) [tetralin skeleton] + $Nu^{\ominus}$ (RLi, RMgBr) ⟶ [*Si(X)(1-Naph)(Nu)]$^{\ominus}$ ⟶ *Si(1-Naph)(Nu) + $X^{\ominus}$

X = F, Cl, OMe; 8-Si-4 → 10-Si-5 → 8-Si-4 (6.50)

Let us consider the ring size effect on stereochemistry of the reaction of alkoxyphosphonium salt and sodium hydroxide to yield phosphine oxide. In the case of an acyclic alkoxyphosphonium salt (**56***), inversion takes place to give **57*** by the apical entry–apical departure through an intermediate phosphorane (**K’**) (Eq. 6.51).

TABLE 6.1 Stereochemistry of Reaction Between $R^1R^2R^3SiX$ and RLi

	X	MeLi	PhLi	$PhCH_2Li$
	F	Ret	Ret	Ret
	Cl	Ret (95)	Ret (95)	Inv (85)
	OMe	Ret	Ret	Ret
		Ret = 90–100%		

TABLE 6.2 Stereochemistry of Reaction Between $R^1R^2R^3SiX$ and RMgBr

	X	MeMgBr	PhMgBr	$PhCH_2MgBr$
	F	Inv (94)	Ret (95)	Inv (95)
	Cl	Inv (80)	Ret (56)	Inv (89)
	OMe	Ret (94)	Ret (95)	Inv (51)

56* + $^{\ominus}$OH ⟶ K′ ⟶ 57* (Inversion) + EtOH

β-Naph = β-naphthyl

(6.51)

In the case of alkoxyphosphonium salts of the six-member ring (**58**), *trans*-**58** yields *cis*-**59** and *cis*-**58** gives *trans*-**59**, quantitatively (Eq. 6.52). Hence, the stereochemistry at phosphorus proceeds by complete inversion. This is due to long enough lifetime of intermediate **L′** for departure of the methoxy group, in which the six-member ring occupies diequatorial positions without any steric strain.

trans-**58** / *cis*-**58** + $^{\ominus}$OH ⟶ L′ ⟶ *cis*-**59** + *trans*-**59** (6.52)

	cis-59	*trans*-59
trans-58	100%	0%
cis-58	0%	100%

trans-60 cis-60 Both are racemic M' cis-61 trans-61

	cis-61	trans-61
trans-60	49%	51%
cis-60	58%	42%

(6.53)

In the case of alkoxyphosphonium salts of the five-member ring **60**, both *trans*-**60** and *cis*-**60** (both are racemic) yield almost a 1:1 mixture of *cis*-**61** and *trans*-**61** (Eq. 6.53). Hydroxide ion attacks *trans*-**60** by apical entry to generate a phosphorane intermediate (**M′**) first. The five-member ring occupies diequatorial positions in **M′** and the ring strain is much larger in **M′** than in **L′**. Thus competition between the departure of the methoxy group to give inverted *cis*-**61** and the BPR giving retained *trans*-**61** takes place as a result of the shorter lifetime and considerable instability of **M′** compared to **L′** (Eq. 6.54). Thus the ratio of *cis*-**61** to *trans*-**61** becomes almost 1:1. The result of *cis*-**60** can be understood similarly [14].

trans-60 M' BPR BPR cis-61 Inversion trans-61 Retention (6.54)

We can now realize that the balance among the leaving ability of substituents, ring strain, and steric hindrance on five-coordinate hypervalent intermediate (10-M-5) are influential in controlling the reactivity of starting materials.

In S_N2 reaction at carbon, the five-coordinate hypervalent species (10-C-5) is the transition state and inversion occurs at carbon completely. In the case of main group elements below the third period, five-coordinate hypervalent species (10-M-5) can have lifetime long enough to effect BPR and even can be isolated stably. In 10-M-5, two apical bonds consist of the electron-rich three-center four-electron bond and the three equatorial bonds consist of the sp^2 bond. This situation results in the coexistence of bonds with different character at one atom, and hence BPR takes place to make them equivalent and a variety of apparent reactions can emerge.

6.9 FREEZING BPR AND ITS EFFECT

Pseudorotation and the unique reactivity of five-coordinate hypervalent phosphorane (10-P-5) are at the origin of the apparently complex phenomena of this class of compounds and these were explained in detail in Section 6.8. Based on the rationalization, two approaches have been made: (i) to freeze BPR and (ii) to place a carbon substituent on an apical position, which is anti-apicophic.

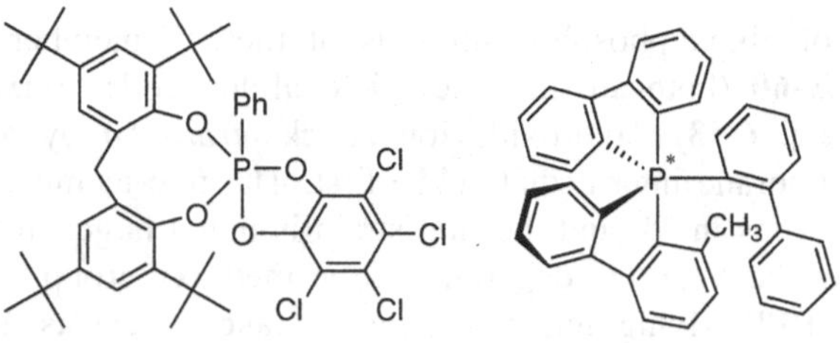

Ph P O O O O Cl Cl Cl Cl CH3 P*

62 63*

It is established that the five-member ring prefers to occupy the apical–equatorial positions in order to avoid ring strain. In compound **62**, the eight-member ring is forced to occupy diequatorial positions and *t*-butyl groups are placed to inhibit ring inversion and pseudorotation, and thus a carbon substituent (phenyl group) can locate at an apical position [15a]. Optically active **63*** is separated stably because BPR is frozen by steric hindrance between the *ortho*-methyl and the biphenyl groups [15b].

Diastereomeric **64**, that is, *exo*-**64** and *endo*-**64**, are prepared by chlorination of phosphoranide anion (**N′**) followed by substitution of the chloro group by alcohol (**64a** and **b**) and by thiol (**64c**). They are separated stably as the exo and endo isomers (Eq. 6.55). This means that BPR of**64** is completely frozen at ambient temperature by the two five-member rings (Martin ligand).

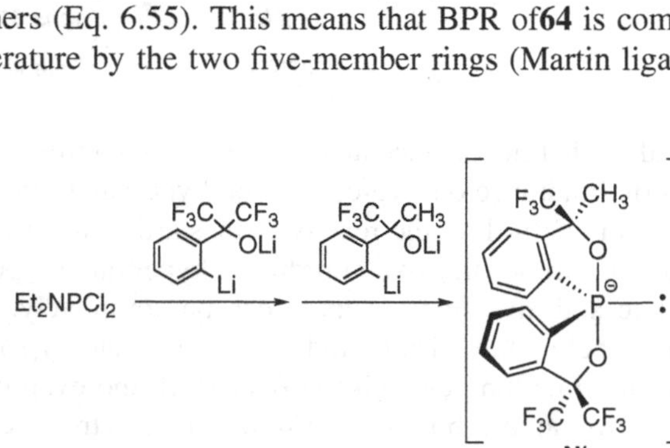

Et_2NPCl_2 → (F_3C, CF_3, OLi, Li) → (F_3C, CH_3, OLi, Li) → [F_3C CH_3 O P⊖ : O F_3C CF_3] **N′** (6.55)

$SOCl_2$ XH

64 X
(a) OMe
(b) $OCH_2CH_2NMe_2$
(c) SMe

exo-**64** > *endo*-**64**

Reactions of pure *exo*-**64a** and *exo*-**64b** containing 10% *endo*-**64b** with methyllithium afford *exo*-**65** completely after recrystallization, and *exo*-**64c** gives *endo*-**65** quantitatively (Eq. 6.56). In the former, methyllithium attacks the phosphorus from the front side keeping the stereochemistry at phosphorus, because methyllithium is coordinated with the oxygen of equatorial substituents. However, methyllithium does not coordinate with sulfur so strongly as oxygen, and therefore attacks the phosphorus from the rear side, inverting the stereochemistry. The result illustrates the first example of stereochemistry of nucleophilic substitution of phosphorane through six-coordinate hypervalent phosphoranate anion (**O′**, **P′**).

exo-**64**
(a) X= OMe
(b) X= $OCH_2CH_2NMe_2$
(c) X= SMe

MeLi; (a), (b) → **O′** → Retention → *exo*-**65**; (c) → **P′** → Inversion → *endo*-**65**

(6.56)

By addition of 2 equivalents of methyllithium to P–H phosphorane (**66**), the phosphoranide anion just like **N′** is generated in situ and the second mole of methyllithium attacks the anion (the anion is still positively charged, see Chapter 2) to afford a ring-opened oxide anion (**Q′**). By iodine oxidation of **Q′**, **R′** is formed and then **R′** cyclizes to give *O-cis* **67** by extrusion of an iodide anion. *O-cis* **67** has a benzene ring at an apical position and it is one of unstable positional isomers of BPR. *O-cis* **67** isomerizes to the most stable isomer of *O-trans* **67** by heating in solution (Eq. 6.57) [16a, b] (see these structures in Fig. 2.11).

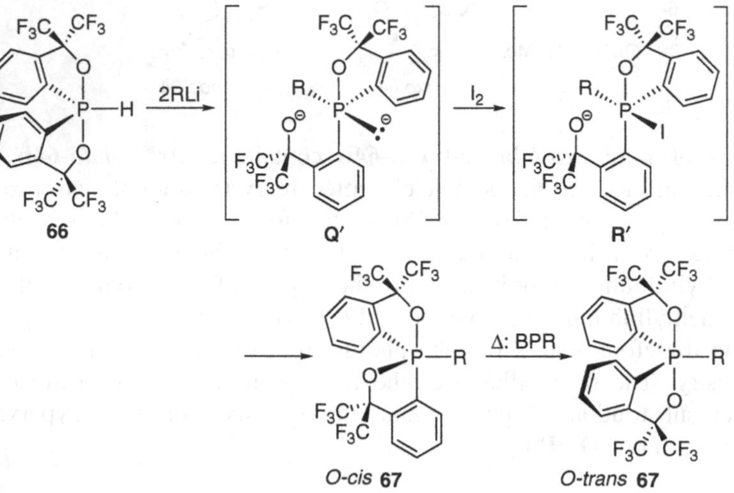

(6.57)

(a) R = Me, (b) R = Bu, (c) R = CH_2Ph, (d) R = *t*-Bu

(e) R = 2-Methylphenyl, (f) R = Mes, (g) R = 2,4,6-Triisopropylphenyl

X-ray structure determination of *O-cis* **67** and *O-trans* **67** revealed the difference of character of the same bond (P–O or P–C), depending on whether it is at apical or equatorial position. For example, the following facts illustrate the stability and inertness of *O-trans* **67**: (i) benzyl proton of *O-trans* **67c** is inert for deprotonation with KHMDS; (ii) *O-trans* **67b** is stable for attack by a fluoride anion; (iii) *O-trans* **67b** is also stable for attack by methyllithium, and the starting material is recovered quantitatively in the three cases mentioned above (Eq. 6.58).

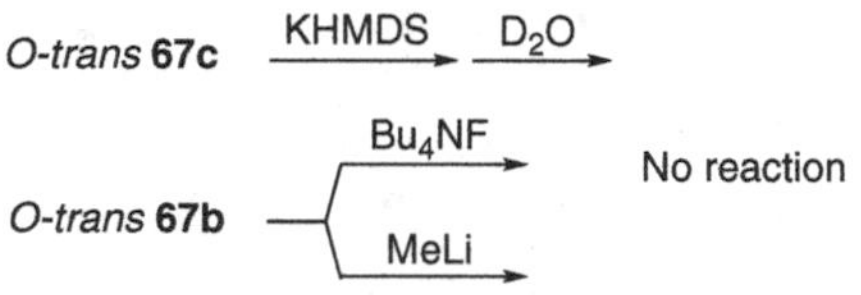

O-trans **67c** —KHMDS→ —D_2O→

O-trans **67b** —Bu_4NF / MeLi→ No reaction (6.58)

On the other hand, *O-cis* **67** is susceptible for all three reactions: (i) benzyl proton of *O-cis* **67c** is deprotonated by KHMDS and deuterated with D_2O (Eq. 6.59); (ii) a fluoride anion adds to *O-cis* **67b** to form **68**; (iii) methyllithium also adds to *O-cis* **67b** to give **U′**, and **U′** affords **69** by acidic hydrolysis (Eq. 6.60).

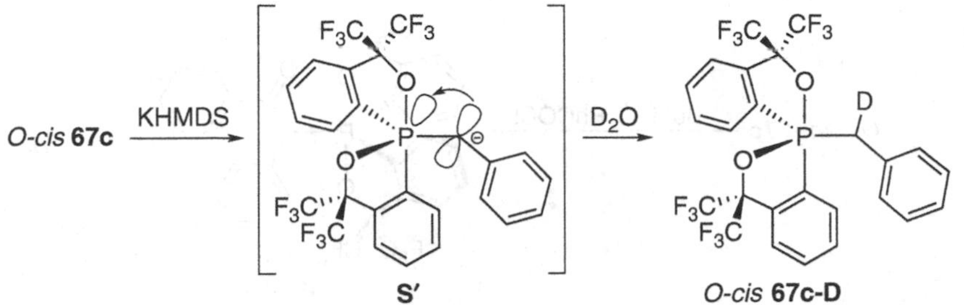

(6.59)

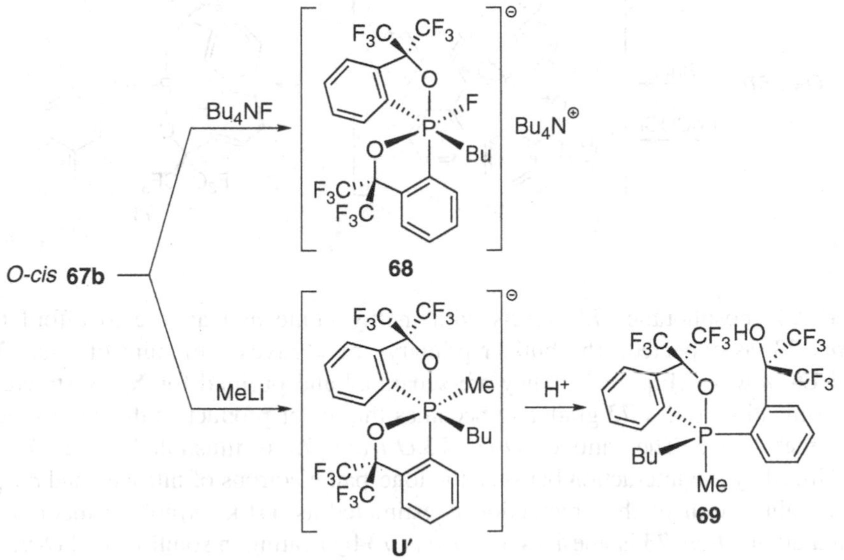

(6.60)

The result contrasts the difference of reactivity between *O-trans* **67** and *O-cis* **67**. This can be realized on the basis of the concept that the orbital energy of the equatorial bond of σ^*_{P-O} is considerably lower than that of σ^*_{P-C}. α-Carbanion of *O-cis* **67** can be stabilized by electron-donating interaction to σ^*_{P-O} (**S′**) but this kind of interaction is not possible with σ^*_{P-C} which has a much higher energy. Moreover, the fluoride anion and methyllithium can interact with σ^*_{P-O} and add to the phosphorus. A couple of examples are presented, which can be understood by the same concept.

By deprotonation of a benzyl proton of *O-trans* **67c** with a strong base such as BuLi and successive addition of benzoyl chloride, α-benzoyl product (**70**) is obtained as expected (Eq. 6.61). Under the same processes, *O-cis* **67c** yields vinyl ether (**71**) (Eq. 6.62). This is rationalized by the fact that α-benzoyl product (**V′**) is first formed but the carbonyl oxygen of **V′** attacks σ^*_{P-O} to rearrange to give **71** accompanied by BPR [17].

O-trans **67c** $\xrightarrow{\text{BuLi}}$ $\xrightarrow{\text{PhCOCl}}$ **70** (6.61)

O-cis **67c** $\xrightarrow[\text{PhCOCl}]{\text{BuLi}}$ [**V′**] $\longrightarrow$ **71** (6.62)

P–Cl phosphorane (**72**) reacts with methylamine and aniline to afford *O-trans* **73**, as expected. The bulkier primary amine gives a mixture of *O-cis* **73** and *O-trans* **73** (Eq. 6.63). They are separated and purified for X-ray structure determination. *O-cis* **73** gradually becomes the major product as the alkyl group becomes larger. The ratio of *O-cis* **73**:*O-trans* **73** is illustrated. *O-cis* **73** is stabilized by the interaction between the lone-pair electrons of nitrogen and $\sigma^*_{\text{P-O}}$. The stabilization of this interaction is estimated as 4.0 kcal/mol by theoretical calculation. *O-cis* **73** isomerizes to *O-trans* **73** by heating in solution, and *O-trans* **73a** is more stable than *O-cis* **73a** by 10.1 kcal/mol [18]. This transformation is similar to that of *O*-cis **67** to *O*-trans **67**.

72 + RNH_2 $\longrightarrow$ *O-cis* **73** + *O-trans* **73** (Δ: BPR)

	O-cis **73** : *O-trans* **73**
(a) R= Pr	36 : 64
(b) R= $PhCH_2$	62 : 38
(c) R= *i*-Pr	80 : 20
(d) R= *t*-Bu	98 : 2

(6.63)

The difference of character between *O-cis* and *O-trans* species is schematically illustrated as **S′** and **S′ ′** and **W′** and **W′ ′**. The σ^*_{P-O} orbital is visualized as a larger ellipse at the rear side of the P–O bond, which is parallel to the carbanion of the benzyl group, than the smaller ellipse of the σ^*_{P-C} orbital in order to qualitatively illustrate the easier interaction of the former than the latter with carbanion and nucleophile. For **67a**, the orbital energy of σ^*_{P-O} (1.21 eV) is calculated to be lower than that of σ^*_{P-C} (2.02 eV) and the difference in energy of the two is in accordance with experimental results.

Optically pure P–H phosphorane (**66***) and P–C phosphorane (**74***) have been synthesized on the basis of the fact that the BPR of phosphorane could be frozen [19].

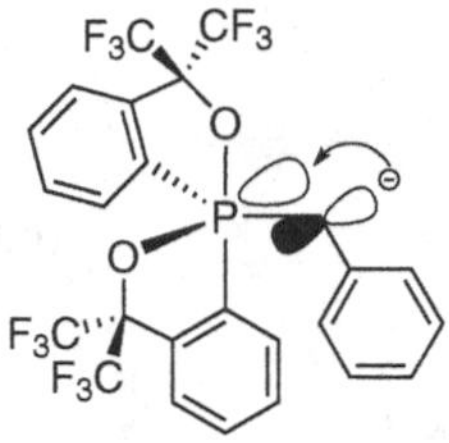

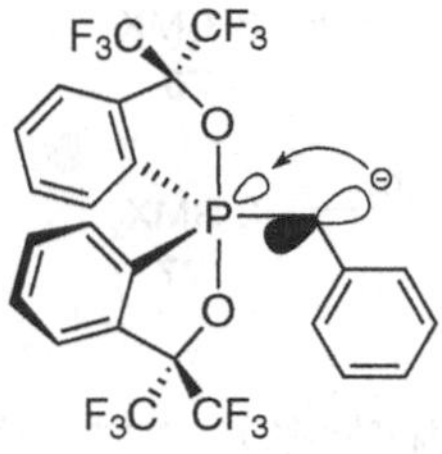

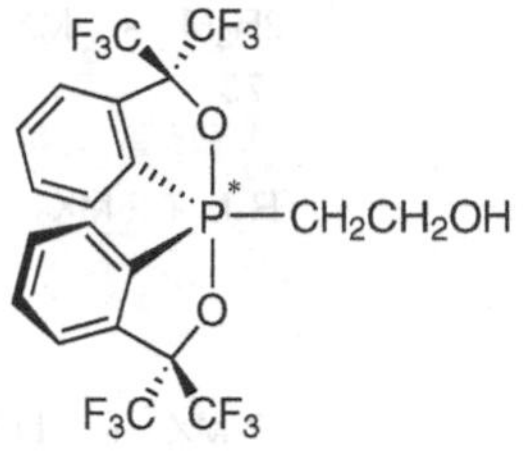

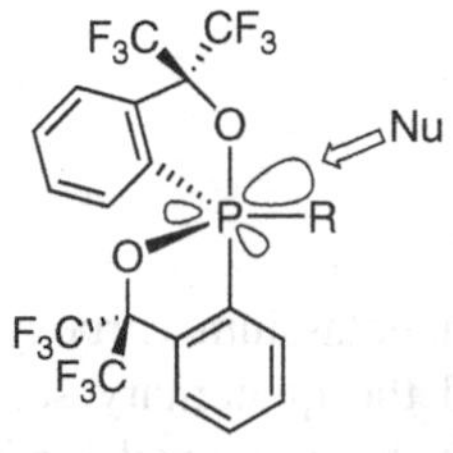

A couple of examples of organometallic complexes of tertiary phosphines and their synthetic utility were illustrated in Figs. 6.1–6.3. Research related to the area has been quite active and fruitful, and there are well-written books and articles which the readers can consult.

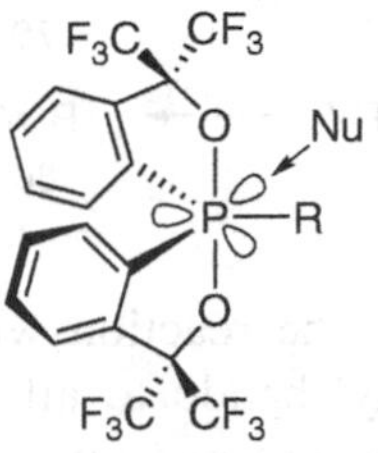

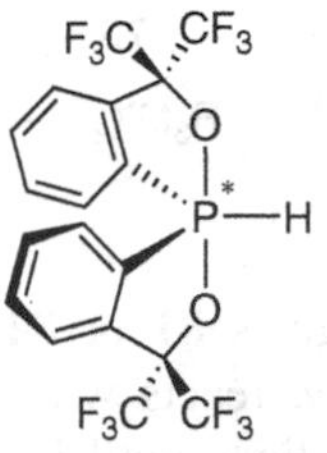

6.10 ANTIMONY AND BISMUTH COMPOUNDS

Trivalent organoantimony and bismuth compounds (R_3M:**75**) are prepared by the reaction of metallic chloride (MX_3) and organometallic reagent (RY) (Eq. 6.64). Lower alkyl compounds are strongly toxic, and the butyl group is usually employed as an alkyl group. Aromatic compounds are more stable and easy for handling.

$$3RY + MX_3 \longrightarrow \underset{\mathbf{75}}{R_3M} + 3\,XY \qquad (6.64)$$

(Y = Li, MgX) (M = Sb, Bi; X = Cl, Br) (R= alkyl, aryl)

Heating metallic chloride (MX_3) with **75** yields the monohalogen compound **76** (Eq. 6.65a) by disproportionation, and **76** is reduced to the corresponding hydride (**78**). Dihalogen compound (**77**) can also be obtained by disproportionation (Eq. 6.65b). R_3M (**75**) reacts with halogen to afford pentavalent dihalide (**79**), and **79** yields the pentavalent organoantimony and bismuth compound **80** by reaction with organometallic reagent (RY).

$$\underset{\mathbf{75}}{2R_3M} + MX_3 \xrightarrow{(a)} \underset{\mathbf{76}}{3R_2MX}$$

$$R_3M + 2\,MX_3 \xrightarrow{(b)} \underset{\mathbf{77}}{3\,RMX_2} \qquad (6.65)$$

$$R_2MX + LiAlH_4\ (NaBH_4) \longrightarrow \underset{\mathbf{78}}{R_2MH}$$

$$\underset{\mathbf{75}}{R_3M} + X_2 \longrightarrow \underset{\mathbf{79}}{R_3MX_2}$$

$$R_3MX_2 + 2RY \longrightarrow \underset{\mathbf{80}}{R_5M} \qquad (6.66)$$

Pentavalent **79** gives **81** or **82** by the reaction with potassium carboxylate (Eq. 6.67), and R_3M reacts with alkyl halide to afford the quaternary salt **83** (Eq. 6.68). The reactions until Eq. (6.68) can commonly be employed for both antimony and bismuth compounds [20a,b].

$$\underset{\mathbf{79}}{R_3MX_2} \xrightarrow{AcOK} \underset{\mathbf{81}}{R_3M(OAc)_2}$$

$$\underset{\mathbf{79}}{R_3MX_2} \xrightarrow{K_2CO_3} \underset{\mathbf{82}}{R_3M(CO_3)} \qquad (6.67)$$

$$R_3M + R'X \longrightarrow \underset{\mathbf{83}}{R'R_3M^+X^-} \qquad (6.68)$$

The quaternary stibonium salt bearing an electron-withdrawing group (**83a**) yields the ylide **X′a** by reaction with base, and the ylide can undergo carbonyl olefination (Wittig reaction) (Eq. 6.69). Tetraphenylstibonium triflate activates the epoxide as Lewis acid forming a complex, and the complex reacts with amine to give aminoalcohol (**84a**) regioselectively (selectivity is 100%) (Eq. 6.70).

$$[Bu_3SbCH_2E]^+Br^- \xrightarrow{KOt\text{-}Bu} [Bu_3Sb{=}CHE] \xrightarrow{R^1COR^2} R^1R^2C{=}CHE + Bu_3SbO \quad (6.69)$$

83a → **X′a**

$E = CO_2Me, CONMe_2, CN$

$$\text{R-epoxide} + R'_2NH \xrightarrow{Ph_4SbOTf} \mathbf{84a} + \mathbf{84b} \quad (6.70)$$

$R = Me, Ph, MeOCH_2$
$NR'_2 = NEt_2, NHPh$

84a : **84b** = 100 : 0

Dicarboxylates of pentavalent bismuth (**81-Bi, 82-Bi**) oxidize alcohol to the carbonyl compound in the presence of a base (Eq. 6.71). In this reaction, α-hydrogen (deuterium) is eliminated to give ketone **86** (the main path) and the hydrogen replaces the bismuth moiety to give a deuterated benzene ring (side path).

$$\mathbf{85} \xrightarrow[Base]{An_3BiCO_3} [\mathbf{Y'a}] \rightarrow An_3Bi + DOCO_2^{\ominus} + \mathbf{86}; \quad [\mathbf{Y'b}] \rightarrow \mathbf{86} + \text{D-C}_6\text{H}_4\text{-OMe} + An_2BiOCO_2^{\ominus} \quad (6.71)$$

$An = p\text{-}MeOC_6H_4$

1,2-Glycol (**87**) and dicarboxylates of pentavalent bismuth (**81-Bi, 82-Bi**) afford two molecules of carbonyl compound through cyclic transition state (**Z′**), where the central carbon–carbon bond is cleaved (Eq. 6.72). Quaternary salt (**83b**) of triphenylbismuth bearing an acetonyl or a phenacyl group gives the ylide **X′b**, and the ylide reacts with aldehyde or imine to afford epoxide (**88**) or aziridine (**89**), differently from Wittig-type olefination (Eq. 6.73) [21].

$$R^1R^2C(OH)\text{-}C(OH)R^3R^4\ (\mathbf{87}) + Ph_3BiCO_3 \rightarrow [\mathbf{Z'}] \rightarrow R^1R^2C{=}O + R^3R^4C{=}O + Ph_3Bi \quad (6.72)$$

83b R=Me, Ph; $Br^{\ominus}$; KOt-Bu; X'b; R'CHO → 88 + Ph_3Bi; R'CH=NTs → 89 + Ph_3Bi

(6.73)

The quaternary salt (**83c**) of triphenylbismuth bearing a vinyl group generates vinylidene carbene and the carbene is trapped with olefin to give methylenecyclopropane **90** in the presence of a base (Eq. 6.74). This is the same type of reaction with iodonium salt, which is explained in Chapter 8, and is due to the enormous ability of the LG of triphenylbismuth and iodobenzene.

83c $BF_4^{\ominus}$; KOt-Bu; Ph; 90 + Ph_3Bi

(6.74)

REFERENCES

1. Prim D, Campagne J-M, Joseph D, Andrioletti B. Tetrahedron 2002;58:2041.
2. (a) Rouch AM. C & EN News, 2002 Dec 23, p. 29, 34; (b) Schwab P, Grubbs RH, Ziller JW. J Am Chem Soc 1996;118:100; (c) Sanfold MS, Day MW, Grubbs RH. J Am Chem Soc 2003;125:10103; (d) Deiters A, Martin SF. Chem Rev 2004;104:2199.
3. (a) Horner L, Winkler H, Rapp A, Mentrup A, Hoffmann H, Beck P. Tetrahedron Lett 1961; 161; (b) Horner L, Winkler H. Tetrahedron Lett 1964;3275.
4. Donohue J, Mandel N, Farnham WD, Murray RK, Mislow K, Benschop HP. J Am Chem Soc 1971;93:3792.
5. Nauman K, Zon G, Mislow K. J Am Chem Soc 1969;91:7012.
6. Ohkuma T, Kitamura M, Noyori R. In: Ojima I, editor. Catalytic asymmetric synthesis. 2nd ed. New York: John Wiley & Sons, Inc.; 2000, Chapter 7.1.
7. (a) Trippet S. Quart Rev (London) 1963;17:406; (b) Bestmann H. J Angew Chem 1965;77:609, 651, 850; (c) Maercker A. Volume 14, Organic reactions. New York: John Wiley & Sons, Inc.; 1965. p. 270; (d) Maryanoff BE, Reitz AB, Mutter MS, Inners RR, Almond HR Jr, Whittle RR, Olofson RA. J Am Chem Soc 1986;108:7664.
8. Boutagy J, Thomas R. Chem Rev 1974;74:87.
9. Denmark SE, Chen C-T. J Org Chem 1994;59:2922.
10. Hannesian S, Gomysyan A, Pane A, Hervé Y, Beaudoin S. J Org Chem 1993;58:5032.
11. (a) Wittig G, Rieber M. Justus Liebigs Ann Chem 1949;562:187; (b) Wittig G, Torssell K. Acta Chem Scand 1953;7:1293.

12. Westheimer FW. Acc Chem Res 1968;1:70.
13. DeBruin KB, Johnson DM. J Am Chem Soc 1973;95:4675.
14. Marsi KM, Burns FB, Clark RT. J Org Chem 1975;40:1779.
15. (a) Holmes RR. Volumes I, II, Pentacoordinate phosphorus-structure and spectroscopy, ACS Monograph 175, 176. Washington (DC): ACS; 1980; (b) Hellwinkel D, Lindner W. Chem Ber 1976;109:1497.
16. (a) Kajiyama K, Yoshimune M, Nakamoto M, Matsukawa S, Kojima S, Akiba K.-y. Org Lett 2001;3:1873; (b) Nakamoto M, Kojima S, Matsukawa S, Yamamoto Y, Akiba K.-y. J Organometal Chem 2002;643–644:441.
17. Matsukawa S, Kojima S, Kajiyama K, Yamamoto Y, Akiba K.-y., Re SY, Nagase S. J Am Chem Soc 2002;124:13154.
18. Adachi T, Mastukawa S, Nakamoto M, Kajiyama K, Kojima S, Yamamoto Y, Akiba K.-y., Re SY, Nagase S. Inorg Chem 2006;45:7269.
19. Kojima S, Kajiyama K, Akiba K.-y. Bull Chem Soc Jpn 1995;68:1785.
20. (a) Finet J-P. Chem Rev 1989;89:1487; (b) Huang YZ. Acc Chem Res 1992;25:182.
21. Matano Y, Yoshimune M, Suzuki H. J Org Chem 1995;60:4663.

NOTES 6

DREAMS OF STAUDINGER AND WITTIG

Staudinger wrote in 1919:

(i) (Nach der Werner'schen Auffassung hat der Stickstoff die Koordinationszahl 4; nur 4 Atome oder Atomgruppen koenen direct an Stickstoff gebunden sein; die fuenfte Valenz ist prinzipiell von den anderen Valenzen verschieden. Es ist deshalb eine theoretisch interessante Frage, ob man nicht doch Stickstoffverbindungen herstellen kann, bei denen 5 Atome oder Atomgruppen gleichartig an Stickstoffgruppen gebunden sind; derartige Verbindungen sollten besonders stabil sein, wenn Stickstoffatom 5 C-Atome gelargert ist, weil das Kohlenstoffatom indifferent ist und weil es sich mit dem Stickstoffatom besonders fest bindet.) (ii) (In der vorigen Mitteilung sind einige resultatlose Versuche zur Gewinnung von neuen organischen Stickstoffverbindungen mit fuenfwertigem Stickstoff beschrieben. Man konnte hoffen, bestaendige Verbindungen mit fuenfwertigem Phosphor oder Arsen leicher gewinnen zu koennen.) [1].

According to Werner's view, the coordination number of nitrogen is 4 and there cannot be any nitrogen compounds bearing five homopolar (σ type) bonds. It is interesting theoretically to challenge the basic idea and to try to synthesize pentavalent nitrogen, phosphorus, and arsenic compounds]. He started to investigate the above possibility.

He tried to react diethylzinc (organometallic reagent of that time) with ammonium and phosphonium salts. Reaction with ammonium salts was unsuccessful, and a reductive reaction took place with tetraethylphosphonium iodide to give triethylphosphine, butane, and zinc dichloride (Eq. N6.1).

Organo Main Group Chemistry, First Edition. Kin-ya Akiba.

$$R_4NI + Et_2Zn \not\longrightarrow R_4NEt + EtZnI$$

$$2Et_4PI + Et_2Zn \not\longrightarrow Et_4PEt + EtZnI$$
$$\longrightarrow 2Et_3P + Et{-}Et + ZnI_2 \qquad \text{(N6.1)}$$

Then, diphenyldiazomethane and phenyl azide were reacted with triethylamine or triphenylphosphine. No reaction occurred with triethylamine. Triphenylphosphine reacted with diphenyldiazomethane to yield phosphazine (**1**) and the latter gave phosphine-alkylene (**2**) by thermolysis. In the case with phenyl azide, the triazene first formed was thermally unstable and decomposed to give phosphine-imine (**3**) (Eq. N6.2).

$$Ph_3P + Ph_2C{=}\overset{+}{N}{=}\overset{-}{N} \longrightarrow \underset{\mathbf{1}}{Ph_3P{=}N{-}N{=}PPh_3} \xrightarrow{\Delta} \underset{\mathbf{2}}{Ph_3P{=}CPh_2} + N_2$$

$$Ph_3P + Ph{-}\overset{+}{N}{=}\overset{-}{N}{=}N \longrightarrow [Ph_3P{=}N{-}N{=}N{-}Ph] \longrightarrow \underset{\mathbf{3}}{Ph_3P{=}NPh} + N_2 \qquad \text{(N6.2)}$$

$$\underset{\mathbf{2}}{Ph_3P{=}CPh_2} + O{=}C{=}CPh_2 \xrightarrow{(a)} Ph_2C{=}C{=}CPh_2 + Ph_3P{=}O$$
$$\underset{\mathbf{2}}{Ph_3P{=}CPh_2} + O{=}C{=}NPh \xrightarrow{(a)} Ph_2C{=}C{=}NPh + Ph_3P{=}O$$
$$\underset{\mathbf{3}}{Ph_3P{=}NPh} + O{=}C{=}CPh_2 \xrightarrow{(b)} PhN{=}C{=}CPh_2 + Ph_3P{=}O$$
$$\underset{\mathbf{3}}{Ph_3P{=}NPh} + O{=}C{=}NPh \xrightarrow{(b)} PhN{=}C{=}NPh + Ph_3P{=}O \qquad \text{(N6.3)}$$

A: $Ph_3P{-}CPh_2$ / $O{-}C$ (four-membered ring)　B: $Ph_3P{-}NPh$ / $O{-}C$ (four-membered ring)

Phosphine-alkylene (**2**) reacted with ketene and isocyanate to afford allene and keteneimine and triphenylphosphine oxide (Eq. N6.3a). Phosphine-imine (**3**) reacted similarly with two reagents and yielded keteneimine and iminoisonitrile and triphenylphosphine oxide (Eq. N6.3b).

Reaction intermediates (**A, B**) of these are four-member rings containing the pentavalent phosphorus moiety. Had these been isolated, homopolar pentavalent phosphorus compounds would have been successfully prepared, but they were thermally unstable and decomposed in situ to yield phosphine oxide and the corresponding cumulenes. These reactions are certainly the same type as the Wittig reaction. Hence, Staudinger established the prototype of Wittig reaction in pursuit of preparing homopolar pentavalent phosphorus compounds.

On the other hand, in 1947 Wittig started the research on organic heteroatom compounds with the same objective as Staudinger (i.e., to synthesize homopolar pentavalent nitrogen and phosphorus compounds) [2]. He tried reactions of ammonium and phosphonium salts with lithium compounds (mainly phenyllithium) instead of diethylzinc (Eq. N6.4). Wittig expected to obtain pentavalent nitrogen (**C**) and phosphorus (**D**) compounds, but ylides (**E, F**) were generated by deprotonation because of the strong basicity of phenyllithium. The ylides reacted with benzophenone, and ammonium (**4**) and phosphonium (**5**) salts of the adducts were obtained after acidic treatment.

$$Me_4NBr + PhLi \longrightarrow \underset{\mathbf{E}}{[Me_3N{=}CH_2]} + Ph_2C{=}O \xrightarrow{HBr} \underset{\mathbf{4}}{\left[Me_3NCH_2C(OH)Ph_2\right]} Br$$

$$Me_4PI + PhLi \longrightarrow \underset{\mathbf{F}}{[Me_3P{=}CH_2]} + Ph_2C{=}O \xrightarrow{HI} \underset{\mathbf{5}}{\left[Me_3PCH_2C(OH)Ph_2\right]} I$$

$$\underset{\mathbf{C}}{Me_4NPh} \qquad \underset{\mathbf{D}}{Me_4PPh} \qquad \text{(N6.4)}$$

After these attempts, Wittig employed alkyltriphenylphosphonium salt and generated the ylide (phosphine-alkylene: **G**) by deprotonation with base. The ylide reacted with carbonyl compound to give alkene, and thus Wittig reaction was generally established (Eq. N6.5). The intermediate of a Wittig reaction is a four-member ring (**H**: oxaphosphetane) containing the pentavalent phosphorus moiety.

$$[Ph_3PCHR^1R^2]Br + B{:} \longrightarrow \underset{\mathbf{G}}{[Ph_3P{=}CR^1R^2]} + R^3R^4C{=}O$$

$$\underset{\mathbf{H}}{\left[\begin{array}{l} Ph_3P-CR^1R^2 \\ \;\;\;|\qquad | \\ \;\;\;O-CR^3R^4 \end{array}\right]} \longrightarrow R^1R^2C{=}CR^3R^4 + Ph_3P{=}O \qquad \text{(N6.5)}$$

The ylide (**G**), however, is not a homopolar pentavalent phosphorus compound. Wittig further tried to react tetraphenylphosphonium salt with phenyllithium and succeeded to obtain pentaphenylphosphorane (**6**) in 1949 (Eq. N6.6). Pentaphenylphosphorane (**6**) is the first organo-hypervalent compound. Using similar reactions, Wittig prepared homopolar pentavalent arsenic, antimony, and bismuth compounds (**7**) [3].

$$Ph_4PI + PhLi \longrightarrow \underset{\mathbf{6}}{Ph_5P} + LiI$$

$$\underset{M\,=\,As,\ Sb,\ Bi}{Ph_4MI + PhLi} \longrightarrow \underset{\mathbf{7}}{Ph_5M} + LiI \qquad \text{(N6.6)}$$

It is interesting and suggestive that Wittig pursued the same dream as Staudinger, without knowing it, to synthesize homopolar pentavalent phosphorus compounds and opened the door to the chemistry of organo-hypervalent compounds. More interesting was that the Wittig reaction was found as the byproduct during the above effort [2, 4]. In addition, without X-ray analysis, Wittig did not know that the structure of pentaphenylphosphorane was a (distorted) trigonal bipyramid and apical bonds consist of 3c–4e bond and equatorial ones are sp^2 and their bonding character is different.

Recently, oxaphosphetane (**8**), an intermediate of Wittig reaction, was isolated using a bidentate ligand (Martin ligand) and structure determined by X-ray analysis (Eq. N6.7) [5]. It is of value that **8c** bearing electron-withdrawing ethoxycarbonyl group was isolated because oxaphosphetane with electron-withdrawing group is quite unstable and cannot be detected during a regular Wittig reaction.

8 → **9** + **10** (N6.7)

(a) E = H, $R^1 = R^2 =$ Ph
(b) E = H, $R^1 = R^2 = CF_3$
(c) E = CO_2Me, $R^1 = R^2 = CF_3$

Anti-apicophilic oxaphosphetane (*O-cis* **11**) was also isolated and structure determined by X-ray analysis (Eq. N6.8). The presence of *O-cis* type oxaphosphetane was predicted theoretically, but *O-cis* **11** isomerized to *O-trans* **11** by heating and then afforded alkene in solution [6]. With azaphosphetane, equilibration was observed between *O-trans* **12** and *O-cis* **12** (Eq. N6.9) [5].

O-*cis* **11** → O-*trans* **11** → $Ar_2C{=}CH_2$ + (N6.8)

O-*trans* **12** ⇌ O-*cis* **12** (N6.9)

REFERENCES

1. Staudinger H, Meyer J. Helv Chim Acta 1919;2:608, 612, 619, 635.
2. Wittig G. Variationen zu einem Thema von Staudinger; Ein Beitrag zur Geschichte der Phosphororganischen Carbonyl-Olefinierung. In: IUPAC Symposium on Organophosphorus Compounds, Heidelberg. London; Butterworths; 1964. pp. 245–254.
3. (a) Wittig G, Rieber M. Justus Liebigs Ann Chem 1949;562:187; (b) Wittig G, Tossell K. Acta Chem Scand 1953;7:1293.
4. Akiba K.-y. Chemistry of phospnine-alkylene, Kagaku no Ryouiki 1966; 20: 492, (Japanese).
5. Kawashima T. Bull Chem Soc Jpn 2003;76:471.
6. Kojima S, Sugino M, Matsukawa S, Nakamoto M, Akiba K.-y. J Am Chem Soc 2002;124:7674.

NOTES 7

STEREOCHEMISTRY IN NUCLEOPHILIC SUBSTITUTION OF MX_4-TYPE COMPOUNDS: INVERSION OR RETENTION

When a nucleophile attacks an MX_4-type compound, MX_5-type intermediate (or transition state) should be generated. An MX_5-type compound is hypervalent and BPR (Berry pseudorotation) would take place in case it has a certain lifetime (stability). Owing to this, the stereochemistry of substitution becomes complex, that is, inversion and retention can be competitive. Incidentally, it is common and fundamental knowledge that MX_5-type species of S_N2 reaction on carbon is unstable and is a transition state to effect complete inversion.

Let us designate substituents of tetrahedron (MX_4) as LG and 1, 2, and 3. There are two kinds of mechanisms to explain the stereochemistry of the substitution, that is, (i) face attack and (ii) edge attack.

Face Attack (Fig. N7.1):

Path (a): nucleophilic attack (Nu^-) of 1–2–3-plane from the opposite side of LG and path (b): nucleophilic attack (Nu^-) of a plane containing LG and two substituents from the opposite side of one substituent. In both cases, LG leaves M from an apical position.

Path (a): **A** is generated by attack of Nu^-. In **A**, LG and Nu^- are at apical positions and the configuration of M inverts according to the departure of LG^-.

Path (b): **B** is generated by attack of Nu^-. There are three faces including LG (2, 3, LG; 1, 2, LG; 1, 3, LG). In **B**, LG is at an equatorial position;

Organo Main Group Chemistry, First Edition. Kin-ya Akiba.

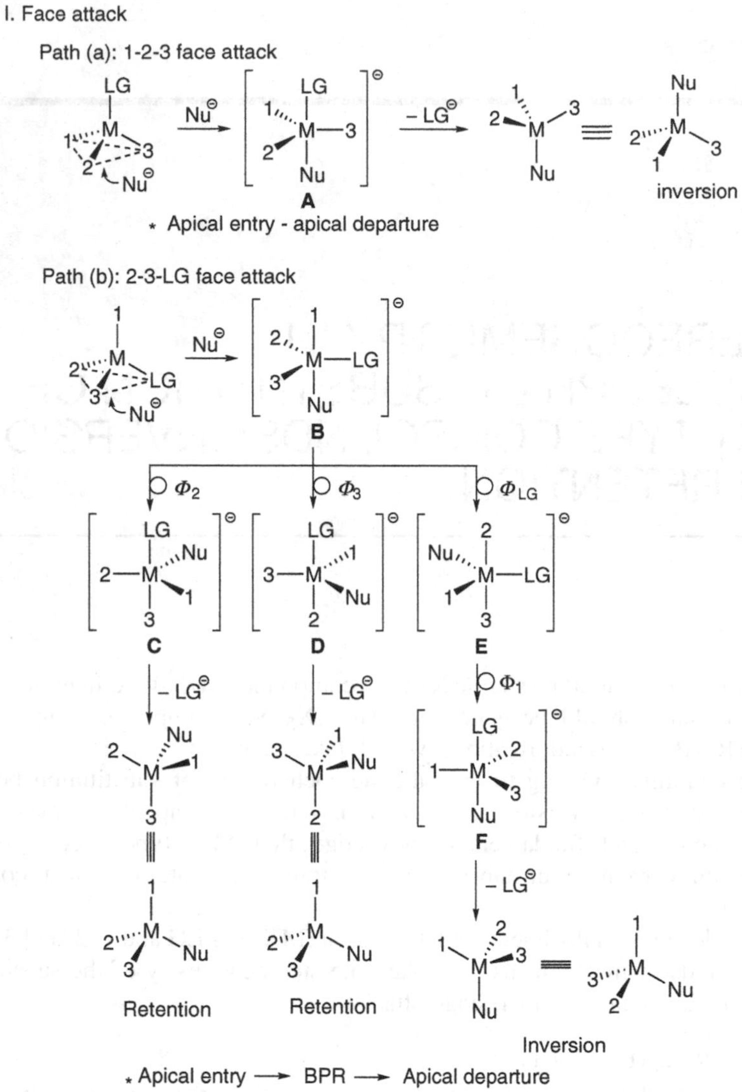

Figure N7.1 Mechanism of face attack and the resulting stereochemistry.

thus, it cannot leave M. Therefore, BPR takes place using one of equatorial substituents as pivot. By BPR using substituent 2 (ϕ_2) or 3 (ϕ_3) as pivot, **C** or **D** is generated to let LG come to an apical position. Retention of configuration of M results through the departure of LG from **C** or **D**. **E** results from BPR using LG (ϕ_{LG}) as the pivot, and the succeeding BPR employing **1** (ϕ_1) as the pivot yields **F** and LG can leave M. In **F**, LG and Nu^- are both at apical positions; thus, inversion occurs.

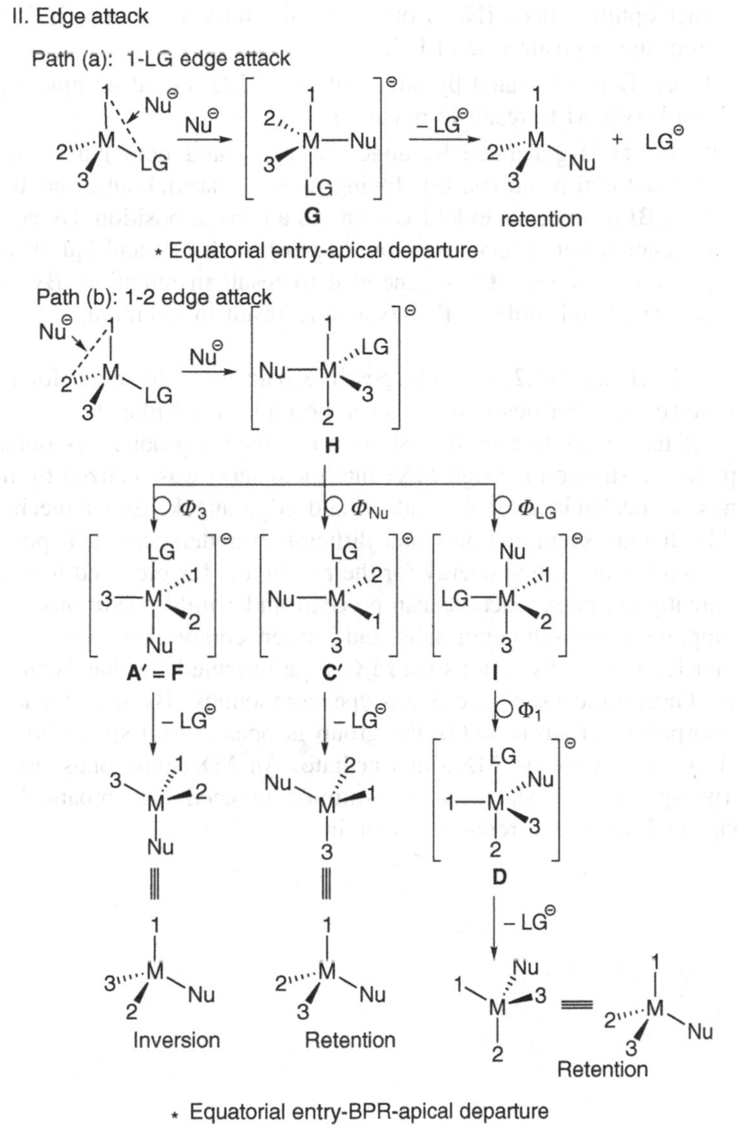

II. Edge attack

Path (a): 1-LG edge attack

G

retention

* Equatorial entry-apical departure

Path (b): 1-2 edge attack

H

A′ = F

C′

I

D

Inversion

Retention

Retention

* Equatorial entry-BPR-apical departure

Figure N7.2 Mechanism of edge attack and the resulting stereochemistry.

Paths (a) and (b) compete each other and BPR takes place because MX_5 has a certain lifetime; therefore, a mixture of inversion and retention products is obtained.

Edge Attack (Fig. N7.2):

Path (a): nucleophilic attack (Nu^-) of an edge consisting of LG and one of the substituents (1, 2, and 3) from the opposite side of M and path (b):

nucleophilic attack (Nu^-) of one of the three edges (1–2; 2–3; 3–1) from the opposite side of LG.

Path (a): **G** is generated by attack of Nu^-: LG sits at an apical position and leaves M to result in retention.

Path (b): **H** is generated by attack of Nu^-, and both Nu^- and LG sit at equatorial positions. Employing three equatorial substituents (3, Nu, LG), BPR proceeds to let LG come to an apical position. By ϕ_3, $\mathbf{A}' = \mathbf{F}$ is generated to effect inversion, because both LG and Nu sit at apical positions. By ϕ_{Nu}, $\mathbf{C}'$ is generated to result in retention. By ϕ_{LG}, **I** is generated and further BPR gives **D** to result in retention.

In Figs. N7.1 and N7.2, the only possible path was illustrated for paths (a) and (b); however, other possible paths can be understood likewise.

The fact that a mixture of inversion and retention products is obtained by nucleophilic substitution through MX_5-intermediate(s) was realized by invoking two kinds of mechanisms of face attack and edge attack. Either mechanism is acceptable. It may seem complex and difficult to understand such possibilities for a reaction but activation energy for the reaction is lowered and it is accepted that apparently complex reactions can proceed under mild conditions.

By employing phosphonium salts and silicon compounds, the stereochemistry of nucleophilic substitutions via MX_5-type intermediates has been explored in detail. Phosphoric esters are MX_4-type compounds. By the attack of Nu^- to the phosphoryl group (P = O), the group is opened to a single bond (oxide group: $P{-}O^-$) to yield an MX_5-intermediate. An MX_5-type phosphorus compound (phosphorane) is stable and substitution proceeds accompanied by BPR (cf. Chapters 2 and 6 and references therein).

CHAPTER 7

SULFUR, SELENIUM, AND TELLURIUM COMPOUNDS

Group / (n) / Period	15 5B	16 6B	17 7B
1 1s			
2 [He] 2s2p	0.70 1402 $^{14}_{7}$**N** Nitrogen 3.0 292	0.66 1314 (1.40^{2-}) $^{16}_{8}$**O** Oxygen 3.5 351	0.58 1681 (1.33^{-}) $^{19}_{9}$**F** Fluorine 4.0 441
3 [Ne] 3s3p	1.10 1012 $^{31}_{15}$**P** Phosphorus 2.1 264	1.04 1000 (1.74^{2-}) $^{32}_{16}$**S** Sulfur 2.5 272	0.99 1251 (1.81^{-}) $^{35}_{17}$**Cl** Chlorine 3.0 352
4 [Ar:3d^{10}] 4s4p	1.21 947 $^{75}_{33}$**As** Arsenic 2.0 200	1.17 941 (1.91^{2-}) $^{80}_{34}$**Se** Selenium 2.4 245	1.14 1140 (1.96^{-}) $^{79}_{35}$**Br** Bromine 2.8 293
5 [Kr:4d^{10}] 5s5p	1.41 834 $^{121}_{51}$**Sb** Antimony 1.9 215	1.37 869 (2.24^{2-}) $^{130}_{52}$**Te** Tellurium 2.1	1.33 1008 (2.20^{-}) $^{127}_{53}$**I** Iodine 2.5 213
6 [Xe:4f$_{14}$5d^{10}] 6s6p	1.52 703 $^{209}_{83}$**Bi** Bismuth 1.9 143	1.53 812 $^{209}_{84}$**Po*** Polonium 2.0	(0.57^{5+})930 (2.27^{-}) $^{216}_{85}$**At*** Astatine 2.2

Organo Main Group Chemistry, First Edition. Kin-ya Akiba.

7.1 SULFUR COMPOUNDS

A simple substance of sulfur exists as a variety of polymers, and crown-shaped S_8 is the most stable of them and the substance is commercially available as cheap powder. Simple substances of selenium and tellurium are sold commercially as powder or metallic rod. The three elements can bear a variety of valences from -2 to $+6$; hence, there are a variety of functional groups centered at the element. Fundamental characters of the element have been summarized in Table 1.1 of Chapter 1; in this chapter, acidity constant (pK_a), bond angle, and bond length of dihydride (H_2M) are summarized in Table 7.1.

There are big differences in character between H_2O (second period) and other three H_2M (third period and heavier). This is mainly due to the difference in hybridization, that is, oxygen is fundamentally sp^3 hybridized but the s orbital of the other three is essentially not hybridized with the p orbitals. Bond angle of H_2O is 104.5° but that of the three H_2M is close to 90°. Actually, bond angle of H_2Te is 89.5° and is the closest to the ideal angle of 90°. H_2O is neutral but all the three H_2M are considerably strong acids and H_2Te is the strongest of the three and its pK_a is 2.6. The acidity of alkyl thiol (RSH) is 10.3–10.5, and those of alcohol, thiophenol, and phenol are 16–18, 7.0, and 10.0, respectively.

By the reduction of a simple substance (M) with $NaBH_4$, Na_2M is obtained and the latter is alkylated to afford RMNa by the reaction with alkyl halide. On the other hand, the reaction of simple substance with RLi (MgX) gives RMLi (MgX). These (RMgX, RMNa, and RMLi) afford RMH when treated with water, and RMH usually gives RM–MR by oxidation.

Among the three simple substances, sulfur has been utilized from ancient times and investigated to show colorful chemistry. Therefore, as an example, fundamental structure and nomenclature of sulfur compounds bearing a variety of oxidation (valence) states are listed in Fig. 7.1.

In Fig. 7.1, compounds from thiolate to iminosulfonic acid and their derivatives are classified as organosulfur compounds and they are arranged according to the valence of sulfur. Sulfinic acid and sulfonic acid have their halogeno-derivatives [1].

Electrons left on the sulfur are formally shown as dot(s). Chlorosulfuric acid, thionyl chloride, and sulfonyl chloride are commonly used in organic synthesis. Monopersulfuric acid is utilized as an oxidizing and also as an epoxidizing reagent just like metachloroperbenzoic acid. They are derivatives of hydrogen peroxide.

7.2 SYNTHESIS OF ORGANOSULFUR COMPOUNDS

In this section, syntheses of popular and fundamental organosulfur compounds are described. Alkyl thiol (**1**) is prepared by the reaction of alkyl halide with thiolating reagents (Eq. 7.1). As thiolating reagents, sodium hydrogen sulfide is the most

TABLE 7.1 Characteristics of H_2M (M = O, S, Se, Te)

	H_2O	H_2S	H_2Se	H_2Te
p*K*a	14.0	7.0	3.7	2.6
∠HMH(°)	104.5	92.1	91.0	89.5
Length(MH) (Å)	0.96	1.34	1.46	1.69

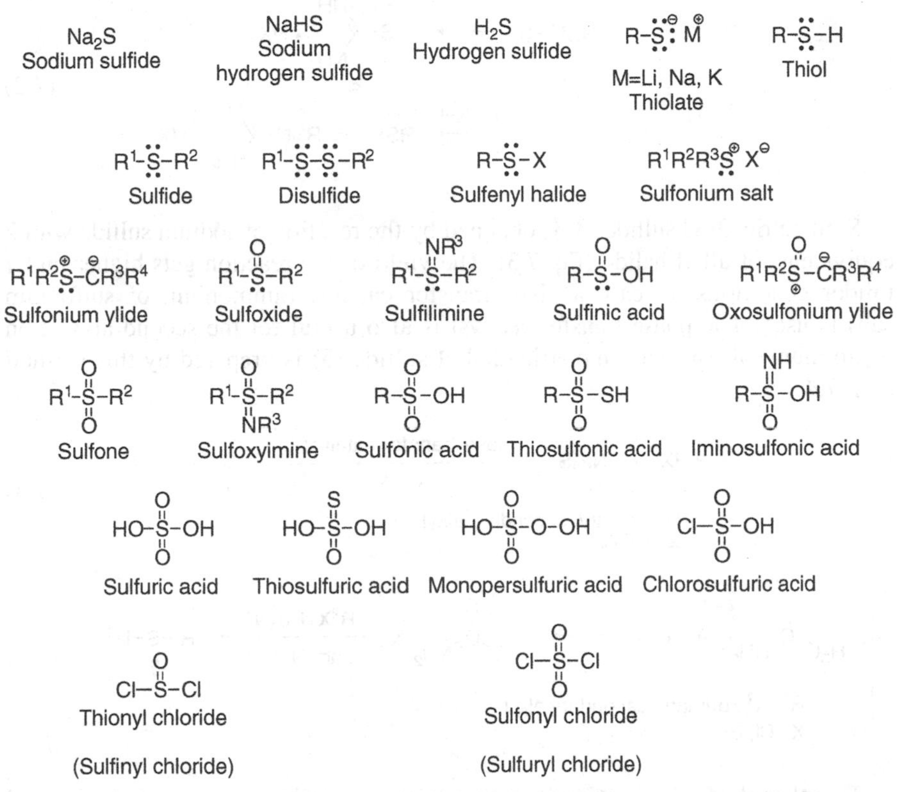

Na_2S
Sodium sulfide

NaHS
Sodium hydrogen sulfide

H_2S
Hydrogen sulfide

$R\text{-}S^{\ominus}\ M^{\oplus}$
M=Li, Na, K
Thiolate

R-S-H
Thiol

$R^1\text{-}S\text{-}R^2$
Sulfide

$R^1\text{-}S\text{-}S\text{-}R^2$
Disulfide

R-S-X
Sulfenyl halide

$R^1R^2R^3S^{\oplus}\ X^{\ominus}$
Sulfonium salt

$R^1R^2S^{\oplus}\text{-}C^{\ominus}R^3R^4$
Sulfonium ylide

$R^1\text{-}S(=O)\text{-}R^2$
Sulfoxide

$R^1\text{-}S(=NR^3)\text{-}R^2$
Sulfilimine

R-S(=O)-OH
Sulfinic acid

$R^1R^2S^{\oplus}(=O)\text{-}C^{\ominus}R^3R^4$
Oxosulfonium ylide

$R^1\text{-}S(=O)_2\text{-}R^2$
Sulfone

$R^1\text{-}S(=O)(=NR^3)\text{-}R^2$
Sulfoxyimine

$R\text{-}S(=O)_2\text{-}OH$
Sulfonic acid

$R\text{-}S(=O)_2\text{-}SH$
Thiosulfonic acid

R-S(=NH)(=O)-OH
Iminosulfonic acid

$HO\text{-}S(=O)_2\text{-}OH$
Sulfuric acid

HO-S(=S)(=O)-OH
Thiosulfuric acid

$HO\text{-}S(=O)_2\text{-}O\text{-}OH$
Monopersulfuric acid

$Cl\text{-}S(=O)_2\text{-}OH$
Chlorosulfuric acid

Cl-S(=O)-Cl
Thionyl chloride
(Sulfinyl chloride)

$Cl\text{-}S(=O)_2\text{-}Cl$
Sulfonyl chloride
(Sulfuryl chloride)

Figure 7.1 Sulfur compounds with a variety of valences: each acid has esters, and so on.

commonly used (i). Derivatives of thiolate anion such as sodium thiocyanate (ii) and sodium dithiocarbamate (iii), in which the thiolate anion is stabilized for easy use, are alkylated to give stable isolable derivatives as intermediates. These intermediates are reduced or hydrolyzed to afford the corresponding thiols. Isothiouronium salt (**2**) is obtained in high yield by the alkylation of thiourea, and thiol is obtained by the addition of secondary amine to **2** accompanied by guanidine (Eq. 7.2).

$$R{-}X \begin{cases} \text{(a) } NaHS \xrightarrow{H^+,\ H_2O} \\ \text{(b) } NaSCN \longrightarrow RSCN \xrightarrow{LiAlH_4} \\ \text{(c) } Me_2NC(S)SNa \longrightarrow Me_2NC(S)SR \xrightarrow{H^+,\ H_2O} \end{cases} R{-}SH\ \mathbf{1} \tag{7.1}$$

R : primary, secondary alkyl
X : Cl, Br, I, OTs

$$R{-}X + (H_2N)_2C{=}S \longrightarrow RS{-}C({=}NH)NH_2 \cdot HX\ (\mathbf{2}) \xrightarrow{R'_2NH} RSH + R'_2N{-}C({=}NH)NH_2 \cdot HX \tag{7.2}$$

Symmetric alkyl sulfide (**3**) is obtained by the reaction of sodium sulfide with 2 equivalents of alkyl halide (Eq. 7.3). The yield of the reaction gets higher under milder conditions, in case a phase-transfer catalyst (ammonium or sulfonium salt) is used. The phase-transfer catalyst is also useful for the second alkylation of iminium salt (**4**). Unsymmetric dialkyl sulfide (**5**) is prepared by this method (Eq. 7.4).

$$2\,RX + Na_2S \xrightarrow[H_2O,\ \Delta]{\text{Phase-transfer catalyst}} R{-}S{-}R\ (\mathbf{3}) \tag{7.3}$$

R : primary, secondary alkyl
X : Cl, Br

$$H_3C{-}C({=}S)NH_2 + R^1X \longrightarrow H_3C{-}C(SR^1){=}NH_2^+\ X^-\ (\mathbf{4}) \xrightarrow[aq.\ NaOH]{R^2X,\ Bu_4NBr} R^1{-}S{-}R^2\ (\mathbf{5}) \tag{7.4}$$

R^1, R^2 : primary, secondary alkyl
X : Cl, Br

Thiophenol (**6**: Ar = phenyl) is prepared by heating benzene sulfonic acid with iodine and triphenylphosphine through a one-pot process. This is a rather new and useful method. In general, aryl thiol is obtained by the reaction of organometallic reagent with a simple substance of sulfur. This is also applicable for the preparation of alkyl thiol.

$$\left.\begin{array}{l} ArSO_3H \xrightarrow{I_2,\ Ph_3P} \\ ArLi(MgBr) + S_8 \end{array}\right\} \longrightarrow ArSH\ (\mathbf{6}) \tag{7.5}$$

1,2-(Diisopropylthio)benzene (**7**) is obtained by heating 1,2-dichlorobenzene with sodium isopropyl thiolate in DMF (*N*,*N*-dimethylformamide). 1,2-Benzenedithol (**8**) is afforded by the reduction of **7** with sodium metal. This shows that thiolate can substitute aromatic halogen. Aryl sulfide (**9**) is prepared by heating iodobenzene and sodium sulfide in DMF in the presence of copper(I) iodide (Eq. 7.7).

$$o\text{-}C_6H_4Cl_2 + 2\ i\text{-PrSNa} \xrightarrow[\Delta]{\text{DMF}} o\text{-}C_6H_4(S\text{-}i\text{-Pr})_2\ (\mathbf{7}) \xrightarrow[\text{HCl}]{\text{Na}} o\text{-}C_6H_4(SH)_2\ (\mathbf{8}) \tag{7.6}$$

$$R\text{-}C_6H_4\text{-}I + Na_2S \xrightarrow{\text{Cu(I), DMF}} R\text{-}C_6H_4\text{-}S\text{-}C_6H_4\text{-}R\ (\mathbf{9}) \tag{7.7}$$

Sulfoxide (**10**) is obtained by the oxidation of sulfide (**5**), but it is important to suppress the formation of sulfone (**15**) as a by-product (Eq. 7.8). To suppress overoxidation, a variety of oxidants and reaction conditions have been examined. Oxidants such as hydrogen peroxide, MCPBA (*meta*-chloroperoxybenzoic acid), sodium periodate, and bromine are used and the yield is usually 80–100%.

$$\underset{\mathbf{5}}{R^1\text{-}S\text{-}R^2} + \text{Oxidizing reagents} \longrightarrow \underset{\mathbf{10}}{R^1\text{-}S(=O)\text{-}R^2} \tag{7.8}$$

Sulfoxide (**10**) can also be prepared by the reaction of sulfinic acid chloride or its ester with Grignard reagent (Eq. 7.9). The yield of this method is usually excellent and the process is easy to handle. The sulfinyl group [R(S = O)] is stereochemically stable even though it has lone pair electrons. Thus, a variety of asymmetric sulfoxides have been synthesized and utilized to investigate the stereochemistry of reactions on sulfur.

$$\underset{\mathbf{11}}{R^1SOCl} + R^2MgBr \longrightarrow \underset{\mathbf{10}}{R^1\text{-}S(=O)\text{-}R^2} \tag{7.9}$$

$$p\text{-Tol}\text{-}S(=O)\text{-}Cl + \text{HO-menthyl}\ (\mathbf{12}) \xrightarrow[\text{Resolution}]{\text{Recrystallization}} (S)\text{-}(-)\text{-}p\text{-Tol}\text{-}S^*(=O)\text{-O-menthyl}\ (\mathbf{13^*}) \xrightarrow[\text{Complete inversion}]{\text{PhMgBr}} (R)\text{-}(+)\text{-}p\text{-Tol}\text{-}S^*(=O)\text{-}Ph\ (\mathbf{14^*}) \tag{7.10}$$

For example, a mixture of diastereomers of sulfinic ester (**13**) is obtained by the reaction of *p*-toluenesulfinyl chloride and menthol (**12**); then, the optically active ester (**13***) can be obtained by recrystallization (Eq. 7.10). The reaction of **13*** with Grignard reagent affords an optically active sulfoxide (**14***); this proceeds with complete optical inversion [2]. This is the same type of reaction as the synthesis of the optically active phosphine oxide described in Eq. (6.25) of Chapter 6.

Sulfone (**15**) is obtained by the oxidation of sulfide and/or sulfoxide (Eq. 7.11). Hydrogen peroxide, peroxyacetic acid, potassium permanganate, monopersulfuric acid, and monoperboronic acid are commonly used. Conditions for oxidation of sulfide are stronger than those for sulfoxide. Sulfone is much more stable than sulfoxide and is hardly overoxidized.

$$\underset{\mathbf{5}}{R^1\text{-}S\text{-}R^2} + \text{Oxidizing reagents} \longrightarrow \underset{\mathbf{15}}{R^1\text{-}S(=O)_2\text{-}R^2} \qquad (7.11)$$

Sulfone (**16**) is also prepared by the alkylation of sodium sulfinate in the presence of a phase-transfer catalyst (Eq. 7.12). The same type of sulfone (**16**) results from the Friedel–Crafts reaction of sulfonyl chloride with aromatic hydrocarbon in the presence of aluminum trichloride (Eq. 7.13).

$$Ar\text{-}S(=O)\text{-}ONa + R\text{-}X \xrightarrow{\text{Phase-transfer catalyst}} \underset{\mathbf{16}}{Ar\text{-}S(=O)_2\text{-}R} \qquad (7.12)$$

Ar: phenyl, *p*-tolyl
R: primary alkyl, benzyl, phenacyl

$$R\text{-}S(=O)_2\text{-}Cl + Ar\text{-}H \xrightarrow{AlCl_3} \underset{\mathbf{16}}{Ar\text{-}S(=O)_2\text{-}R} \qquad (7.13)$$

R: alkyl, phenyl
Ar–H: benzene, xylene, mesitylene

Sulfinic acid adds to the activated conjugate alkene by Michael-type reaction to afford sulfone (**17**) (Eq. 7.14). Moreover, sulfinic acid can add to a carbonyl group and an imino group to give the corresponding sulfone. Sulfonyl chloride adds to double and triple bonds catalyzed by copper(I) chloride in a radical-type reaction. The former affords **19** and the latter yields a mixture of *trans*-**18** and *cis*-**18** (Eq. 7.15).

$$Ar\text{-}S(=O)\text{-}OH + R^1CH{=}CH\text{-}C(=O)R^2 \longrightarrow \underset{\mathbf{17}}{Ar\text{-}S(=O)_2\text{-}CH(R^1)\text{-}CH_2\text{-}C(=O)\text{-}R^2} \qquad (7.14)$$

Ar: phenyl, *p*-tolyl; R^1, R^2: H, Me, Ph

$$RSO_2Cl \xrightarrow[\text{(a) } HC{\equiv}CPh]{CuCl} \underset{trans\text{-}\mathbf{18}}{RSO_2CH{=}C(Cl)Ph} + \underset{cis\text{-}\mathbf{18}}{RSO_2CH{=}C(Cl)Ph}$$
$$RSO_2Cl \xrightarrow[\text{(b) } H_2C{=}CHPh]{CuCl} \underset{\mathbf{19}}{RSO_2CH_2CHClPh} \quad (7.15)$$

Formation of sulfonium salt by the reaction of alkyl sulfide and alkyl halide is quite general (Eq. 7.16). Trifluoromethane sulfonic acid ester and oxonium salt (Meerwein reagent) are quite strong alkylating reagents, and the reaction proceeds cleanly because their resulting anions are quite stable (**21**; Eq. 7.17). A combination of alkyl iodide and silver salt is generally used for alkylation.

$$R^1SR^2 + R^3X \longrightarrow \underset{\mathbf{20}}{R^1R^2\overset{+}{S}{-}R^3\ \overset{-}{X}} \quad (7.16)$$

R^1, R^2 : alkyl
R^3 : primary, secondary alkyl
X : Br, I

$$R^1SR^2 \xrightarrow{CF_3SO_3CH_3\ (Me_3O^+BF_4^-)} \underset{\mathbf{21}}{R^1R^2\overset{+}{S}{-}CH_3\ CF_3SO_3^-}$$
$$R^1SR^2 \xrightarrow{R^3I,\ AgBF_4} \xrightarrow{NH_4PF_6} \underset{\mathbf{20}}{R^1R^2\overset{+}{S}{-}R^3\ PF_6^-} \quad (7.17)$$

R^1 : alkyl, aryl
R^2, R^3 : alkyl

Alkoxysulfonium salt (**22**) results from the alkylation of sulfoxide (Eq. 7.18). The salt (**22**) is converted to sulfonium salt (**20**) by nucleophilic substitution with Grignard reagent. Reaction of dimethyl sulfoxide (DMSO) and methyl iodide gives trimethyl oxosulfonium salt (**23**); this is a lucky exception in which alkylation takes place at the sulfur of sulfoxide (Eq. 7.19). Oxosulfonium salt (**24**) is generally prepared by the reaction of sulfonium salt with sodium peroxybenzoate (Eq. 7.20). Sulfonium salt and oxosulfonium salt are useful reagents for organic synthesis, as mentioned later.

$$R^1{-}\overset{O}{\overset{\|}{S}}{-}R^2 \xrightarrow{Me_3O^+BF_4^-} R^1{-}\overset{OMe}{\overset{|}{\underset{+}{S}}}{-}R^2\ BF_4^- \xrightarrow{R^3MgBr} \underset{\mathbf{20}}{R^1R^2\overset{+}{S}{-}R^3\ BF_4^-}$$
$$R^1{-}\overset{O}{\overset{\|}{S}}{-}R^2 \xrightarrow{RI,\ AgBF_4} \underset{\mathbf{22}}{R^1{-}\overset{OR}{\overset{|}{\underset{+}{S}}}{-}R^2\ BF_4^-} \xrightarrow{R^3MgBr} \mathbf{20} \quad (7.18)$$

R^1 : alkyl, aryl
R^2 : alkyl

$$H_3C{-}\overset{O}{\overset{\|}{S}}{-}CH_3 + CH_3I \longrightarrow \underset{\mathbf{23}}{(CH_3)_2\overset{O}{\overset{\|}{\underset{+}{S}}}{-}CH_3\ I^-} \quad (7.19)$$

$$R^1R^2R^3S^+ BF_4^- + \text{(3-ClC}_6\text{H}_4\text{)C(=O)–O–O–Na} \longrightarrow R^1R^2R^3\text{–}\overset{+}{S}(=O)\ BF_4^- \ \ \mathbf{24} \tag{7.20}$$

7.3 REACTIONS OF ORGANOSULFUR COMPOUNDS

There are two categories of reactions of organosulfur compounds, that is, (i) reaction at the central sulfur and (ii) reaction at the α-carbon of sulfur. The two categories of reactions take place in a similar manner for both organosilicon and organophosphorus compounds. Here, organosulfur compounds are taken as an example of the three.

1. Reaction at the central sulfur
   - nucleophilic substitution by thiolate (RS^-);
   - addition of sulfenyl cation (RS^+) to alkene and conjugate addition of sulfinate (RSO_2^-) to activated alkene;
   - addition of thiyl radical ($RS\bullet$) and sulfonyl radical ($RSO_2\bullet$) to alkene;
   - nucleophilic addition of organometallic reagents to sulfinyl and sulfonium groups.
2. Reaction at the α-carbon of sulfur
   - stabilization of α-carbocation (unshared electron pair effect);
   - stabilization of α-carboradical (unshared electron pair effect);
   - stabilization of α-carbanion (effect of σ^*_{X-C});
   - stabilization of β-carbocation (effect of σ_{X-C}).

These directly correspond to the main group element effect mentioned in Table 2.1 of Chapter 2.

These reactivities are naturally used for the synthesis of organosulfur compounds. Here, several synthetic reactions are mentioned, in which the above-mentioned characteristics of organosulfur compounds are evidently used.

$$H_3C\text{–}S(=O)\text{–}CH_3 + Cl\text{–}C(=O)\text{–}C(=O)\text{–}Cl \longrightarrow \left[H_3C\text{–}\overset{+}{S}(\text{–O–C(=O)–C(=O)–Cl})\text{–}CH_3\ \ Cl^- \right]_{\mathbf{A}} \longrightarrow \left[H_3C\text{–}\overset{+}{S}(Cl)\text{–}CH_3\ Cl^- \ \ + CO_2 + CO \right]_{\mathbf{B}}$$

$$\xrightarrow[Et_3N]{RCH_2OH} \left[H_3C\text{–}\overset{}{S}(Cl)(\text{–O–CH}_2R)\text{–}CH_3 \ \ :NEt_3 \right]_{\mathbf{C}} \longrightarrow CH_3SCH_3 + O{=}C(H)R + HCl + Et_3N{\cdot}HCl \tag{7.21}$$

By the addition of oxalyl dichloride to DMSO, sulfonium ion (**A**) is generated by O-acylation and successive attack by the resulting chloride ion on the sulfonium salt gives chlorosulfonium salt (**B**) by the elimination of carbon monoxide and carbon dioxide (Eq. 7.21). By the addition of alcohol and a base (Et_3N) to **B**, chlorooxysulfurane (**C**) is generated and following deprotonation from a methyl group by the base affords an aldehyde and dimethyl sulfide, induced by the intramolecular electronic shifts written for **C**. This is generally used for the synthesis of an aldehyde or a ketone by the dehydrogenation of alcohol and is recognized as the *Swern oxidation*.

By heating sulfoxide in the presence of acetic anhydride, O-acetylation takes place to generate sulfonium salt (**D**) and the resulting acetate ion deprotonates to give sulfenium ylide (**E**) and acetic acid and another acetate ion. The resulting acetate ion subsequently adds to **E** to yield thiolacetate (**26**) (Eq. 7.22). During the reaction, the oxygen of sulfoxide migrates to the α-carbon. This is called the *Pummerer rearrangement*.

$R^1CH_2-S(=O)-R^2$ (**25**) + $(CH_3CO)_2O$ ⟶ [**D**: $R^1CH(H)-S^+(R)-O-C(=O)-CH_3$, $^-O(O)CH_3CH_3$] ⟶ [**E**: $R^1CH=S^+-R^2$ + CH_3CO_2H, $CH_3CO_2^-$] ⟶ $R^1CH(OC(=O)CH^3)-S-R^2$ (**26**)

(7.22)

α-Hydrogen of sulfone (**27**) is acidic and easily deprotonated by the base. When a leaving group (LG) such as halogen is present on the α-carbon, a three-member ring is produced by nucleophilic attack of the carbanion. A mixture of alkenes is obtained by the succeeding elimination of sulfur dioxide (Eq. 7.23). This is called the *Ramberg–Baeckland reaction*.

B: + $R^1CH(H)-SO_2-CH(R^2)X$ (**27**) ⟶ [**F**: episulfone of R^1 and R^2] $\xrightarrow{-SO_2}$ $R^1CH=CHR^2$ (**28**) (7.23)

X = Halogen, leaving group

Dithiane (**29**) can be used as a carbanion twice successively because 2-C with two hydrogens are connected to two sulfur atoms which stabilize the carbanion. Hence, **29** gives **30** and then cyclizes to afford **31**. The resulting **31** is hydrolyzed to ketone, activated by Lewis acid (Eq. 7.24). Therefore, 2-C carbon is converted to carbonyl carbon. By the oxidation of one of the sulfur of dithioacetal to sulfoxide (**32**: FAMSO) or sulfone (**35**: MT-sulfone), acidic hydrolysis becomes much easier and the central carbons of **32** and **35** are converted to the carbonyl carbons of the products (**34**).

$$\mathbf{29} \xrightarrow{\text{BuLi}} \xrightarrow{\text{Cl(CH}_2)_4\text{Br}} \mathbf{30} \xrightarrow{\text{BuLi}} \mathbf{31} \xrightarrow[\text{H}_2\text{O}]{\text{HgCl}_2} \text{O}= \text{cyclopentanone} \tag{7.24}$$

$$\underset{\substack{\mathbf{32}\\ \text{FAMSO}}}{\text{H}_2\text{C(S(O)Me)(SMe)}} \xrightarrow{\text{BuLi}} \xrightarrow{\text{R–X}} \underset{\mathbf{33}}{\text{RHC(S(O)Me)(SMe)}} \xrightarrow{\text{H}^+/\text{H}_2\text{O}} \underset{\mathbf{34}}{\text{RCHO}} \tag{7.25}$$

$$\underset{\substack{\mathbf{35}\\ \text{MT-sulfone}}}{\text{H}_2\text{C(SO}_2\text{–C}_6\text{H}_4\text{–CH}_3\text{)(SMe)}}$$

Dimethylsulfonium methylide (**36**) can transfer the methylene group to double bond and affords cyclopropane, aziridine, and epoxide, respectively (Eq. 7.26). Dimethyloxosulfonium methylide (**37**) also transfers the methylene group to the carbonyl group to yield an epoxide ring. Here it is noted, with conjugated double bond, that **36** forms an epoxide ring, while **37** affords a cyclopropane ring (Eq. 7.27) [3].

$$\underset{\mathbf{21}}{\text{Me}_2\overset{+}{\text{S}}\text{–CH}_3\ \text{I}^-} \rightarrow \underset{\mathbf{36}}{\text{Me}_2\text{S=CH}_2}$$

(a) $\text{R}^1\text{R}^2\text{C=CR}^3\text{R}^4 \rightarrow \text{R}^1\text{R}^2\text{C–CR}^3\text{R}^4$ (bridged by CH_2)

(b) $\text{PhCH=NPh} \rightarrow \text{PhHC–NPh}$ (bridged by CH_2)

(c) $\text{R}^1\text{R}^2\text{C=O} \rightarrow \text{R}^1\text{R}^2\text{C–O}$ (bridged by CH_2)

(d) $\text{PhCH=CHC(=O)Ph} \rightarrow \text{PhCH=CH–C(Ph)–O}$ (bridged by CH_2) (7.26)

$$\underset{\mathbf{23}}{\text{Me}_2\overset{+}{\text{S}}(\text{O})\text{–CH}_3\ \text{I}^-} \rightarrow \underset{\mathbf{37}}{\text{Me}_2\text{S(O)=CH}_2}$$

(a) $\text{R}^1\text{R}^2\text{C=O} \rightarrow \text{R}^1\text{R}^2\text{C–O}$ (bridged by CH_2)

(b) $\text{PhCH=CHC(=O)Ph} \rightarrow \text{PhHC–CH–C(=O)Ph}$ (bridged by CH_2) (7.27)

Corey investigated successfully in order to improve the yield and reaction conditions of the Wittig reaction and used strongly polarized DMSO as the solvent and dimsyl anion (MeSO–CH_2Na) as the base. The product alkene was obtained in higher yield under milder conditions. Then, the methylene transfer reaction was found. It is interesting to note here that there is a unique difference in reactivity between the ylides of phosphorus and sulfur.

The sulfur of allyl sulfide (**38**) can trap carbene much more efficiently than a double bond and yields S-ylide (**G**) *in situ* (Eq. 7.28). The resulting S-ylide (**G**) undergoes [2, 3] sigmatropy with the S-allyl group to give a mixture of alkenes (**40**). By deprotonation of the allyl methylene of sulfonium salt (**41**), S-ylide is generated as an intermediate; then ring enlargement to yield **42** proceeds by [2, 3] sigmatropy in a high yield (Eq. 7.29). Furthermore, 12-member ring containing diene (**44**) is obtained by the second [2, 3] sigmatropy after deprotonation of the sulfonium ion (**43**) prepared by alkylation with bromoacetate [4].

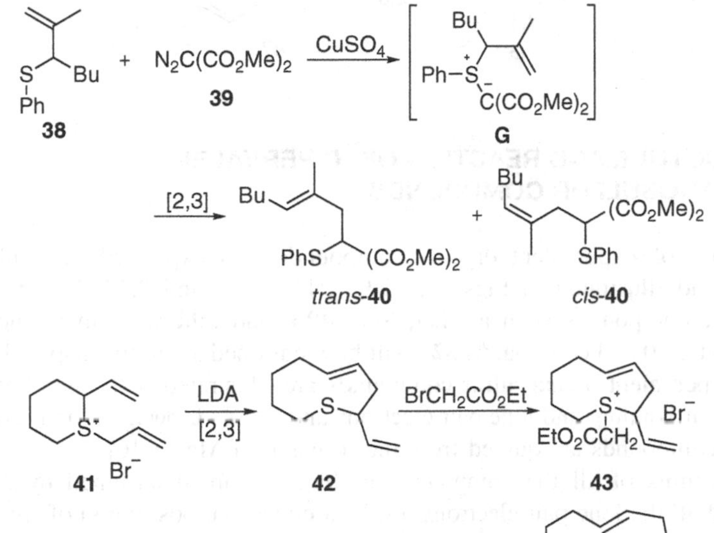

(7.28)

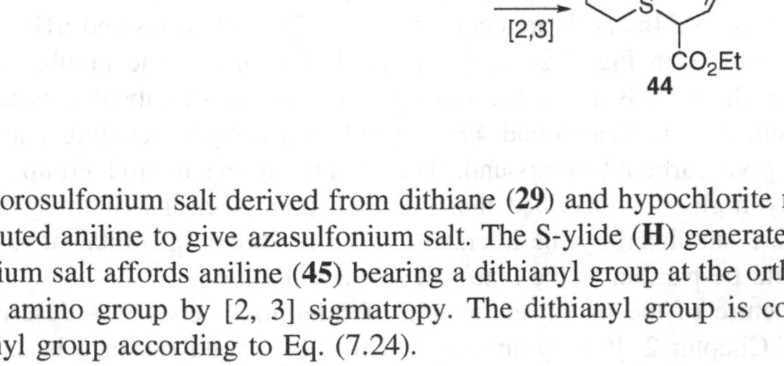

(7.29)

Chlorosulfonium salt derived from dithiane (**29**) and hypochlorite reacts with substituted aniline to give azasulfonium salt. The S-ylide (**H**) generated from the sulfonium salt affords aniline (**45**) bearing a dithianyl group at the ortho position of the amino group by [2, 3] sigmatropy. The dithianyl group is converted to carbonyl group according to Eq. (7.24).

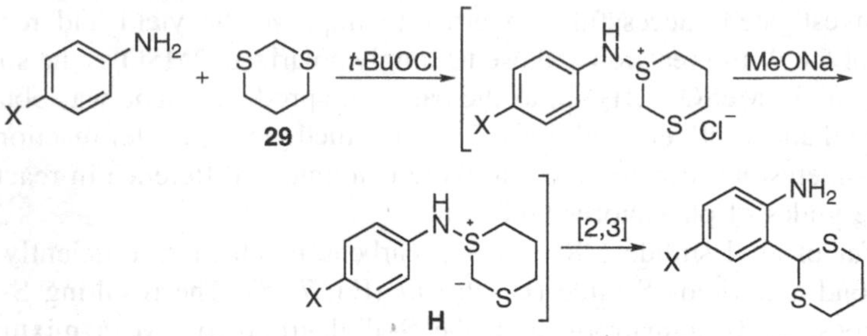

(7.30)

Optically active sulfonium salt (**46**) affords sulfide (**47**) by [2, 3] sigmatropy, in which optical activity is transferred from the sulfur to the carbon in high optical yield (94%). Thus, it is illustrated that [2, 3] sigmatropy can be a useful tool for organic synthesis [5].

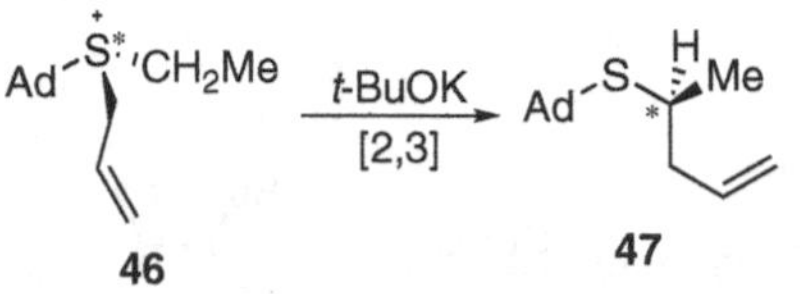

(7.31)

7.4 STRUCTURE AND REACTION OF HYPERVALENT ORGANOSULFUR COMPOUNDS

The structure of hypervalent organic compounds was explained generally in Chapter 2 and illustrated in Figs. 2.5, 2.9–2.11, 2.13, and 2.14. Hypervalent organosulfur compounds such as SF_4, SF_6, **49b**, and **50d** were mentioned in Figs. 2.5 and 2.10 and compounds **52a** will be mentioned again in Chapter 11. In Fig. 7.2, hypervalent organosulfur compounds are illustrated with due attention to hypervalent bonding and lone pair electrons and N–X–L designation is added. The major compounds are quoted from the research of Martin [6].

The structures of all the compounds in Fig. 7.2 are determined by X-ray analysis and all the lone pair electrons reside at equatorial position(s) of sp^2 configuration. The imidazole ring of **48** is perpendicular to the hypervalent Br–S–Br bond. Compounds **51a** and **b** are octahedron and **51a** is O-trans and **51b** is O-cis isomer and others in Fig. 7.2 are all trigonal bipyramid. The number of formal valence electrons is 10 or 12, although the number of substituents (ligand) changes from 3 to 6. Compound **49c** is used as a dehydrogenating reagent of alcohol to give carbonyl compound. The acidity of the methyl group of **50d** is unusually high. It is ascribed that the lone pair electrons of the generated methylide are stabilized by the interaction with antibonding orbital of O–S–O, which should be parallel to the lone pair of the carbanion.

The structure of the parent compound of thiathiophthenes was described in Fig. 2.9 of Chapter 2. It is symmetric and the S–S bond distance is 2.363 Å. The distance changes to 2.499 Å and 2.218 Å when there are phenyl groups

48: 10-S-3 **49a**: 10-S-4 **49b**: 10-S-4 **49c**: 10-S-4

50a: 10-S-3 **50b**: 10-S-4 **50c**: 10-S-5 **50d**: 10-S-4

51a: 12-S-6 **51b**: 12-S-6 **52a**: 10-S-4 **52b**: 10-S-4

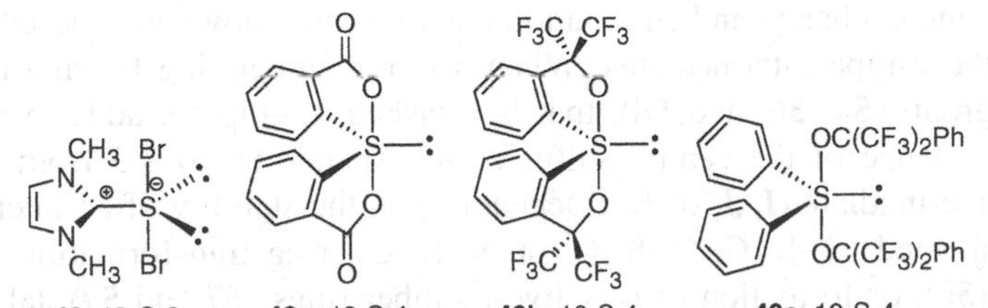

Figure 7.2 Hypervalent sulfur compounds: illustrated with unshared electron pairs and N–X–L designation.

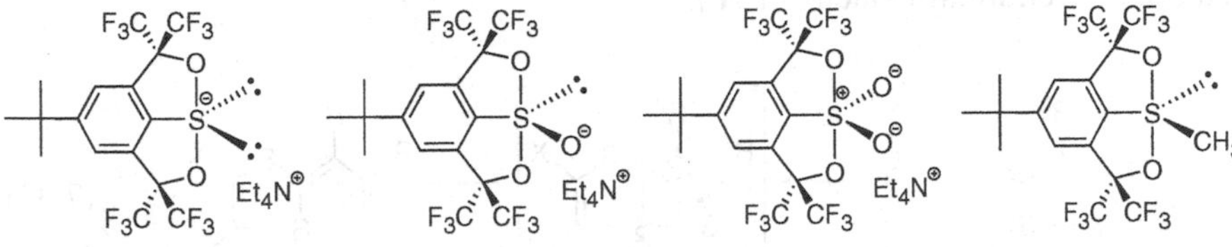

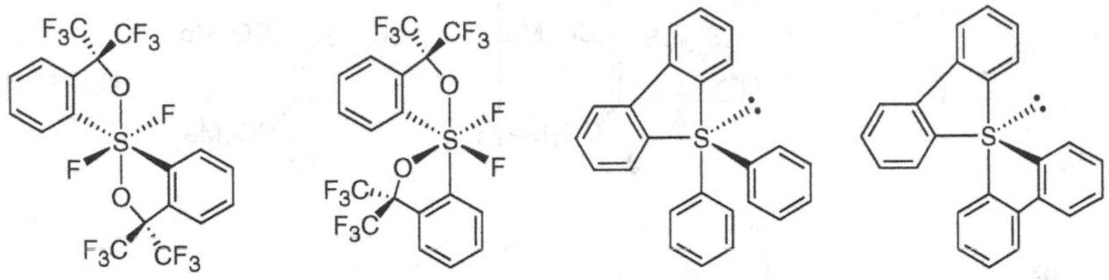

at 2 and 4 positions ($\Delta = 0.28$ Å). It is apparent that the distance is quite susceptible to substituent effect. When the electronegativity of the atom at 6 position increases according to the change in the kind of atom from sulfur to nitrogen and to oxygen, the interaction between the central sulfur (S^{6a}) and the atom at 6 position decreases; therefore S^1–S^{6a} distance shortens to become closer to a covalent bond (2.08 Å, cf. **53b** and **53c**). The structure of **53d** is symmetric and the N–S distance is 1.904 Å (covalent bond length: 1.74 Å) (Fig. 7.3). Now, it is clearly illustrated that the hypervalent bond is longer and weaker than the corresponding covalent bond.

53a **53b** **53c** **53d**

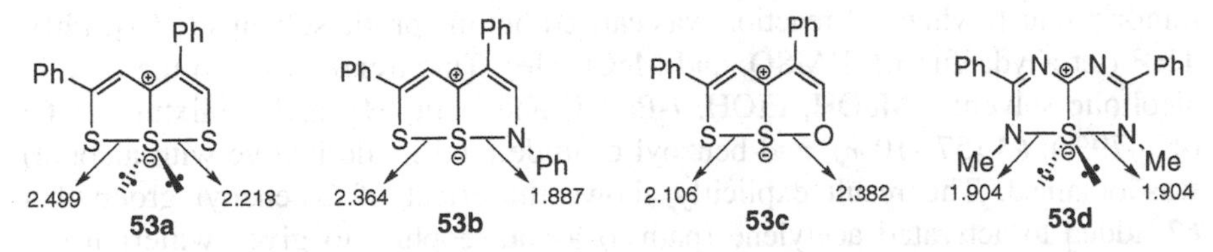

Figure 7.3 Distance of hypervalent bond of thiathiophthene derivatives cf.: Eq. (2.1) and Fig. 2.9.

Several ring exchange and elimination reactions are known to proceed through thiathiophthene-type intermediates. When a five-member ring bearing a thio or an imino group (**54, 56**, and **58**), that is, masked 1,3-dipole, adds to activated acetylene, valence of the central sulfur atom expands to 10-S-3 from 8-S-2 to afford an intermediate (**I, J**, or **K**). Depending on the stability of an intermediate hypervalent bond (S–S–C, C–S–C, or N–S–C), ring transformation by bond switching (**55**), or formation of new five-member rings (**57** and **59**), takes place by the elimination of ethylene or nitrile. Compound **57** is useful as a synthetic material of tetrathiafulvalene (TTF).

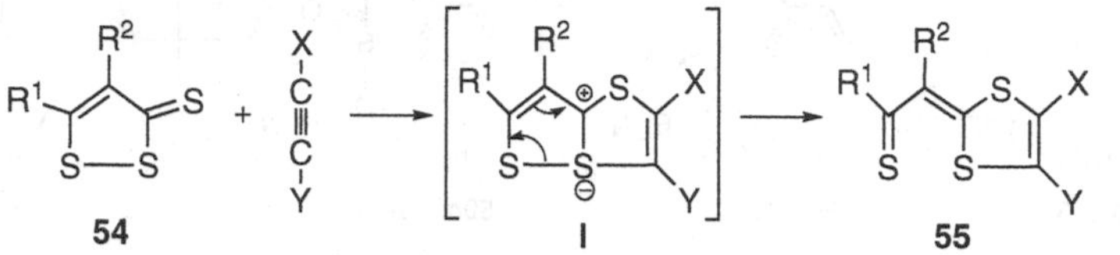

(7.32)

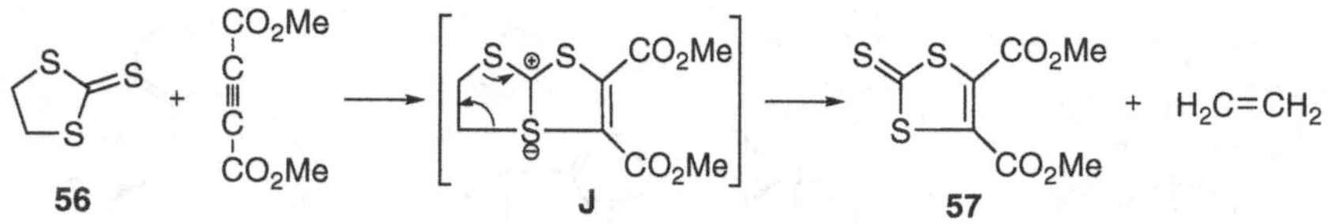

(7.33)

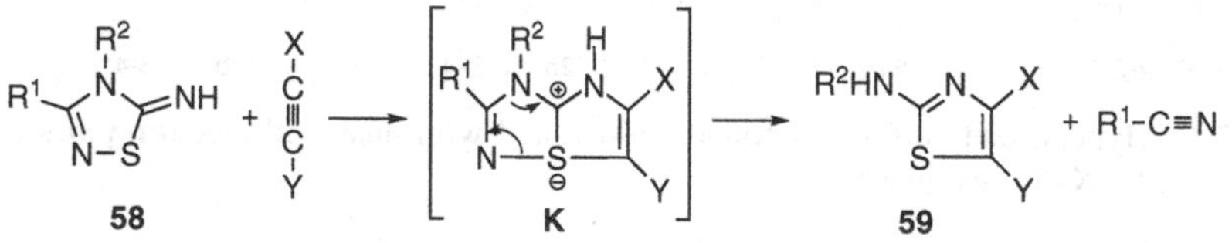

(7.34)

A five-member ring bearing an imino group (**60**: regioisomer of **58**) reacts with activated acetylene to give a simply added vinyl product (**61**: a mixture of cis and trans isomers), in which the elimination of nitrile was not observed (Eq. 7.35). This may result from the failure of formation of an intermediate hypervalent bond (**M′**: C–S–C) due to the weakness of the bond compared to that of **K** (N–S–C). Then, by using **62** having a benzoyl group at the 2 position to increase the electronegativity of the central sulfur, simply added vinyl product (**63**) and five-member ring with ring transformation (**64**) and benzoyl cyanide were obtained (Eq. 7.36). Compound **64** was the major product (73–92%) and **63** was the minor product, when the reaction was carried out in aprotic solvents (CCl_4, PhH, THF (tetrahydrofuran), DMSO, and MeCOMe). This trend was also the case in alcoholic solvents (MeOH, EtOH, *i*-PrOH, and *t*-BuOH), and a mixture of **64** (42–79%), **63** (57–10%), and benzoyl cyanide (and its derivative with alcohol) was obtained. The result explicitly shows the effect of 2-benzoyl group that **62** added to activated acetylene (path a) as nucleophile to give zwitterion (**L**) and (path b) as masked 1,3-dipole to generate hypervalent bond (**M**: C–S–C), competitively.

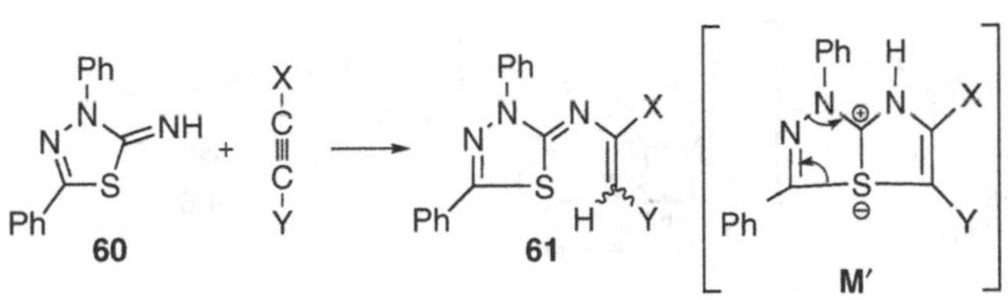

(7.35)

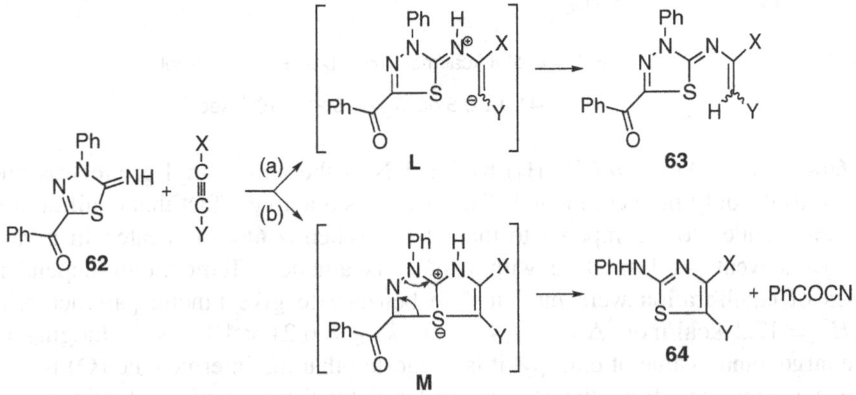

(7.36)

5-Amidino-1,2,4-thiadiazole derivatives (α-**65** and β-**65**) were prepared separately by the addition of the corresponding nitrile to the 5-amino group of 1,2,4-thiadiazoles catalyzed by $AlCl_3$. Surprisingly, the same solid was obtained by the two different processes and it was disclosed that the two compounds (α-**65** and β-**65**) are in equilibrium in solution (Eq. 7.37). The equilibrium was firmly established by the presence of 1:1 equilibrium mixture of α-**65a*** and β-**65a*** in solution when symmetric **65a*** was prepared by employing ^{15}N-acetonitrile. Compound β-**65** bearing substituent (R) at the 3-position of thiadiazole becomes predominant when electronegativity and steric hindrance of R increase. The equilibrium indicates the formation of symmetric hypervalent bond (N: N–S–N), probably by 1,5-sigmatropic shift of hydrogen(s). The 1,6-dimethylated compound (**53d**) of the intermediate **N** was synthesized and the structure was determined by X-ray analysis and S–N bond length was shown to be 1.904 Å [7].

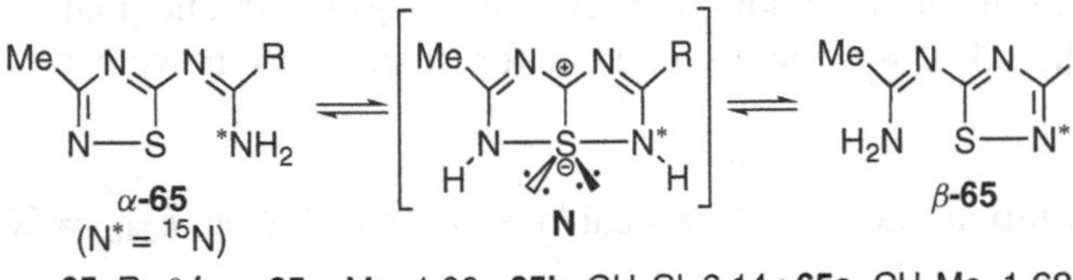

(7.37)

65, R, β / α ; **65a**, Me, 1.00 ; **65b**, CH_2Cl, 2.14 ; **65c**, CH_2Me, 1.68

65d, *t*-Bu, 8.57 ; **65e**, Ph, 14.5 ; **65f**, *p*-ClC_6H_4, ~50

5-Aminovinylisothiazole derivatives (**66**) were synthesized. Equilibration between α-**66** and β-**66** was observed when it was heated in solution (Eq. 7.38).

α-66 ($N^* = {}^{15}N$) ⇌ [O] ⇌ β-66 (7.38)

66a, $Ar^1 = Ar^2 = p\text{-}ClC_6H_4$

$\Delta G^{\ddagger}_{298} = 24.4$ kcal/mol, $\Delta H^{\ddagger} = 12.2 \pm 0.2$ kcal/mol

$\Delta S^{\ddagger} = -41.0 \pm 0.5$ eu, $k_{298} = 6.21 \times 10^{-6}\ \text{sec}^{-1}$

α-**66a*** ($Ar^1 = Ar^2 = p\text{-}ClC_6H_4$) having ^{15}N at the aminovinyl group was prepared as the only pure compound. The success is due to the fact that equilibration of **66** is quite slow compared to that of **65**. When α-**66a*** is heated in a variety of solvents, 1:1 mixture with β-**66a*** is attained. Temperature-dependent rates of equilibration were measured in benzene to give kinetic parameters of $\Delta H^{\ddagger} = 12.2$ kcal/mol, $\Delta S^{\ddagger} = -41.0$ eu, $k_{298} = 6.21 \times 10^{-6}\ s^{-1}$. Judging by the large minus value of entropy, it is suggested that the intermediate (**O**) is very much compressed from the starting material by the formation of hypervalent bond (N–S–N) [8].

Now, we can take it for granted that the structure of intermediate (**O**) should be similar (almost the same) to that of **53d**. As S–N single bond distance in a five-member ring is 1.67 Å, the central sulfur atom moves back and forth eternally on the hypervalent bond (N–S–N: 3.80 Å = 1.90 × 2) in the range of 0.46 Å (3.80 − 2 × 1.67). The two pairs of lone pair electrons on the sulfur swing left to right just like a wiper on the front glass of a car according to the movement of the sulfur (bell-clapper rearrangement). This is the idea of bond switching and is described in more detail in Fig. 12.4 of Chapter 12.

How can bond energy of a hypervalent bond (N–S–N) be estimated experimentally? For the purpose, a model compound (**67**-H) bearing two fused 4,6-dimethylpyrimidine rings on the parent skeleton was synthesized. Only two kinds of the methyl groups (Me^1 and Me^2) could be observed for **67**, but four kinds of methyl groups were observed for α-**67**-Br with a bromine group on the pyrimidine ring (**B**). Temperature-dependent measurement of ^1H NMR (nuclear magnetic resonance) of α-**67**-Br in CD_2Cl_2 enabled the separate observation of the rotation of **A** ring to γ-**67**-Br and that of **B** ring to β-**67**-Br (Eq. 7.39). Kinetic parameters for each rotation were obtained as follows:

$$\text{For } \mathbf{A} \text{ ring rotation}: \Delta H^{\ddagger} = 18.8\ \text{kcal/mol},\ \Delta S^{\ddagger} = 1.8\ \text{eu},\ k_{348} = 26.6\ s^{-1}$$

$$\text{For } \mathbf{B} \text{ ring rotation}: \Delta H^{\ddagger} = 16.1\ \text{kcal/mol},\ \Delta S^{\ddagger} = 2.2\ \text{eu},\ k_{297} = 18.9\ s^{-1}$$

The result says that the rotation energy of **B** ring with bromine is smaller than that of **A** ring without bromine by 2.7 kcal/mol. An electronegative substituent (Br) elongates and weakens the hypervalent S–N bond (X-ray result) and this is in accordance with the character of apicophilicity. A symmetric molecule (**67**) has rotation energy close to the average of **A** ring and **B** ring. The activation energy obtained by this method necessarily contains the rotation energy of pyrimidine ring; thus, hypervalent bond energy of S–N should be smaller by that amount and would be about 15 kcal/mol [9].

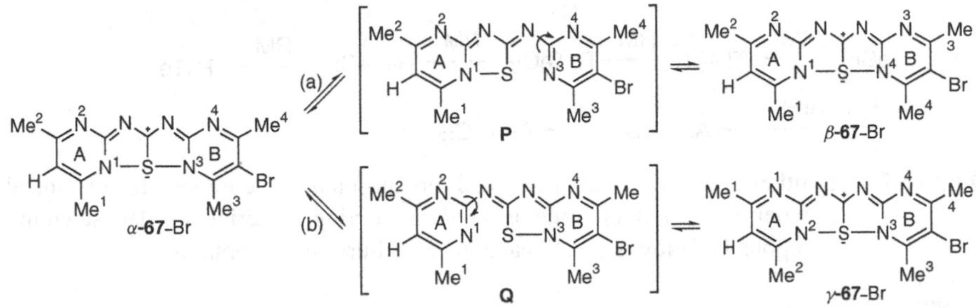

	T_c (°C)	$\Delta G^{\ddagger}_{298}$ (kcal/mol)	$\Delta H^{\ddagger}$ (kcal/mol)	$\Delta S^{\ddagger}$ (eu)
(a) Br-side	24	15.5	16.1 ± 1.3	2.2 ± 0.9
(b) H-side	75	18.3	18.8 ± 0.3	1.8 ± 1.0

(7.39)

7.5 SELENIUM AND TELLURIUM COMPOUNDS

Selenium and tellurium, the same as sulfur, can have valences from −2 to +6 hence there are a variety of compounds. Figure 7.4 illustrates the general synthetic methods for tellurium compounds. Selenium compounds are prepared in a similar manner. They are, however, not so well investigated as the sulfur compounds. Here, useful reactions characteristic of them are selectively explained [10, 11].

Arylselenenyl halide (**69**), prepared from diaryl diselenide (**68**) and a halogen, rapidly adds to an alkene. When a nucleophile such as alcohol exists in solution, *trans*-adduct (**70**) is obtained stereo selectively (Eq. 7.40).

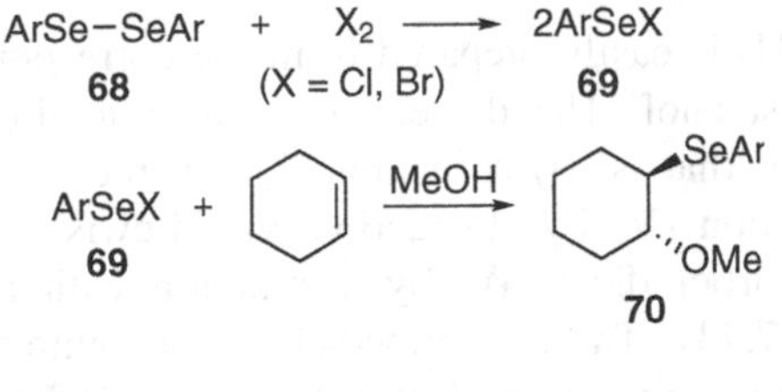

ArSe−SeAr (**68**) + X_2 (X = Cl, Br) ⟶ 2ArSeX (**69**)

ArSeX (**69**) + cyclohexene $\xrightarrow{\text{MeOH}}$ **70** (SeAr, OMe)

(7.40)

Alkyl aryl selenide with *α*-hydrogen (**71**) yields an alkene and aryl selenenic acid by oxidation with a weak oxidant such as hydrogen peroxide, in which selenoxide (**R**) is generated *in situ* and proceeds to the product without isolation. Stereochemistry of the resulting alkene is controlled

$$\text{Te (or Se)} \xrightarrow[\text{(M = Li, MgX)}]{\text{RM}} \text{RTeM} \xrightarrow[(\text{O}_2,\ \text{Br}_2)]{\text{Oxidation}} \text{RTe–TeR} \xrightarrow[\text{(X = Cl, Br)}]{\text{X}_2} \text{RTeX}$$

$$\text{Te (or Se)} \xrightarrow[(\text{NaBH}_4,\ \text{etc.})]{\text{Reduction}} \text{Na}_2\text{Te} \xrightarrow{\text{RX}} \text{RTeR} \xrightarrow{\text{X}_2} \text{R}_2\text{TeX}_2$$

$$\text{RTeR} \xrightarrow{\text{Oxidation}} \text{R}_2\text{TeO} \xrightarrow{\text{Oxidation}} \text{R}_2\text{TeO}_2$$

$$\text{TeCl}_4 \xrightarrow{\text{RM}} \text{RTeCl}_3 \xrightarrow{\text{RM}} \text{R}_2\text{TeCl}_2 \xrightarrow{\text{RM}} \text{R}_3\text{TeCl} \xrightarrow{\text{RM}} \text{R}_4\text{Te}$$

$$\text{TeCl}_4 \xrightarrow{\text{ArH}} \text{ArTeCl}_3 \xrightarrow{\text{ArH}} \text{Ar}_2\text{TeCl}_2$$

Figure 7.4 Synthesis of organotellurium and organoselenium compounds: (a) metal powder or its tetrachloride is used as starting material and (b) selenium compounds follow similar reactions as tellurium compounds.

because selenenic acid is eliminated concertedly through five-member ring as shown in **R**.

$$\mathbf{71} \xrightarrow[\text{THF}]{\text{H}_2\text{O}_2} [\mathbf{R}] \xrightarrow{-\text{ArSeOH}} \text{alkene} \qquad (7.41)$$

When allyl aryl selenide (**72**) is similarly oxidized, an intermediate selenoxide (**S**) subsequently undergoes [2, 3] sigmatropy to give **T**, and **T** is hydrolyzed to allyl alcohol (**73**) by the elimination of aryl selenenic acid.

$$\mathbf{72} \xrightarrow{\text{H}_2\text{O}_2} [\mathbf{S}] \xrightarrow{[2,3]} [\mathbf{T}] \xrightarrow[-\text{ArSeOH}]{\text{H}_2\text{O}} \mathbf{73} \qquad (7.42)$$

Diselenoacetal (**74**) is easily prepared from the corresponding carbonyl compound and benzeneselenol. The diacetal can generate three types of reactive intermediates *in situ*, that is, (i) carbanion (**U**) by metal exchange with butyllithium, (ii) carbocation (**V**) by abstraction with Lewis acid such as tin tetrachloride, and (iii) carboradical (**W**) by abstraction with metallic radical such as tin radical (Eq. 7.43). Three intermediates react with selected reagents as expected and the remaining phenylselenyl group can be utilized for further transformation; thus **74** is quite useful. These reactions are also applied for ditelluroacetal (**74′**).

$$R^1R^2C{=}O \xrightarrow{PhSeH / H^+} R^1R^2C(SePh)_2 \ (\mathbf{74}) \longrightarrow \begin{cases} \text{(a) BuLi} \rightarrow [R^1R^2C^{-}(SePh)] \ \mathbf{U} \rightarrow \\ \text{(b) } SnCl_4 \rightarrow [R^1R^2C^{+}{-}SePh] \ \mathbf{V} \rightarrow \\ \text{(c) } Bu_3Sn^{\bullet} \rightarrow [R^1R^2C^{\bullet}{-}SePh] \ \mathbf{W} \rightarrow \end{cases} \tag{7.43}$$

$R^1R^2C(TePh)_2$ **74′**

The most important characteristics of organotelluride (**75**) are to generate a new type of organometallic reagent (**76′**: R^1M) *in situ* very rapidly at a low temperature by metal exchange with organolithium and organometallic reagent (R^3M) (Eq. 7.44) [12].

$$\underset{\mathbf{75}}{R^1R^2Te} + \underset{\mathbf{76}}{R^3M} \overset{(a)}{\rightleftharpoons} \underset{\mathbf{X}\text{: 10-Te-3}}{[R^1R^2R^3Te]^- M^+} \overset{(a')}{\rightleftharpoons} \underset{\mathbf{76'}}{R^1M} + R^2R^3Te$$

M = Li, Na, K, MgX, AlX2, ZnX, Cu (including ate complexes) (7.44)

$$\underset{\mathbf{75}}{R^1R^2Te} + \underset{\mathbf{77}}{R^3\bullet} \overset{(b)}{\rightleftharpoons} \underset{\mathbf{Y}\text{: 9-Te-3}}{[R^1R^2R^3Te]^\bullet} \overset{(b')}{\rightleftharpoons} \underset{\mathbf{77'}}{R^1\bullet} + R^2R^3Te$$

The intermediate of this exchange is the so-called T-shape intermediate (**X**: 10-Te-3) in which R^1 and R^2 are apical substituents and R^3 is an equatorial substituent of the hypervalent compound bearing two pairs of lone pair electrons at equatorial positions. Organotelluride (**75**) easily reacts with radical species also to generate a new radical ($R^1\cdot$, **77′**) through an intermediate radical (**Y**: 9-Te-3). R^1 may sit at the apical position of the hypervalent radical (**Y**) bearing two pairs of lone pair electrons at equatorial positions.

Relative reaction rates of main group compounds with butyllithium at −70°C in THF are as follows: PhI:PhTeBu:$PhSnMe_3$:$PhSnBu_3$:PhSeBu = 4:1:0.9:0.05:0.003; hence, organotelluride is next to organoiodide in reactivity. By this tellurium–lithium exchange, the following lithium reagents are generated *in situ* and they react as expected; hence, organotelluride is useful as synthetic reagents (Eq. 7.45). In the examples written as Eqs. 7.43, 7.44, 7.45, the main group element effect of selenium and tellurium is exhibited more strikingly than that of sulfur which was described in Section 7.3.

$$\text{PhTePr} + \text{BuLi} \xrightarrow{(a)} [\text{PhLi}] + \text{PrTeBu}$$
Phenyl

$$\text{Ph-CH=CH-TePh} + \text{BuLi} \xrightarrow{(b)} [\text{Ph-CH=CH-Li}] + \text{PhTeBu}$$
Vinyl

$$\text{Br-C}_6\text{H}_4\text{-CH}_2\text{TeBu} + \text{BuLi} \xrightarrow{(c)} [\text{Br-C}_6\text{H}_4\text{-CH}_2\text{Li}] + \text{Bu}_2\text{Te}$$
Benzyl

$$\text{RC(O)TeBu} + \text{BuLi} \xrightarrow{(d)} [\text{RC(O)Li}] + \text{Bu}_2\text{Te}$$
Acyl

$$\text{Et}_2\text{NC(O)TeBu} + \text{BuLi} \xrightarrow{(e)} [\text{Et}_2\text{NC(O)Li}] + \text{Bu}_2\text{Te} \quad (7.45)$$
Carbamoyl

It is noteworthy here that telluroalkene (**78**) undergoes metal exchange with triethylaluminum and diethylzinc by rearrangement of intermediate ylide complex. Stereochemistry of **78** is retained by the exchange (Eq. 7.46).

$$\text{Ph(BuTe)C=CH}t\text{-Bu } (\mathbf{78}) \xrightarrow{\text{a) Et}_3\text{Al}} \text{Ph(Et}_2\text{Al)C=CH}t\text{-Bu} + \text{BuTeEt}$$

$$\text{Ph(BuTe)C=CH}t\text{-Bu } (\mathbf{78}) \xrightarrow{\text{(b) Et}_2\text{Zn}} \text{Ph(EtZn)C=CH}t\text{-Bu} + \text{BuTeEt} \quad (7.46)$$

REFERENCES

1. Cremlyn RJ. An introduction to organosulfur chemistry. Chichester: John Wiley & Sons, Ltd.; 1996.
2. (a) Solladie G, Hutt J, Girardin A. Synthesis 1987:173; (b) Oae S, Uchida Y. Acc Chem Res 1991;24:202.
3. (a) Corey EJ, Chaykovsky M. J Am Chem Soc 1962;84:867, 3782; (b) Corey EJ, Chaykovsky M. J Am Chem Soc 1965;87:1353.
4. Vedejs E, Hagen JP, Roach BL, Spear KL. J Org Chem 1987;43:1185, 4831.
5. (a) Trost BM, Hammen RF. J Am Chem Soc 1973;95:962; (b) Trost BM, Melvin LS Jr. Sulfur ylides. New York: Academic Press;1975, Chapter 7.
6. Hayes RA, Martin JC. In: Csizmadia IG, Mangini A, Bernardi F, editors. Organic sulfur chemistry: theoretical and experimental advances. Amsterdam: Elesevier; 1985, Chapter 8.

7. Akiba K.-y., Yamamoto Y. Heteroatom Chem 2007;18:161.
8. Akiba K.-y., Kashiwagi K, Ohyama Y, Yamamoto Y, Ohkata K. J Am Chem Soc 1985;107:2721.
9. Ohkata K, Ohsugi M, Yamamoto K, Ohsawa M, Akiba K.-y. J Am Chem Soc 1996;118:6355.
10. Paulmier C. Selenium reagents and intermediates in organic synthesis. Oxford: Pergamon Press; 1986.
11. Petragnani N. Tellurium in organic synthesis. San Diego (CA): Academic Press; 1994.
12. (a) Hiiro T, Morita Y, Inoue T, Kambe N, Ogawa A, Ryn. I, Sonoda N. J Am Chem Soc 1990;112:455; (b) Hiiro T, Morita Y, Inoue T, Kambe N, Ogawa A, Ryn I, Sonoda N. Organometallics 1990;9:1355.

NOTES 8

INVERSION MECHANISM OF NH_3 AND NF_3: VERTEX INVERSION OR EDGE INVERSION

It is well known and established that there is rapid equilibration by inversion of ammonia (NH_3) and amines ($R_1R_2R_3N$) at room temperature (Eq. N8.1). Phosphine (PH_3) and phosphines ($R_1R_2R_3P$) do not invert at room temperature but do so at higher temperatures. At the transition state of inversion, the molecule becomes planar (**A**) and its stereochemical characteristics are lost. This is generally illustrated as (Eq. N8.2). Here, substituent A is the unshared electron pair and the process is called *vertex inversion*.

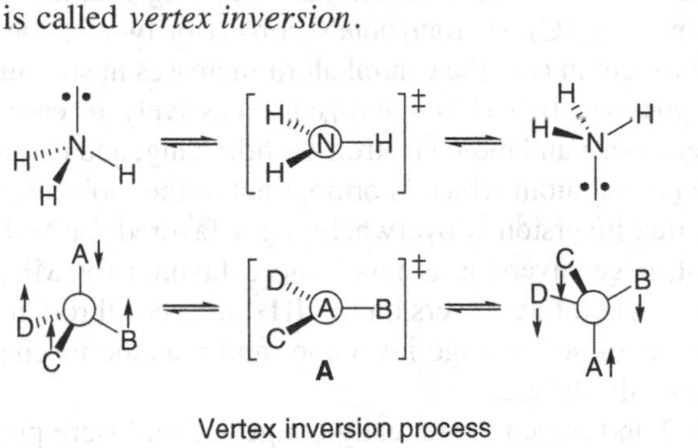

(N8.1)

(N8.2)

Vertex inversion process

Bicyclic phosphine (**1**) has two five-member rings bent like a roof at the ground state, but it is demonstrated to undergo inversion to **1′** (Eq. N8.3). At the transition state (**B**), the O– P– O bond is linear and an unshared electron pair lies

Organo Main Group Chemistry, First Edition. Kin-ya Akiba.

on the N– P line linearly, and the molecular frame is planar; thus it is called the *T-shape transition state* [1]. At the transition state, it is theoretically expected that a vacant p orbital appears on the phosphorus atom perpendicularly to the molecular plane. The essence of the motion is generally shown in Eq. (N8.4), where A can be an unshared electron pair. Among the four substituents (atoms) at the vertices of a tetrahedron, possible combinations of two substituents, for example edges of A–B and C–D, become two perpendicular and linear lines on a plane by the bending motion and proceed to inversion by continued motion (**C**: transition state). The stereochemistry of A is retained as a substituent and at the same time a vacant p orbital appears on the central atom perpendicularly to the molecular plane. This new inversion mechanism is called *edge inversion* [2].

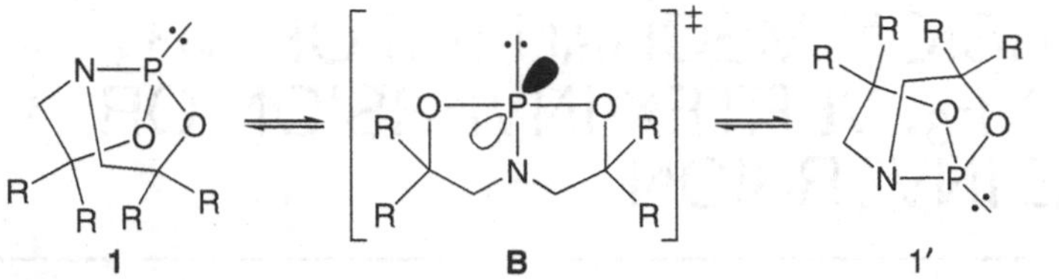

(N8.3)

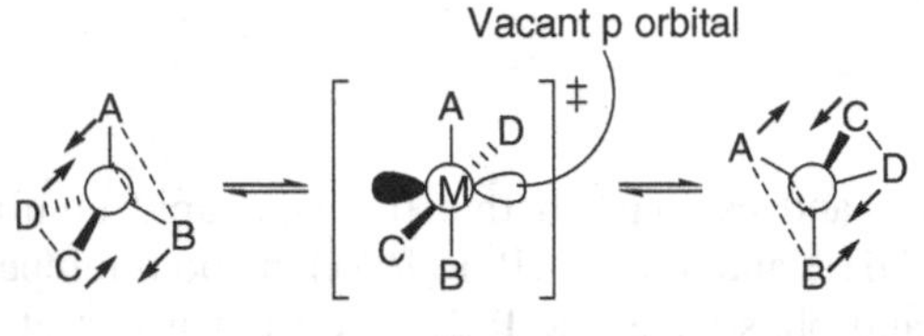

Vacant p orbital

(N8.4)

C

Edge inversion process

On the basis of *ab initio* calculation, the following characteristics are derived on the transition state (**C**): (i) four bonds consist of two 3c–4e bonds; (ii) **C** is stabilized excessively in case the central atom increases in size and becomes more electronically positive; (iii) **C** is stabilized excessively in case the substituents become smaller in size and more electron-withdrawing; and (iv) a vacant p orbital appears on the central atom which is orthogonal to the molecular plane [2c, 3]. It is clear that vertex inversion is overwhelmingly favored for MH_3 (M: P, As, Sb, Bi); in contrast, edge inversion is much more favored for MF_3 (M: P, As, Sb, Bi) (Table N8.1). Therefore, inversion of NH_3 proceeds through vertex inversion and that of NF_3 does so via edge inversion, and thus the mechanism of the two should be essentially different.

Compounds **2** and **3** are not sterically congested and were prepared as models in which vertex inversion and edge inversion can take place competitively. The two trifluoromethyl groups are not equivalent to the central atom stereochemically hence they interconvert according to the inversion of the central atom. The inversion energy can be estimated by observing the rate of interconversion of the trifluoromethyl group. The rate of inversion of **2** (M = Bi) should be faster than that of **3** (M = Sb) (cf. ii), but the rate was too low to be observed in a nonpolar solvent (*o*-dichlorobenzene) even at 175°C. The rate could be observed

TABLE N8.1 Activation Energy of Vertex Inversion and Edge Inversion by Ab Initio Calculation (kcal/mol)

Edge inversion	Vertex inversion		Edge inversion	
PH_3	35.0[a]	34.7[b]	159.9[a]	
AsH_3	41.3[a]	39.7[b]	142.2[a]	
SbH_3	42.8[a]	44.9[b]	112.2[a]	
BiH_3		60.5[b]		
PF_3	85.3[a]		53.8[a]	52.4[b]
AsF_3	66.3[a]		46.3[a]	45.7[b]
SbF_3	57.9[a]		38.7[a]	37.6[b]
BiF_3				33.5[b]

[a]Ref. [2c].
[b]Ref. [3].

TABLE N8.2 Activation Parameters of Inversion of Compounds 2, 4, 5, 7

Compound	Solvent	T_c(°c)	$\Delta G^{\ddagger}_{T_c}$ (kcal/mol)	$\Delta H^{\ddagger}$ (kcal/mol)	$\Delta S^{\ddagger}$ (eu)
2	*o*-Dichlorobenzene	175[a]	21		
	Pyridine-d_5	110[a]	18.0	9.0 ± 0.1	-23.5 ± 0.4
4	Toluene-d_8	125[b]	20.5	12.8 ± 0.4	-18.8 ± 1.0
	Pyridine-d_5	40[a]	14.6	7.1 ± 0.2	-23.6 ± 0.9
	2,6-Dimethylpyridine	170[a]	20.6	12.4 ± 0.2	-18.6 ± 0.5
5	Dichloromethane-d_2	< -90[a]	<8		
7	Dichloromethane-d_2	-55[a]	9.4	9.3	-1.7

[a]By exchange of CF_3.
[b]By exchange of CH_2.

in pyridine however, and the kinetic parameters were obtained: the activation enthalpy is quite small (9.0 kcal/mol) and the activation entropy was a very large negative value (−23.5 eu). This is realized by the coordination effect of pyridine to an emerging vacant p orbital in the transition state **C**, stabilizing the state (**C**, cf. iv). Hence, compounds **4–7** were prepared in which the dimethylaminomethyl group (s) was attached at 2 (and 6) position(s) of a benzene ring. All the structures were determined by X-ray analysis and the dimethylaminomethyl group(s) was shown to coordinate tightly to the central atom, which is illustrated by structures **2–7** [4a, b].

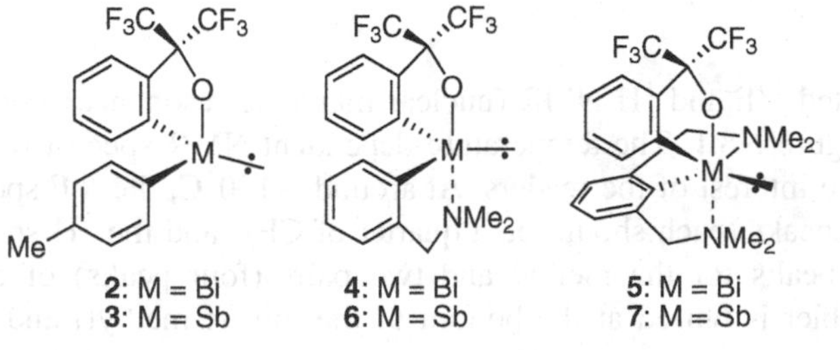

2: M = Bi
3: M = Sb

4: M = Bi
6: M = Sb

5: M = Bi
7: M = Sb

The rate of inversion of **4** (M = Bi) was much higher than that of **2** (M = Bi) in nonpolar solvent (toluene) and the activation enthalpy was 12.8 kcal/mol and the activation entropy was a considerably large negative value (−18.8 eu). This means that the coordination ability of the side arm of the dimethylaminomethyl group to bismuth is quite strong but is weaker than that of pyridine, as mentioned above in the case of **2**. The rate of inversion of **4** in 2,6-dimethylpyridine as solvent was almost the same as in toluene: that is, the activation enthalpy was 12.4 kcal/mol and activation entropy was −18.6 eu. This shows that 2,6-dimethylpyridine cannot coordinate to **4** because of steric hindrance and enhance the rate only slightly in a polar solvent. In pyridine as solvent, the rate of inversion of **4** was considerably accelerated to yield an activation enthalpy of 7.1 kcal/mol and an activation entropy of −23.6 eu. The result demonstrates interestingly that pyridine can coordinate to **4** as the second group at the transition state (**C**) even in the presence of a dimethylaminomethyl group and stabilize it by 5.7 kcal/mol (12.8 − 7.1 = ∼5.7 kcal/mol).

Accordingly, compound **5** (M = Bi) bearing two dimethylaminomethyl groups was prepared. Surprisingly, the rate of inversion of **5** at −90°C was too high to be observed. Then, antimony compound (**7**) with the same skeleton was synthesized and the rate was not too high and was observed to give the kinetic parameters at about −50°C. The activation enthalpy was 9.4 kcal/mol and the activation entropy was almost zero (−1.7 eu). The result means that, during the inversion of **7** to **7′**, two dimethylaminomethyl groups are kept coordinated, forming the N–Sb–N bond (3c–4e) and the five-member ring with two trifluoromethyl groups sweeps widely like a wiper on the windshield of a car (**D**). Pyridine added to the solvent (CD_2Cl_2) did not affect the rate at all. These facts support the conclusion that the N–Sb–N bond is kept tightly in the solution (Eq. N8.5) [4].

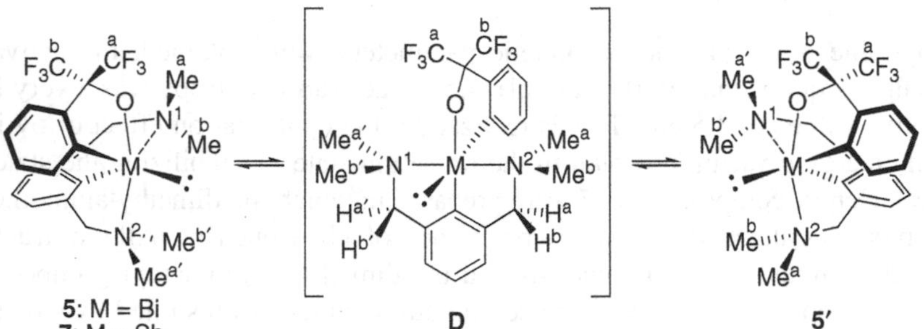

5: M = Bi
7: M = Sb

D

5′
7′

(N8.5)

The related ^{19}F and ^{1}H NMR (nuclear magnetic resonance) spectra are illustrated in Figure N8.1. The temperature-dependent NMR spectra would certainly stimulate the interest of the readers. At around −100°C, the ^{19}F spectrum shows two broad speaks (each should be a quartet of CF_3) and the ^{1}H spectrum shows four sharp peaks for the methyl and two pairs (four peaks) of doublets (one half of doublet is buried at the bottom of the lowest methyl) and benzene part

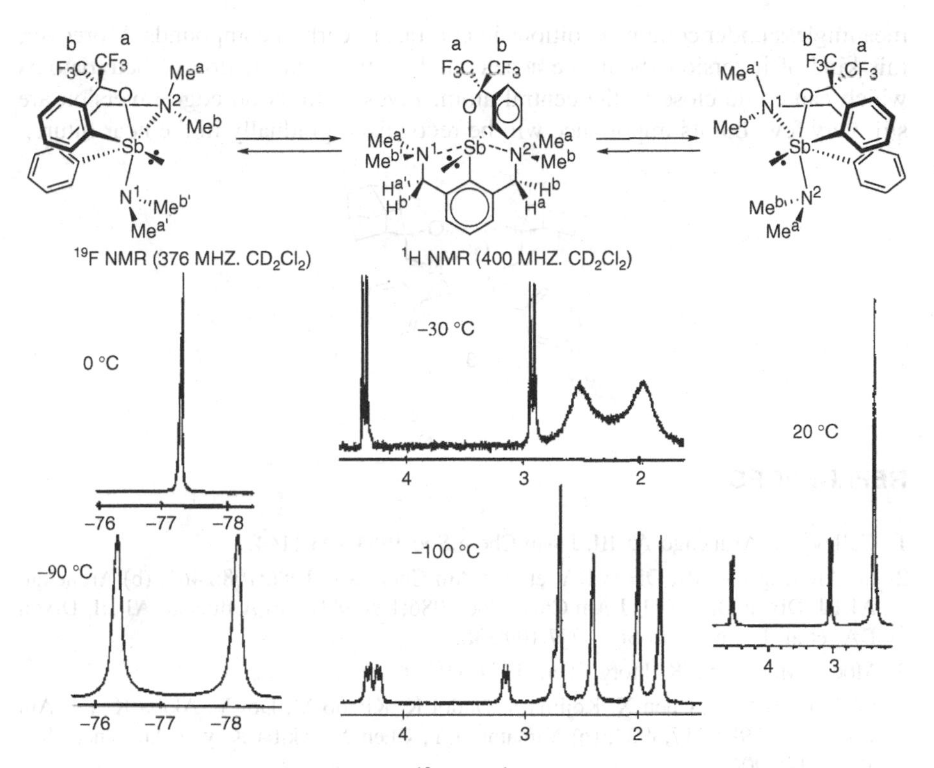

Figure N8.1 Temperature dependent ^{19}F and ^{1}H NMR spectra of antimony compound **7** to show the edge inversion mechanism.

(omitted). The spectra correspond to the structure of **7** in the solid state and the inversion is completely frozen. Above 0°C, spectra of both ^{19}F and four methyls appear as a sharp singlet and the methylene as a pair of doublets. This shows the inversion is quite rapid and **7** and **7**′ cannot be discriminated. At −30°C, the methylene appears as a quartet (doublets of doublet) and the methyls give two broad peaks to show that the methyls of a–a′ and b–b′ are paired and they are the intermediate state to become a singlet.

On the basis of these observations, the transition state of the inversion of **7** in solution is illustrated as **D**. The N1– Sb– N2 bond and the skeleton with a lower benzene ring lie in the plane of the paper and the five-member ring fused with a benzene ring is perpendicular to this plane. It is realized that, at the transition state (**C**), a vacant p orbital is generated on Sb and trapped by two dimethylamino groups to form a 3c–4e bond to effect edge inversion (cf., iv).

The more important contribution of edge inversion is to predict the possibility of inversion without cleaving any one of the four σ bonds. This is strictly proved in the case of the germanium compound (**8**) with two bidentate O,N ligands [5]. That is, heavier main group element compounds generally can invert without breaking a σ bond, and thus asymmetric synthesis of these compounds may be meaningless under certain conditions in contrast to carbon compounds. Moreover, this kind of inversion should be accelerated by the coordination of heteroatoms which can come close to the central atom. Investigations on edge inversion are still very few, but its importance will be recognized gradually in the near future.

8

REFERENCES

1. Culley SA, Arduengo AJ III. J Am Chem Soc 1984;106:1164.
2. (a) Arduengo AJ III, Dixon DA, et al. J Am Chem Soc 1986;108:2461; (b) Arduengo AJ III, Dixon DA, et al. J Am Chem Soc 1986;108:6821; (c) Arduengo AJ III, Dixon DA, et al. J Am Chem Soc 1987;109:338.
3. Moc J, Morokuma K. Inorg Chem 1994;33:551.
4. (a) Yamamoto Y, Chen X, Kojima S, Ohdoi K, Kitano M, Doi Y, Akiba K.-y. J Am Chem Soc 1995;117:3922; (b) Yamamoto Y, Chen X, Akiba K.-y. J Am Chem Soc 1992;114:7906.
5. Arduengo AJ III, Dixon DA, Roe DC, Klein M. J Am Chem Soc 1988;110:4437.

CHAPTER 8

ORGANOHALOGEN COMPOUNDS: FLUORINE AND IODINE COMPOUNDS

Group (*n*) / Period	16 6B	17 7B	18 0
1 1s			(1.28°) 2372 $^{4}_{2}$**He** Helium [5.50]
2 [He] 2s2p	0.66 1314 (1.40^{2-}) $^{16}_{8}$**O** Oxygen 3.5 351	0.58 1681 (1.33^{-}) $^{19}_{9}$**F** Fluorine 4.0 441	2081 $^{20}_{10}$**Ne** Neon [4.84]
3 [Ne] 3s3p	1.04 1000 (1.74^{2-}) $^{32}_{16}$**S** Sulfur 2.5 272	0.99 1251 (1.81^{-}) $^{35}_{17}$**Cl** Chlorine 3.0 352	(1.74°) 1520 $^{40}_{18}$**Ar** Argon [3.20]
4 [Ar:3d^{10}] 4s4p	1.17 941 (1.91^{2-}) $^{80}_{34}$**Se** Selenium 2.4 245	1.14 1140 (1.96^{-}) $^{79}_{35}$**Br** Bromine 2.8 293	1.89 1351 (1.69^{+}) $^{84}_{36}$**Kr** Krypton [2.94]
5 [Kr:4d^{10}] 5s5p	1.37 869 (2.24^{2-}) $^{130}_{52}$**Te** Tellurium 2.1	1.33 1008 (2.20^{-}) $^{127}_{53}$**I** Iodine 2.5 213	2.09 1170 (1.90^{+}) $^{132}_{54}$**Xe** Xenon [2.40]
6 [Xe:4f$_{14}$5d^{10}] 6s6p	1.53 812 $^{209}_{84}$**Po*** Polonium 2.0	1.45 (0.57^{5+})930 (2.27^{-}) $^{216}_{85}$**At*** Astatine 2.2	1040 $^{222}_{86}$**Rn*** Radon [2.06]

Representative and important organohalogen compounds are chlorides and bromides, without argument. Their fundamental character and synthetic methods are described in many chemistry textbooks. Chlorides and bromides for general use are commercially available and their synthesis is briefly summarized. Most of fluorides and iodides are also commercially available but comparatively new and useful reactions have recently been developed for them. Hence, fluorides and iodides are described in more detail.

8.1 SYNTHESIS OF CHLORINE AND BROMINE COMPOUNDS

Alkyl and aryl chlorides and bromides are fundamental and basic compounds of organohalogen compounds. Their synthesis and character are well documented in textbooks and in advanced volumes [1]. Here, several essential synthetic reactions are summarized as a reminder.

Aliphatic alcohols are directly converted to their chlorides and bromides (**1**) with inorganic acids (HX), sulfur ($SOCl_2$), and phosphorus reagents (PX_3, $P(O)X_3$) (path a). Alcohols are sulfonylated to give esters of sulfonic acid (**2**) which are precursors to **1**. The reactivity of **2** with nucleophiles (X^-: halide ions) is controlled depending on the kinds of sulfonyl group, that is, mesyl, trifyl, tosyl, and brosyl, and the yield of **1** is usually excellent (path b).

$$\text{R-OH} \xrightarrow{\text{(a) HX, SOCl}_2\text{, PX}_3\text{, P(O)X}_3} \underset{\mathbf{1}}{\text{R-X}}$$
$$\text{R-OH} \xrightarrow{\text{(b) R}'\text{-SO}_2\text{-Cl}} \underset{\mathbf{2}}{\text{R-O-SO}_2\text{R}'} \xrightarrow{:X^-} \underset{\mathbf{1}}{\text{R-X}} + \text{R}'\text{SO}_3^- \tag{8.1}$$

R′—SO₂– : H_3C-SO_2- : F_3C-SO_2- : $H_3C-C_6H_4-SO_2-$: $Br-C_6H_4-SO_2-$

: Mesyl, Ms; Trifyl, Tf; Tosyl, Ts; Brosyl, Bs

Halogens combine with alkenes easily to afford dichlorides and dibromides (**3**), and bromine reacts faster than chlorine. As bromination reagents dioxane dibromide and pyridine hydrobromide dibromide are conveniently used (path a). Hydrogen chloride and bromide (HX) also add to alkene according to the Markownikow rule (**4**), where Lewis acids ($AlCl_3$, $ZnCl_2$, etc.) accelerate the addition effectively (path b).

$$\text{RCH=CH}_2 \xrightarrow[\text{Addition of halogen}]{\text{(a) Br}_2\text{, Cl}_2\text{, dioxane}\bullet\text{Br}_2\text{, pyridinium }\overset{+}{\text{N}}\text{-HBr}_3^-} \underset{\mathbf{3}}{\text{RCH(X)-CH}_2\text{X}}$$
$$\text{RCH=CH}_2 \xrightarrow[\text{Markovnikov addition}]{\text{(b) HX (X = Cl, Br), AlCl}_3\text{, ZnCl}_2\text{ (catalyst)}} \underset{\mathbf{4}}{\text{RCH(X)-CH}_3} \tag{8.2}$$

HX adds to conjugate alkene bearing electron-withdrawing group (**5**), in which X adds to the methylene group (**6**), induced by the relative stability of an intermediate carbocation (seemingly contrary to the Markownikow rule).

$$H_2C{=}CHY \xrightarrow{HX\ (X:\ Cl,\ Br)} XCH_2{-}CH_2Y \quad (8.3)$$

5 Y : CO_2Et, CN → **6**

Allylic and benzylic C–H bonds are substituted for bromine with *N*-bromosuccinimide (NBS) by radical reaction in good yields. Thus, an aromatic methyl group is converted to a bromomethyl or dibromomethyl group, frequently under irradiation.

Cyclohexene + NBS (AIBN, 80%) → 3-bromocyclohexene (8.4)

1-($CHBr_2$)naphthalene ← (2 NBS, 80%) 1-(CH_3)naphthalene → (NBS, 90%) 1-(CH_2Br)naphthalene (8.5)

The α-hydrogen of a carbonyl group (**7**) is selectively substituted for bromine under irradiation quantitatively, in which the α-hydrogen bound to a substituted carbon is preferred in accordance with the stability of an intermediate radical (**8**). An epoxide is added to trap the resulting HBr in situ.

7 (2-methylcyclohexanone) → (Br_2, cyclohexene oxide, hν; CCl_4, 100%) → **8** (2-bromo-2-methylcyclohexanone) (8.6)

The α-hydrogen of an aldehyde (**9**) is also converted to bromine (**13**) through a series of reactions (Eq. 8.7). Enol acetate (**10**) is obtained by heating **9** in acetic anhydride with potassium carbonate, and bromine is added to the double bond of **10** to yield a dibromide (**11**). The dibromide is converted to an acetal (**12**) with excess methanol and the acetal is hydrolyzed to the product (**13**). The enol acetate (**10**) can also be replaced by the corresponding silyl enol ether.

$$RCH_2CHO \xrightarrow{Ac_2O,\ K_2CO_3} RCH{=}CHOAc \xrightarrow{Br_2} RCH(Br){-}CH(Br)OAc \xrightarrow{CH_3OH} RCH(Br){-}CH(OCH_3)_2 \xrightarrow{aq.\ HCl} RCH(Br){-}CHO \quad (8.7)$$

9 → 10 → 11 → 12 → 13

An aromatic hydrogen can be substituted for bromine (**14**), in which Fe (powder) is used as a catalyst, under heating. An aromatic hydrogen activated by an electron-donating group (phenol or aniline) is easily substituted for bromine at the para position with dioxane dibromide (**15, 16**).

2,3-dimethylnaphthalene $\xrightarrow[90\%]{Fe\ (powder:\ catalyst),\ Br_2}$ 1-bromo-2,3-dimethylnaphthalene (**14**) (8.8)

Phenol / aniline + dioxane $\bullet Br_2$ → (a) Br–C$_6$H$_4$–OH (**15**); (b) Br–C$_6$H$_4$–NH$_2$ (**16**) (8.9)

8.2 FLUORINE COMPOUNDS

The Schiemann reaction (cf. Eq. 8.24: fluorination of aromatic diazo compound) was the only one synthetic method, established a long time ago, to prepare aromatic fluoro compounds. Fluorination of organic compounds had been difficult, and organofluoro compounds had been treated as a special class of compounds for a long time.

As the electronegativity of fluorine is the largest, the X–F bond is necessarily polarized as positive X and negative F. Therefore, X behaves as a strong electrophile in every X–F reagent. Accordingly, Br–F and I–F are strong halogenating reagents, and I–F can substitute the aromatic hydrogen by itself (Eq. 8.10). Fluorine in aqueous acetonitrile generates active species (**19**) as HO–F in solution and the species behaves as an electrophilic hydroxylating reagent (HO^+) (Eq. 8.11).

TABLE 8.1 Relative Rate of Addition of Halogen to Double Bond

Halogen	*I*–I	*I*–Br	*Br*–Br	*I*–Cl	*Br*–Cl
Relative rate	1	3×10^3	10^4	10^5	4×10^6

$$F_2 + I_2 \xrightarrow[-78\,^\circ C]{CFCl_3} 2IF\ (\mathbf{17})$$

$$C_6H_5CO_2Et + IF\ (\mathbf{17}) \longrightarrow m\text{-}IC_6H_4CO_2Et\ (\mathbf{18}) \quad (8.10)$$

$$F_2 + H_2O + CH_3CN \longrightarrow HO\text{–}F\bullet CH_3CN\ (\mathbf{19})$$

$$RC(OSiMe_3)\!=\!CH_2\ (\mathbf{20}) + HO\text{–}F\bullet CH_3CN \longrightarrow RC(O)CH_2OH\ (\mathbf{21}) \quad (8.11)$$

One of the typical reactions of halogen is the electrophilic addition to alkene. This is expected to proceed faster with polar reagents such as Br–F and I–F rather than nonpolar Br–Br. The relative rates of addition of halogen to the alkene clearly support this (Table 8.1). That is, the rates of addition of mixed halogens are much higher than that of the halogens themselves and the differences are quite remarkable.

The trifluoromethyl group can stabilize gem-diol (**22a**), but dehydration takes place subsequently to give enol (**22b** and **c**) when there is an α-hydrogen (Eq. 8.12).

$$F_3CC(O)H + H_2O \xrightarrow{(a)} F_3CCH(OH)_2\ (\mathbf{22a})$$

$$F_3CC(O)CHF_2 + H_2O \xrightarrow{(b)} F_3CC(OH)\!=\!CF_2\ (\mathbf{22b}) \quad (8.12)$$

$$F_2HCC(O)CO_2Me + H_2O \xrightarrow{(c)} F_2C\!=\!C(OH)CO_2Me\ (\mathbf{22c})$$

Let us summarize the effect of fluorine in a general way: (i) fluorine destabilizes β-carbocation due to strong electron-withdrawing effect (**B**: inductive effect); (ii) fluorine stabilizes α-carbocation because of short F–C bond length (**A**: resonance effect); and (iii) fluorine behaves as a leaving group when there is an electron-donating substituent (e.g., lone pair electrons, double bond) at β-position (**C** and **D**).

A B C D

Fluorine gas diluted with nitrogen reacts with the tertiary hydrogen of adamantane (**23**) to give the monofluoro derivative (**24**) selectively at low temperature (Eq. 8.13). Xenon difluoride (**25**: commercially available) affords the monofluoro compound (**26**) from carboxylic acid and the fluorinated aromatic compound (**27**) by electrophilic substitution (Eq. 8.14) [2].

F_2/N_2; $CFCl_3$, −78 °C

23 **24** (8.13)

XeF_2 (**25**) — (a) RCO_2H → $RF + CO_2 + HF$ (**26**)

R: **a** = $PhCH_2$; **b** = $CH_3COCH_2CH_2$; **c** = Ph

(b) C_6H_5R → FC_6H_4R (**27**) (8.14)

R: **a** = H; **b** = F; **c** = Cl; **d** = OCH_3; **e** = CF_3; **f** = NO_2

Fluorine gas diluted with nitrogen also reacts with pyridine to afford *N*-fluoropyridinium salt (**E**), and its triflate (**29**) is a stable reagent for electrophilic fluorination. Aromatic hydrogen and silyl enol ether are converted to fluoride with **29b** (Eq. 8.16).

28 + F_2/N_2 → [**E**, F^-] —TfONa→ **29** (TfO^-) (8.15)

R: **a** = 2,4,6-Me_3; **b** = H; **c** = 3,5-Cl_2; **d** = 2,6-Cl_2

29b (TfO^-) — (a) benzene → PhF; (b) PhMgBr → PhF; (c) $OSiMe_3$, CO_2Et → **30** (F, CO_2Et) (8.16)

Hydrofluoric acid (aq. 70%) gives pyridinium hydrogen fluoride (**31**), and **31** is used as a nucleophilic fluorinating reagent (Eq. 8.17). The fluoride anion of **31** substitutes chlorine of **32a** and opens the aziridine ring to give **33b** and **c**. Furthermore, the 1-β group of glucose is selectively converted to a fluoride (**34b**), retaining the stereochemistry (Eq. 8.18).

Pyridine + aq. HF (70%) ⟶ pyridinium (N^+–H) F^- (HF-Py) **31** (8.17)

31 ⟶ (a) $PhCOCH_2Cl$ **32a**, HgO ⟶ $PhCOCH_2F$ **32b**

(b) aziridine **33a** (Ph, Me, H, Me, N–H) ⟶ **33b** (Ph, Me, F, NH_2, H, Me) + **33c** (Ph, Me, H_2N, F, H, Me) (8.18)

(c) **34a** (OCH_3, OBz, OBz, H_3CO, BzO) ⟶ **34b** (OCH_3, F, OBz, H_3CO, BzO)

TBAF (tetrabutylammonium fluoride) is commonly used as a nucleophilic fluorinating reagent. TBAF substitutes an activated alcoholic group to fluoride (**35b**), activates allylsilane to react with aldehyde (**36b**), and opens an epoxide ring to give **37b**.

$Bu_4N^+F^-$ TBAF ⟶ (a) **35a** (OTs) ⟶ **35b** (F)

(b) Me_3Si (allyl) **36a**, PhCHO ⟶ **36b** (Ph, OH) (8.19)

(c) Ph (epoxide) **37a** ⟶ **37b** (Ph, F, HO)

DAST (diethylaminosulfur trifluoride) and MOST (morphorinosulfur trifluoride) are also useful and commercially available fluorinating reagents.

Stereochemistry of alcohol is inverted by substitution (**38b, 39b**), and difluoride (**40b**) is obtained from a carbonyl compound (Eq. 8.20) [3]. TASF [(tris(dimethylamino)sulfonium trimethylsilyldifluoride)] is a strong nucleophilic fluorinating reagent, which is also commercially available. For instance, it activates silyl groups (**41a, 42a**) and formally generates α-carbanion in situ (Eq. 8.21).

HO OBz (a) BzO O BzO OMe **38a** → F OBz BzO O BzO OMe **38b**

Et_2NSF_3 DAST, O NSF$_3$ MOST; (b) OH **39a** → F **39b**; (c) H O **40a** → H F F **40b** (8.20)

(a) Cl Cl Me $SiMe_3$ **41a**, PhCHO → Cl Cl Me Ph OH **41b**

$(Et_2N)_3S^+ \, Me_3SiF_2^-$ TASF; (b) Ph $SiMe_2F$ (cat) **42a**, PhI → Ph Ph **42b** + Ph Ph **42c** (8.21)

An aromatic amino compound is converted to its fluoro compound via diazonium salt (Schiemann reaction: Eq. 8.22) [4]. Aromatic diazonium tetrafluoroborate is separated first as solid from the corresponding diazonium chloride by the addition of aqueous tetrafluoroboric acid, and the solid is thermally decomposed to give the fluoride. Similar conversion of the amino group to the iodo group in aromatic compounds proceeds smoothly in aqueous solution from diazonium chloride by the addition of potassium iodide.

$$ArNH_2 \xrightarrow[\text{ap. HCl}]{NaNO_2} ArN_2^+ Cl^- \begin{cases} \xrightarrow{\text{(a) aq. } HBF_4} ArN_2^+ BF_4^- \xrightarrow{\Delta} ArF + N_2 + BF_3 \\ \xrightarrow{\text{(b) aq. KI}} ArI \end{cases} \quad (8.22)$$

8.3 IODINE COMPOUNDS

Hydrogen iodide generated in situ from potassium iodide and phosphoric acid adds to an alkene (**43**) according to the Markownikow rule (Eq. 8.23). Chlorine iodide reacts similarly to give **44** in excellent yield (Eq. 8.24).

$$\xrightarrow{KI,\ H_3PO_4} \mathbf{43} \quad (8.23)$$

$$PhHC{=}CH_2 + ClI \longrightarrow \underset{\mathbf{44}}{PhHC(Cl){-}CH_2I} \quad (8.24)$$

An alkene bearing a carboxyl group (**45**) affords a five-member lactone (**46**) in one step when it is treated with iodine under alkaline conditions. An intermediate iodonium salt (**E**) is trapped by the carboxylate ion (Eq. 8.25). This is called *iodolactonization* and is useful for organic syntheses [5]. The catechol borane adduct of acetylene (**F**) gives vinyl iodide (**47**) under similar alkaline conditions with iodine (Eq. 8.26).

$$\mathbf{45} \xrightarrow{I_2,\ \text{aq. } NaHCO_3} [\mathbf{E}] \longrightarrow \mathbf{46} \quad (8.25)$$

$$+ \text{catecholborane (BH)} \longrightarrow [\mathbf{F}] \xrightarrow{I_2,\ \text{aq. NaOH}} \mathbf{47} \quad (8.26)$$

An activated alcohol (**48**) gives the corresponding iodide (**49**) by treatment with potassium iodide (Eq. 8.27) [6]. An alcohol (**50**) is activated with a mixture of diethyl ester of azodicarboxylic acid and triphenylphosphine, and then methyl iodide is added to the mixture. The resulting iodide ion attacks an intermediate alkoxyphosphonium salt (**G**) to afford an alkyl iodide (**51**) (Eq. 8.28). The latter reaction proceeds in an S_N2 manner and the stereochemistry of the central carbon is inverted completely (Mitsunobu reaction).

(8.27)

48 —OTs —NaI→ 49 —I

(8.28)

50: Ph—C*(Me)(OH) —$EtO_2C{-}N{=}N{-}CO_2Et$; Ph_3P; MeI→ [$EtO_2C{-}N(Me){-}N(H){-}CO_2Et$; $Ph_3\overset{+}{P}{-}O{-}C^*(Ph)(Me)$ I^-] **G** —Inversion→ 51: Ph—C*(Me)(I)

In Eq. 8.28, examples of iodine to react as a nucleophile (I^-) or electrophile (I^+) are described. On the other hand, iodine is easily oxidized and can bear a variety of valences (I^{-1}, I^0, I^{+1}, I^{+3}, I^{+5}, I^{+7}). This can be expected from its low ionization energy and large polarizability (cf. Table 1.1 of Chapter 1 and Figure 8.1). Actually, a number of iodine reagents with a variety of oxidation states are known, including iodosobenzene.

When iodobenzene is oxidized with peroxyacetic acid in an acetic acid solution, diacetoxyiodobenzene (**52**) is obtained quantitatively (Eq. 8.29). Compound **52** is converted to iodosobenzene (**53**) by treatment with aqueous sodium hydroxide and to di(trifluoroacetoxy)iodobenzene (**54**) by treatment with trifluoroacetic acid. Furthermore, the stable Dess–Martin reagent (**55**) is prepared from *ortho*-iodobenzoic acid and peroxyacetic acid in acetic acid solution by a one-pot process (Eq. 8.30).

(8.29)

PhI —40% $H_3CC(=O){-}OOH$, CH_3CO_2H→ $PhI(OCOCH_3)_2$ (**52**)

52 —aq. NaOH→ PhI=O (**53**)

52 —CF_3CO_2H→ $PhI(OCOCF_3)_2$ (**54**)

(8.30)

o-iodobenzoic acid (CO_2H) —40% $CH_3C(=O){-}OOH$ / CH_3CO_2H→ AcO, AcO, OAc periodinane

55: Dess–Martin reagent

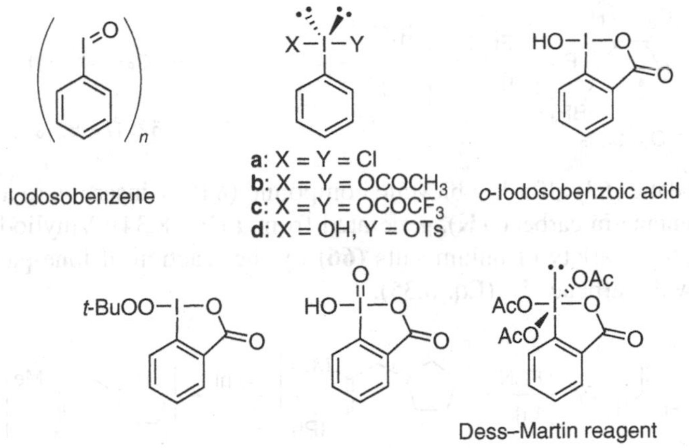

Figure 8.1 Oxidized reagents of iodobenzene.

When vinylsilane (**56**) reacts with iodosobenzene in the presence of boron trifluoride, phenylvinyl-λ^3-iodane (**57**) is obtained through an intermediate silyl-cation (**H**) (Eq. 8.31). The stereochemistry of the α-carbon of the vinyl group is changed completely. The structure of **57** was illustrated in Figure 5.9, showing that three atoms of vinyl carbon–iodine–fluorine of BF_4^- are linear. Hereafter λ^3-iodanes are referred to simply as *substituted iodanes*, because the iodobenzene part is common for them.

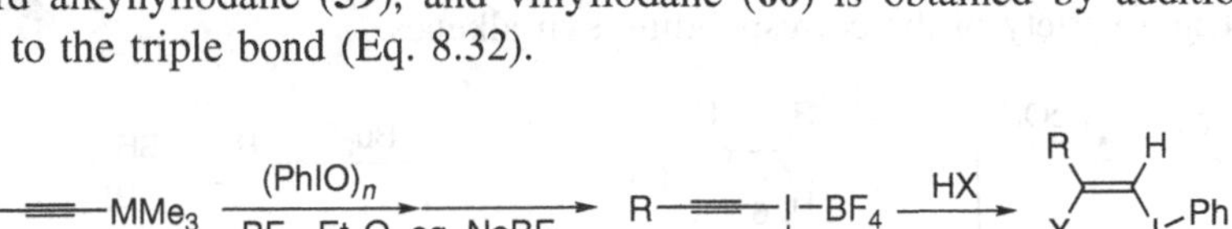

(8.31)

Metallic derivatives of acetylene (**58**) react similarly with iodosobenzene to afford alkynyliodane (**59**), and vinyliodane (**60**) is obtained by addition of an acid to the triple bond (Eq. 8.32).

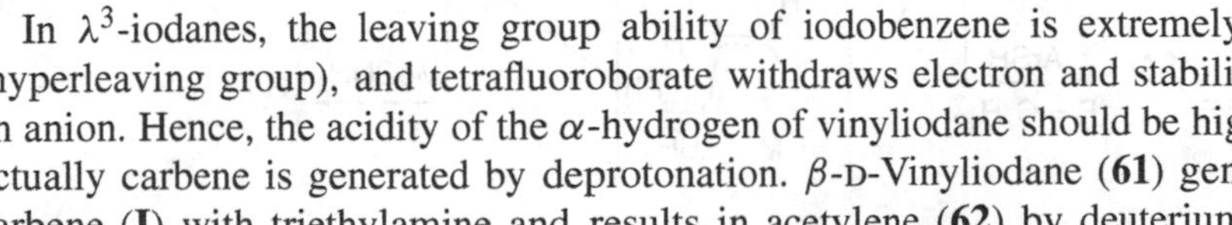

(8.32)

In λ^3-iodanes, the leaving group ability of iodobenzene is extremely high (hyperleaving group), and tetrafluoroborate withdraws electron and stabilizes as an anion. Hence, the acidity of the α-hydrogen of vinyliodane should be high and actually carbene is generated by deprotonation. β-D-Vinyliodane (**61**) generates carbene (**I**) with triethylamine and results in acetylene (**62**) by deuterium shift (Eq. 8.35). Deuterium is retained completely during the rearrangement [7].

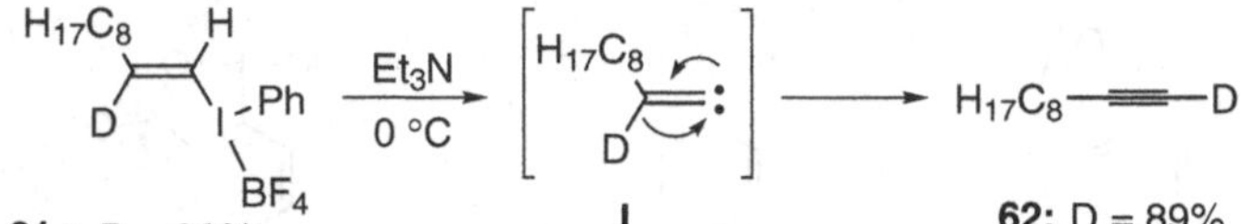

(8.33)

Vinyliodane (**63**) affords a bicyclic compound (**64**) by insertion to a C–H bond of cyclopentane in carbene (**K**), generated from **J** (Eq. 8.34). Vinyliodane (**65**) is converted to a variety of onium salts (**66**) by the reaction of lone-pair electrons of Ph_nX with carbene (**L**) (Eq. 8.35).

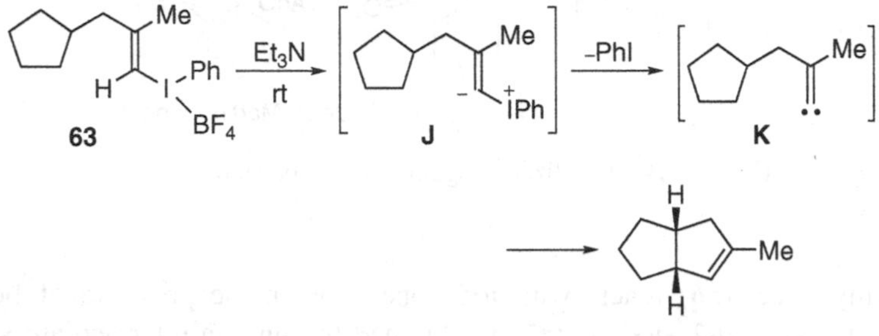

(8.34)

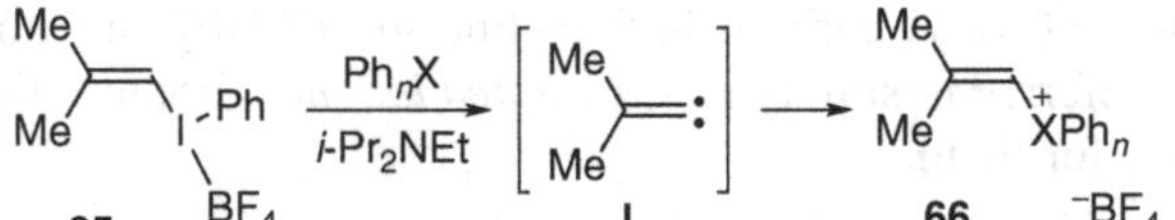

(8.35)

$Ph_nX = Ph_3P, Ph_3Sb, Ph_2S, Ph_2Se, Ph_2Te$

When nucleophiles of low basicity react with vinyliodane (**67**), elimination of iodobenzene takes place through a transition state (**M**) to effect substitution at the α-carbon without deprotonation of the α-hydrogen (Eq. 8.36). At the transition state (**M**), inversion of the sp^2 carbon takes place via an S_N2-type substitution to result in a variety of the corresponding syn alkenes.

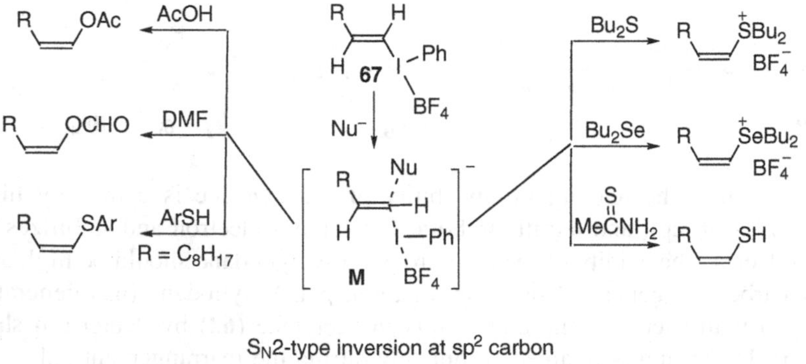

(8.36)

S_N2-type inversion at sp^2 carbon

By treatment of *ortho*(trimethylsilyl)phenyl iodane (**68**) with TBAF, benzyne (**N**) is generated and trapped with furan (and alkenes) in high yield (Eq. 8.37).

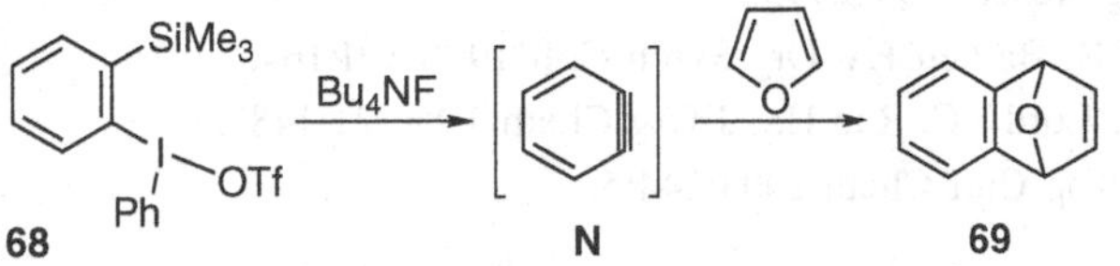

(8.37)

The Dess–Martin reagent (**55**) converts primary alcohol (**70**) to aldehyde (**71**) even in the presence of functional groups, and thus it is used as a selective oxidation reagent of primary alcohol (Eq. 8.38).

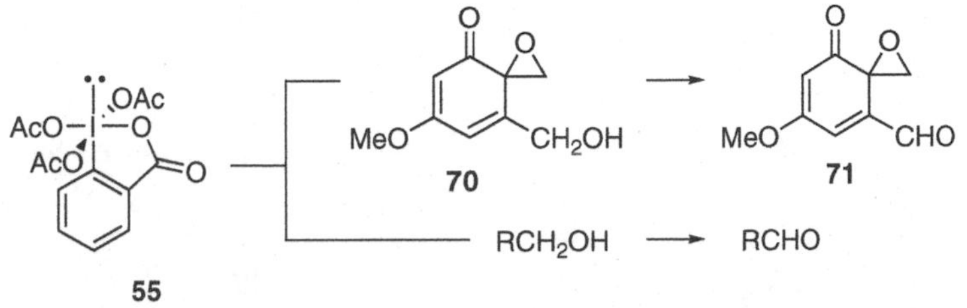

(8.38)

In general, it is believed that the reaction of phenyl iodane (**72**) with nucleophile proceeds through the transition state (**O**) of S_N2-type to give the product (**73**) (Eq. 8.39). On the other hand, there is another possibility for **72** to result in **73** by ligand coupling reaction through a hypervalent intermediate (**P**) (cf. Chapter 11). There are several synthetically useful reactions of **72**, although distinction of the two types of mechanisms is quite difficult. It should be interesting to pay attention to the unique reactivity and possible use of higher valent iodine and bismuth compounds.

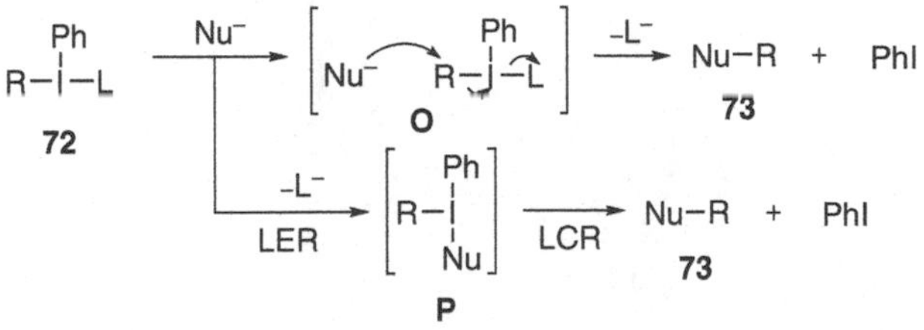

(8.39)

REFERENCES

1. Refer to standard books of organic experiments and organic synthesis. (a) Textbooks of "Organic Chemistry" by McMurry, Morrison and Boyd, etc; (b) Fieser LF, editor. Organic experiments. Boston (MA): D. C. Heath Co.; 1964; (c) Fieser LF, Fieser M, editors. Reagents for organic synthesis. Wiley-Interscience (series); (d) Org. Synth. Coll. Vol. I–VII to present; (e) The Chemical Society of Japan. Organic synthesis I. Volume 19, 4th Jikken Kagaku Kouza (4th Encyclopedia of Experimental Chemistry, in Japanese). Tokyo: Maruzen; 1992, Chapter 2.

2. Rozen S. Acc Chem Res 1988;21:307.
3. Middleton WJ. J Org Chem 1975;40:574.
4. Roe A. Org React 1968;5:193.
5. Gonzalez FB, Bartlett PA. Org Synth Coll 1990;VII:164.
6. House HO, Lord RC, Rao HS. J Org Chem 1956;21:1487.
7. Ochiai M. Top Curr Chem 2003;244:5.

CHAPTER 9

ATRANE AND TRANSANNULAR INTERACTION: FORMATION OF HYPERVALENT BOND

9.1 INTRODUCTION

As fundamental examples of 3-center 4-electron bond, $[I-I-I]^-K^+$ and $[F-F-F]^-K^+$ have been mentioned already. The former is stable at ambient temperature and the latter is observed at −258°C (stable up to −231°C). These are linear molecules and obtained by addition of potassium iodide to iodine and of potassium fluoride to fluorine, respectively. They form 3-center 4-electron bond by the donation of electrons of iodide and fluoride to σ^*X-X (here that of I–I and F–F) and the resulting system is more stabilized as a whole than the original system of two separate molecules [1].

Let us consider the hydrogen bond here. Among hydrogen halides, hydrogen fluoride has the strongest hydrogen bond. The structure of potassium salt of the dimer $[(F-H-F)^-K^+]$ was determined a long time ago and the salt is linear and symmetric with H–F bond distance of 1.13 Å. $(HF)_5$ is a pentagon and the bond distances of H–F and H···F are 1.00 and 1.50 Å, respectively. If we assume that the fluorine stays on the vertex then the hydrogen moves back and forth as a piston within the range of 0.5 Å along an edge of the pentagon of 2.5 Å. The H–F bond distance of $[(F-H-F)^-K^+]$, that is, 1.13 Å, is longer than the single bond (1.00 Å) and shorter than the hydrogen bond (1.50 Å). From the stand point of formal logic, the $[F^-\cdots H^+\cdots F^-]$ bond is formed by the coordination

Organo Main Group Chemistry, First Edition. Kin-ya Akiba.

of two unshared electron pairs of the fluoride anion to a proton. As there is only one orbital (1s) to accept the two unshared electron pairs, the 3-center 4-electron bond should be invoked. The proton contains four valence electrons formally, thereby forming a hypervalent compound 4-H-2.

By the same token, hydrogen bonds (X–H···X) of ammonia and water can also be examples of hypervalent bonds. The nature and essential characters of a hydrogen bond are still actively investigated. Let us consider some examples and characters of hypervalent bonds that form between different atoms.

9.2 SILATRANE AND ATRANE

When triethanolamine reacts with trimethoxysilane, a tricyclic compound bearing three pseudo five-member rings containing a silicon and a nitrogen is obtained, which is accompanied by the elimination of three molecules of methanol (Eq. 9.1). The product formed is the parent compound of silatrane (**1b**: R=H), and a group of compounds with this molecular frame is called *silatrane* (**1**). Silatrane is an efficient rodenticide, which was actively investigated in Soviet Union and a variety of derivatives were synthesized and their structures were determined [2].

$RSi(OMe)_3 + N(CH_2CH_2OH)_3 \longrightarrow$ **1**, **1'**, **1''** (9.1)

1a: R=Ph **1b**: R=H **1c**: 2.116, 1.905, NO_2

The structure of silatrane should be lantern-type (**1′**) and lone pair electrons of nitrogen should be directed to the outside of the molecular skeleton. This is commonly and reasonably expected. On basis of the X-ray analysis results of these compounds, the lone pair electrons are withdrawn into the skeleton and 3-center 4-electron bond is formed as a N–Si–R bond to form ammoniosilicate (**1″**), which is identical to **1**. The structure around the silicon is revealed to be a distorted trigonal bipyramid and the silicon is granted to be pentavalent. The N-Si bond is polar and weak and becomes stronger and shorter with respect to the increase in electronegativity of R. For example, the bond distance N–S of **1c** bearing a *m*-nitrophenyl group is 2.116 Å and that of Si–C is 1.905 Å. The bond distance N–Si is 13% longer than that of the N–Si covalent bond, 1.87 Å (0.70 + 1.17), and considerably shorter than the sum of van der Waals radius, 3.45 Å (1.55 + 1.90). The bond distance of Si–C is shorter than that of Si–C (sp^3), 1.94 Å (1.17 + 0.77), but slightly longer than that of Si–C (sp^2), 1.90 Å (1.17 + 0.73). The N–Si bond of silatrane is weak and labile, but, generally, stays in the range of 2.0–2.2 Å.

The basicity of nitrogen of silatrane is very weak and it cannot be methylated by Meerwein reagent. 1-Methoxysilatrane (**2a**) is methylated at the oxygen to give an oxonium salt (**2′**) (Eq. 9.3). The N–Si distance of **2′** is 1.96 Å and it is the shortest among silatranes until now. The silicon of **2′** is a trigonal bipyramid bearing a nitrogen and an oxygen atoms at the apical positions [3].

$$ROH + Si(OEt)_4 + N(CH_2CH_2OH)_3 \longrightarrow \mathbf{2} \qquad (9.2)$$

2a: R=Me
2b: R=Et

$$\mathbf{2a} + Me_3\overset{\oplus}{O}\overset{\ominus}{B}F_4 \longrightarrow \mathbf{2'} \quad (1.96\ \text{Å};\ Me_2O^{\oplus};\ BF_4^{\ominus}) \qquad (9.3)$$

1-Hydrosilatrane (**1b**) is converted to 1-halosilatrane (**3**), and from 1-iodosilatrane (**3c**) compounds **2** and **1d** are derived, respectively, with alcohol and acetylene (Eq. 9.4).

$$\mathbf{1b} \xrightarrow{Ph_3CX\ (X=Cl,\ Br)} \mathbf{3}; \qquad \mathbf{3c} \xrightarrow{ROH} \mathbf{2}; \qquad \mathbf{3c} \xrightarrow{RC\equiv CH} \mathbf{1d} \qquad (9.4)$$

3a: X=Cl, **3b**: X=Br, **3c**: X=I

The general structure of an atrane is illustrated in Fig. 9.1, and silatrane (E=Si) is one of the examples of atrane. The central atom (E) is usually a main group element of Groups 13, 14, and 15, but it can be substituted with a transition element of Groups 4, 5, and 6. Moreover, it is apparent that the characteristics of E can be modified by substituting nitrogen instead of oxygen and also by Z.

Atrane is conventionally named, which indicates the central atom (E), and aza is added when Y = NR and not O. For example, phosphatrane (E=P, Y=O) corresponds to the basic structure, and oxygen is replaced with NR in the case of azaphosphatrane (E = P, Y = NR). The structures of azatitanatrane (E=Ti,

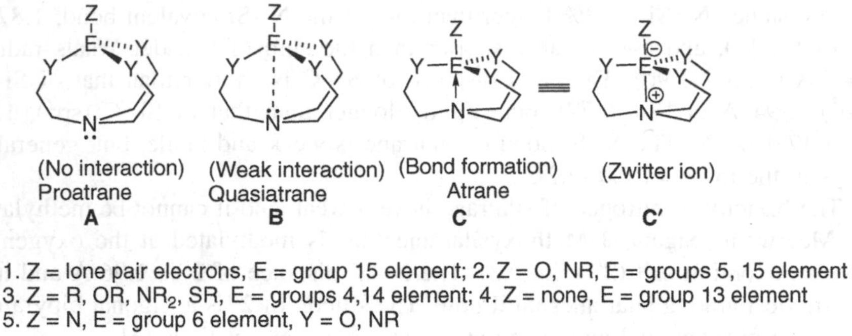

(No interaction) Proatrane **A** | (Weak interaction) Quasiatrane **B** | (Bond formation) Atrane **C** | (Zwitter ion) **C′**

1. Z = lone pair electrons, E = group 15 element; 2. Z = O, NR, E = groups 5, 15 element
3. Z = R, OR, NR_2, SR, E = groups 4,14 element; 4. Z = none, E = group 13 element
5. Z = N, E = group 6 element, Y = O, NR

Figure 9.1 Structure of atrane and formation of a hypervalent bond.

Y=NR) and azavanadatrane (E = Va, Y = NR) have been explicitly written, and Z is substituted at 1-position (1–Z).

The fact that lone pair electrons of nitrogen are withdrawn toward the direction of E (the inside of the molecule) is called transannular interaction. Depending on the strength of transannular interaction, there is a change in the E–N bond distance between the sum of van der Waals radius and that of the covalent radius of the two atoms. The molecule is named proatrane (**A**), when there is no transannular interaction, and hence the lone pair electrons of nitrogen reside outside the molecule which has a lantern-like shape. When E–N distance becomes intermediate as a result of transannular interaction it is quasi-atrane (**B**), and it is atrane (**C**) when the distance becomes short enough to form 3-center 4-electron bond as a result of strong transannular interaction. In **B** and **C**, lone pair electrons of nitrogen are withdrawn to enhance electron density of E or Z, thus E or Z shows exceptionally high basicity and unique reactivity.

Reaction of tris(dimethylamino)phosphine and triethanolamine was expected to yield prophosphatrane (**4**), which, however, resulted in a mixture of polymers. The mixture afforded P–H phosphatrane (**5**) by successive treatment with Meerwein reagent. It is unexpected here that phosphine, and not nitrogen, is protonated [4].

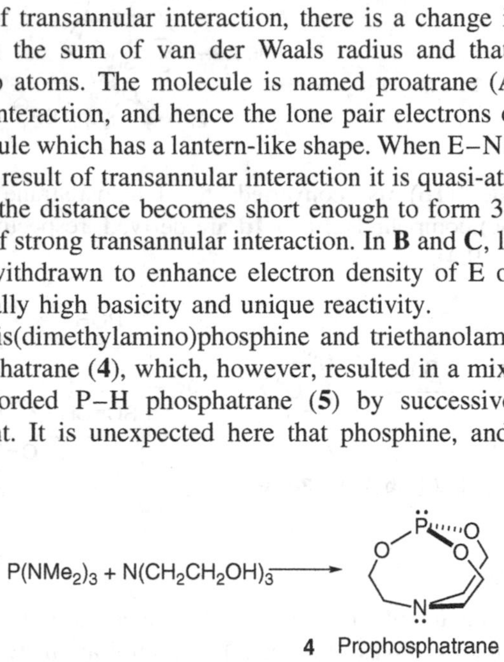

$$P(NMe_2)_3 + N(CH_2CH_2OH)_3 \longrightarrow \mathbf{4} \quad (9.5)$$

4 Prophosphatrane

X-ray analysis of P–H phosphatrane (**5**) revealed that the phosphorus is a trigonal bipyramid and P–N and P–H distances are 1.986 Å and 1.35 Å, respectively. Chemical shift of phosphorus appears at a high field ($\delta_P = -20.9$ ppm) and that of hydrogen δ_H is 6.03 (d) ppm, with a large coupling constant ($J_{PH} = 791$ Hz). Compound **5** is a typical atrane. The P–H bond is strong and its acidity is quite weak. Deprotonation of the P–H does not proceed with sodium methoxide nor even with a strong base of 1,8-dimethylaminonaphthalene. The nitrogen maintains its ammonium structure and is not protonated even in magic acid ($FSO_2OH{\cdot}SbF_3$ in SO_2) [4].

$$4 + Me_3\overset{\oplus}{O}\overset{\ominus}{BF_4} \longrightarrow \mathbf{5}\ (BF_4^{\ominus}) \qquad (9.6)$$

l(P–N) = 1.986 Å ; l(P–H) = 1.35 Å;
∠O–P–O = 120° ; ∠O–P–N = 87° ; ∠H–P–N = 172°
δ_H = 6.03(d) ; J_{PH} = 791 Hz ; $\delta^{31}P$ = – 20.9

Reaction of an ester of phosphoric acid and triethanolamine affords quasi-phosphatrane (**6**) (Eq. 9.7). Silylation of **6** takes place at the oxygen and not at the nitrogen, yielding phosphatrane (**7**) (Eq. 9.8). Chemical shift of phosphorus shifts to a higher field with an increase in electron-withdrawing ability of oxygen.

$$O{=}P(OEt)_3 + N(CH_2CH_2OH)_3 \longrightarrow \mathbf{6} \qquad (9.7)$$

6 Quasi-phosphatrane

$\delta^{31}P$ = – 6.6 ppm

(9.8)

6 → (**a**) $BF_3 \cdot Et_2O$ → **7a** $\delta^{31}P$ = – 18.1

6 → (**b**) Et_3SiClO_4 → **7b** ($ClO_4^{\ominus}$) $\delta^{31}P$ = – 18.7

6 → (**c**) $2Et_3SiClO_4$ → **7c** ($2ClO_4^{\ominus}$) $\delta^{31}P$ = – 25.6

Reaction of bis(dimethylamino)chlorophosphine and tris(aminoethyl)amine affords P–H azaphosphatrane (**8a–c**) easily (Eq. 9.9). Compound **8a** (R=H) is obtained quite easily in the presence of a strong acid (Eq. 9.10). Acidity of azaphosphatrane (**8a–c**) is quite weak and has a pK_a value of 26.8–29.6 with KO-*t*-Bu (pK_a of HO-*t*-Bu is 26.8 and the basicity of potassium salt is quite strong) in dimethyl sulfoxide (DMSO). By deprotonating **8** under similar conditions, **9** is generated and the resulting free phosphorus easily reacts with oxygen and selenium to give **10** (Eq. 9.11).

$ClP(NMe_2)_2 + N(CH_2CH_2NHR)_3 \xrightarrow{-2Me_2NH}$ **8** ($Cl^{\ominus}$) (9.9)

8a: R=H, **8b**: R=Me, **8c**: R=$PhCH_2$

$P(NMe_2)_3 + N(CH_2CH_2NH_2)_3 + CF_3SO_2OH \longrightarrow$ **8a** (TfO^-) (9.10)

8a + *t*-BuOK ⟶ **9** ⟶ **10** (9.11)

9a: R=H,
9b: R=Me
9c: R=$PhCH_2$

10a: Ch=O, R=H
10b: Ch=Se, R=H
10c: Ch=Se, R=Me

The basicity of azaphosphoatrane (**9**) is stronger than DBU and the basicity of a series of compounds is measured by ^{31}P NMR (Fig. 9.2). DBU, **12a**, and **12b** are intentionally designed as strong organic bases; **9b, 11a**, and **11b** are stronger bases than the designed molecules. This illustrates that phosphorus withdraws electrons of nitrogen strongly as a result of transannular interaction.

Compound **9b** (R=Me) reacts as a basic catalyst and gives a cyclic trimer of phenyl isocyanate and also polymerizes ε-caprolactam to 6-nylon (Eq. 9.12).

9b + Ph—N=C=O ⟶ [adduct] $\xrightarrow{Ph—N=C=O}$ ⟶ **9b** + (PhNCO)$_3$ (9.12)

The characteristics of azatrane can be modified by the substituent on nitrogen [5]. Moreover, the central atom (E) can be a transition element, and such derivatives as tantalum (**14**), tungsten (**15**), vanadium (**13d**), and molybdenum (**13e**) are synthesized (Figs. 9.3 and 9.4). Compound **13e** (E=Mo) is an intermediate in nitrogen fixation [6]. Compound **14** is prepared from

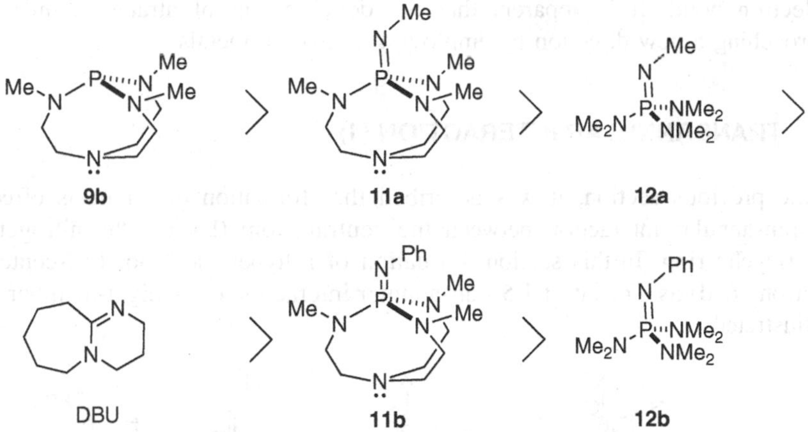

Figure 9.2 Order of strength of a base.

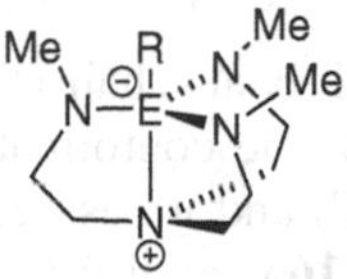

13	E	R
a	Si	Me
b	Sn	NMe_2
c	Ge	Me
d	V	=O
e	Mo	≡N

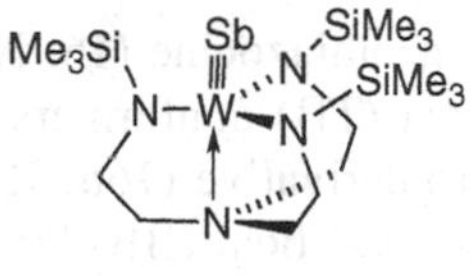

Figure 9.3 Azatranes of several elements.

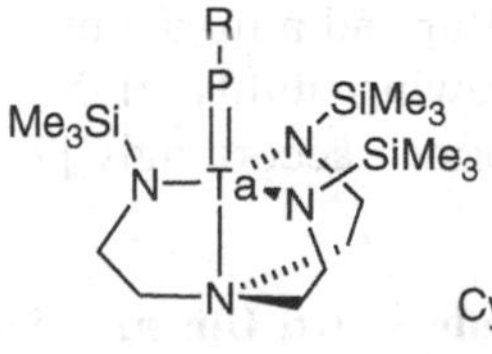

14a: R = Ph
14b: R = Cy
14c: R = Bu

Cy = cyclohexyl

Figure 9.4 Azatrane of tantalum.

$[(Me_3SiNCH_2CH_2)_3NTaCl_2]$ and 2 equivalents of phosphine (RPHLi), and is a complex of Ta(III) and phosphinidene (R–P:). The presence of a triple bond (bond distance of W≡Sb is 2.525 Å) in **15** was determined by X-ray analysis. Transannular interaction (bond formation) is often designated with an arrow as shown in **15**; however, it is often illustrated as atrane (**C′**) of ylide structure in this chapter in order to emphasize the formation (participation) of 3-center

4-electron bond. It is apparent that the development of atrane chemistry is approaching a new direction by employing transition metals.

9.3 TRANSANNULAR INTERACTION (1)

In the previous section, it was described that formation of atrane is effected by transannular interaction between the central atom (E) and the nitrogen of the tricyclic ring. In this section, formation of a hypervalent bond (3-center 4-electron bond) as a result of 1,5-transannular interaction of an eight-member ring is illustrated.

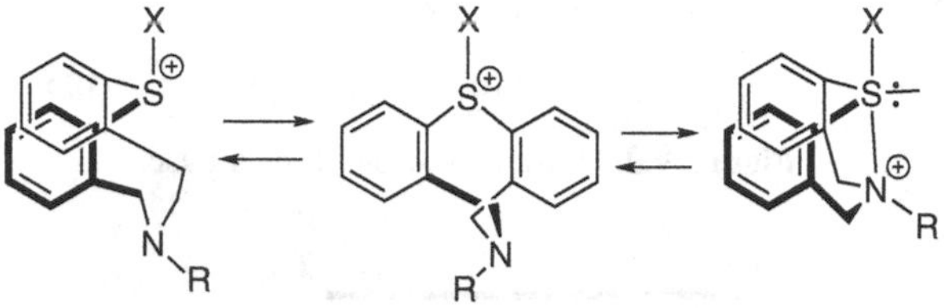

16 BC **16** TB **16** BB (9.13)

Dibenzothiazocine (**16**: X=none) is in equilibrium with boat–chair (BC) and twist-boat (TB) conformers in solution, and there is only one conformation for *S*-chloro derivative (**16d**: X=Cl) in solution (−30 to 43°C) and it has been clarified as boat–boat (BB) by ^{1}H NMR. Other derivatives (**16a–c**) also have only BB conformation (Eq. 9.13). Compounds **16a–d** have been shown to be BB conformers in solid by X-ray analysis. The distance between sulfur and nitrogen is 2.609 Å in sulfoxide (**16a**: X=O) and decreases to 2.091 Å in *S*-Chloro derivative (**16d**: X = Cl) and the configuration of sulfur becomes closer to an ideal trigonal bipyramid with an increase in electron-withdrawing ability of X (Table 9.1). The energy of S–N interaction increases from 4.2 (**16a**) to 39.7 kJ/mol (**16d**). That is, the transannular interaction of sulfur and nitrogen becomes gradually stronger with an increase in electron-withdrawing ability of X, and hence it is clearly shown that a hypervalent bond is formed accordingly [7a].

TABLE 9.1 Structure of S-Substituted Dibenzothiazocine (16) (N–S Bond Distance and Bond Energy)

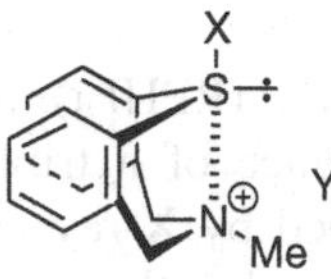

16	X	Y	N–S bond distance (Å)	N–S bond energy (kJ/mol)
a	O	—	2.609	4.2
b	Me	PF_6	2.466	7.5
c	MeO	$SbCl_6$	2.206	24.2
d	Cl	PF_6	2.091	39.7

16

It is theoretically supported that 3-center 4-electron bond is formed as X–S–N bond as a result of transannular interaction. When we deal with the interaction of NH_3 and H_3S^+, hydrogen bond, that is, $H_3N\cdots H{-}S^+H_2$, is considered in the first place. But in this system, the direct interaction between nitrogen and sulfur should be taken into account as shown in Eq. (9.13) and Table 9.1. As a theoretical model, **E** (both species having C_{3v} symmetry) is the first candidate for this. With this model, S–N distance is obtained as 3.33 Å with bond energy of 38.9 kJ/mol [4-31G*(S,N)]. However, because of sp^2 configuration of sulfur in D, S–N distance is shortened to 2.67 Å and bond energy becomes stronger, 92.4 kJ/mol [4-31*G**(S)].

D **E**

Therefore, 3-center 4-electron bond is formed as H–S–N and the configuration of sulfur is sp^2 and its lone pair electrons lie on the plane of the paper [7b]. On the basis of the results, X–S–N bond is the apical bond of a trigonal bipyramid and it is in accordance with the order of apicophilicity that bond distance decreases with an increase in electron-withdrawing ability of X.

In 2-ethylsulfoniumpyridine (**17**), the S–N distance is 2.835 Å and bond formation is not observed, although there is a withdrawing interaction (Fig. 9.5). The structure of sulfur is tetragonal in **17**. The configuration of phosphorus in **18c** is sp^2 and a hypervalent bond is formed as X–P–N. Chemical shift of ^{31}P shifts toward a higher field as a result of an increase in electron-withdrawing ability of X and this trend is similar to that observed in phosphatrane (Eqs. 9.7 and 9.8).

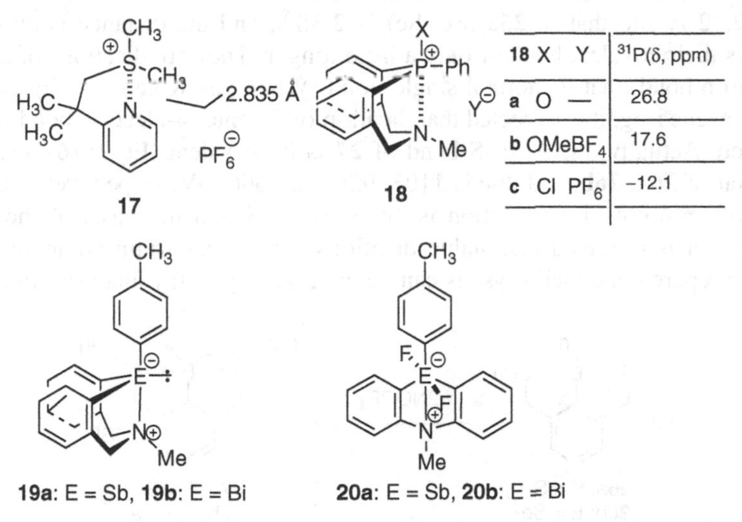

18	X	Y	^{31}P(δ, ppm)
a	O	—	26.8
b	OMe	BF_4	17.6
c	Cl	PF_6	−12.1

Figure 9.5 Transannular interaction between heteroatoms.

When tetravalent phosphorus of **18** is substituted in trivalent antimony and bismuth, a strong interaction takes place between a trivalent element and the nitrogen, and the conformation is kept as BB (**19**), forming a zwitter ion. Compound **19** is oxidized to 12-E-6 by fluorination and the structure becomes octahedral (**20**) [7c].

9.4 TRANSANNULAR INTERACTION (2)

The dication **22** is obtained with transannular bond by oxidation of 1,5-dithiaoctane **21** with anode or $NOBF_4$ (or $NOPF_6$), where a cation radical is supposed to be an intermediate (Eq. 9.14). Dication **22** is also obtained by treatment of monooxide **23** with concentrated sulfuric acid. 1,5-Diseleno-(**24a**) and ditelluro-octane (**24b**) are also oxidized to dications **25a** and **b** (Eq. 9.15). Chemical shifts of selenide and telluride (**24**) shift toward an extremely low field in dications (**25**) [8].

21 —($2NOBF_4$)→ **22** ($2BF_4^{\ominus}$) ←(Conc. H_2SO_4)— **23** (9.14)

24 —($2NOBF_4$)→ **25** ($2BF_4^{\ominus}$) (9.15)

24a: X = Se ^{77}Se δ 141 ppm
24b: X = Te ^{125}Te δ 163 ppm

25a: X = Se ^{77}Se δ 806 ppm
25b: X = Te ^{125}Te δ 1303 ppm

The ortep drawing of dications is quite similar. The bond distance of S–S of **22** is 2.12 Å and that of **25a** (Se–Se) is 2.38Å, and the distance is almost the same as their covalent bond or only a little longer. They are dications of 2-center 2-electron bond, that is, normal single bond. When van Koten-type ligand is set on a benzene ring, it is expected that dication of 3-center 4-electron bond is easily prepared. Actually, the S–E–S bond of **27** is hypervalent (Eq. 9.16). Oxidation potential of **26a, 26b**, and **26c** is 1103, 926, and 666 mV, respectively; thus, the order of feasibility for oxidation is Te > Se < S. On the basis of the results obtained, it is expected that stable dications (**29**) of tellurium compounds (**28**) can be prepared and their crystals can be analyzed by X-ray analysis (Eq. 9.17).

26 —($2NOPF_4$)→ **27** ($2BF_4^{\ominus}$) (9.16)

26a: E = S
26b: E = Se
26c: E = Te

27a: E = S
27b: E = Se
27c: E = Te

28a: RX = SPh
28b: RX = SePh
28c: RX = NMe_2

$\xrightarrow{2NOBF_4}$

29a: RX = SPh
29b: RX = SePh
29c: RX = NMe_2

(9.17)

van Koten-type compound (**30**) bearing a methyl group at the central atom (E) is oxidized with *t*-BuOCl to afford the monocation **31**, during which a dication (**F**) is generated first and methyl chloride is eliminated by the attack of the resulting chloride ion (Eq. 9.18). The monocation **31** is a derivative of Ph–E^+ (**31a**: E=Se; **31b**: E=Te) and E^+ is stabilized by the formation of 3-center 4-electron bond with two dimethylamino groups. Compound **31** has two lone pair electrons, each residing above and below the molecular plane and chemical shifts of Se and Te are shifted toward an extreme low field. The fact that chemical shifts of **31** appear at a much lower field than that of dication **25** is in accordance with the characteristics of 3-center 4-electron bond (N–E–N).

30a: E = Se ^{77}Se δ 90.0 ppm
30b: E = Te ^{125}Te δ 287.2 ppm

$\xrightarrow{t\text{-BuOCl}}$ F $\xrightarrow{KPF_6}$ + CH_3Cl

31a: E = Se ^{77}Se δ 1208.3 ppm
31b: E = Te ^{125}Te δ 7949.4 ppm

(9.18)

REFERENCES

1. Cahill PA, Dykstra CE, Martin JC. J Am Chem Soc 1995;107:6359.
2. Verkade JG. Acc Chem Res 1993;26:483.
3. Garant RJ, Daniels LM, Das SK, Janakiraman MN, Jacobson RA, Verkade JG. J Am Chem Soc 1991;113:5728.
4. Carpenter LE, Verkade JG. J Am Chem Soc 1985;107:7084.
5. Gudat D, Daniels LM, Verkade JG. J Am Chem Soc 1989;111:8520.
6. Kal M, Schrock RR, Kempe R, Davis WM. J Am Chem Soc 1994;116:4382.
7. (a) Akiba K.-y., Takee K, Shimizu Y, Ohkata K. J Am Chem Soc 1986;108:6320; (b) Morokuma K, Hanamura M, Akiba K.-y. Chem Lett 1984:1557; (c) Akiba K.-y., Okada K, Ohkata K. Tetrahedron Lett 1986;27:5221.
8. Furukawa N. Bull Chem Soc Jpn 1997;70:2571.

CHAPTER 10

UNSATURATED COMPOUNDS OF MAIN GROUP ELEMENTS OF THIRD PERIOD AND HEAVIER

10.1 INTRODUCTION

The double bond rule, that is, the double bond cannot exist between main group elements below the third period, has been one of the dogmas in modern organic chemistry since 1965. Historically, diphosphene (**1**) bearing a double bond between phosphorus and phosphorus was reported in 1877, and sarbarsane 606 (**2**) having a double bond between arsenic and arsenic was reported in 1909. The double bond in these compounds looked quite similar to that of the nitrogen–nitrogen double bond of azobenzene (**3**) and azo dyes; hence, the times connived them as it was written.

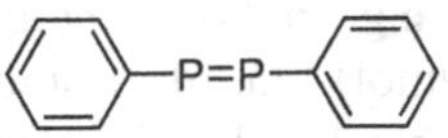

Ph–P=P–Ph

1

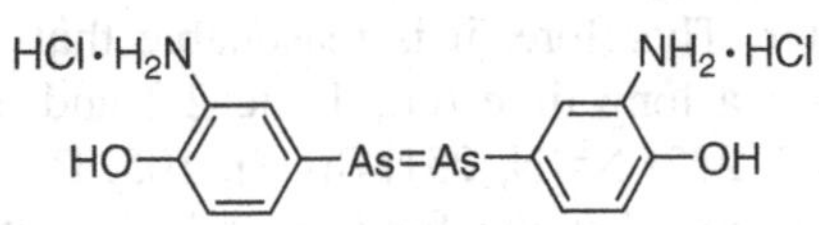

HCl·H₂N / HO–C₆H₃–As=As–C₆H₃–OH / NH₂·HCl

2

Ph–N=N–Ph

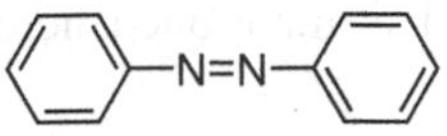

3

Organo Main Group Chemistry, First Edition. Kin-ya Akiba.

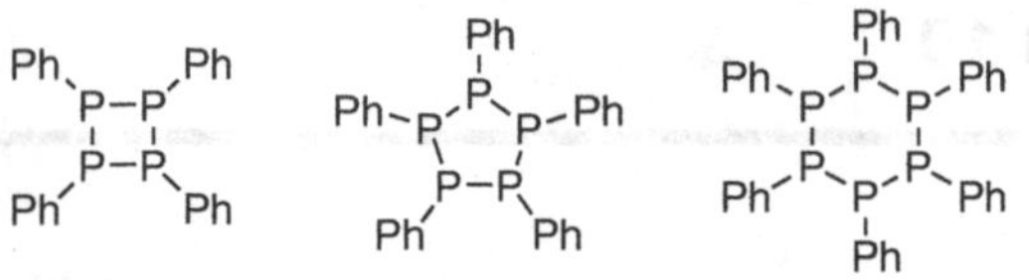
Ph
P
P
Ph
four member ring
five member ring
six member ring

In 1965, however, the molecular weight of **1** was measured, disclosing that **1** should be a mixture of tetramer and pentamer of phosphinidene (Ph–P:, **A**), and possible cyclic structures were proposed for **1**. In 1966, X-ray analysis showed that they are a mixture of pentamer and hexamer in crystals. It was clearly illustrated that there is no double bond in them. The result was the origin of the double bond rule mentioned above.

Surprisingly, in 1981, diphosphene (**4**) [1a, b] and disilene (**5**) [2a,b], which contained a double bond between phosphorus and phosphorus and silicon and silicon, were synthesized, and structures were determined. In both compounds, sterically bulky substituents such as 2,4,6-tri-*t*-butylphenyl (Mes*) and 2,4,6-trimethylphenyl (Mes) were intentionally used to protect the unsaturated bond. Based on these, "the double bond rule" mentioned at the beginning was denied and corrected.

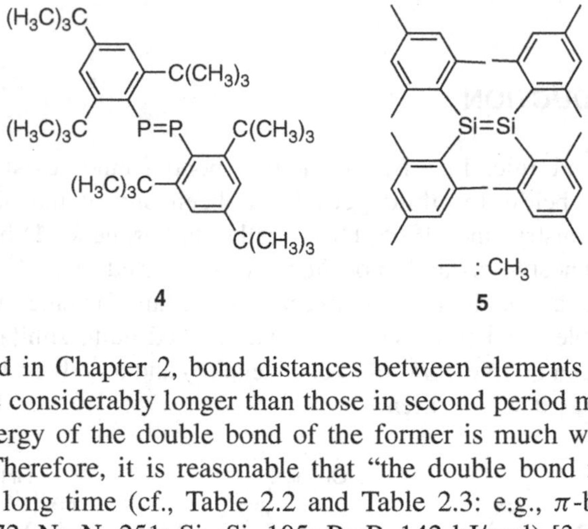
(H3C)3C
C(CH3)3
(H3C)3C
P=P
C(CH3)3
(H3C)3C
C(CH3)3
Si=Si
— : CH3

4 **5**

As explained in Chapter 2, bond distances between elements in third period main group are considerably longer than those in second period main group, and the π-bond energy of the double bond of the former is much weaker than that of the latter. Therefore, it is reasonable that "the double bond rule" had been believed for a long time (cf., Table 2.2 and Table 2.3: e.g., π-bond energy of E=E; C=C, 272; N=N, 251; Si=Si, 105; P=P, 142 kJ/mol) [3a, b]. Moreover, it is understood that the fundamental character of double bonds of group 14 and group 15 elements is different, considering their electronic configurations. The schematic presentation was given in Chapter 2 and is mentioned here for emphasis.

A compound is said to be unstable when

1. Reactivity of a compound is very high with other molecules: the compound reacts with water or oxygen during usual treatment and changes or decomposes.

2. Reactivity of a compound itself is very high, and even in the absence of other molecules, the lifetime is very short and it decomposes or polymerizes by itself.

In order to prevent these, a method was devised to insert or place bulky substituents in a molecule, that is, bulky enough to protect reactive functional group(s) from other molecules. This is called *steric protection*. This is also called *kinetic protection* because the reactivity of functional group(s) is retained. On the other hand, there is another method to stabilize reactive functional group(s) by inserting or placing a substituent, that is, able enough to change the electronic state of the molecule or include reactive functional group(s) into a heterocycle for stabilization. This is called *thermodynamic stabilization* because the reactivity of the molecule is changed (lowered). In this chapter, recent accomplishment by kinetic stabilization of unsaturated bond is mentioned.

10.2 UNSATURATED BONDS OF GROUP 15 ELEMENTS OF THIRD PERIOD AND HEAVIER

The most important method to protect reactive functional group(s) is to devise a sterically bulky group suitable for a certain molecule. 2,4,6-Tri-*t*-butylphenyl group (Mes*) was successfully used for diphosphene (**4**) and 2,4,6-tris[(bistrimethylsilyl)methyl]phenyl (Tbt), which is a revised model of Mes*, are utilized according to the character of the element [4a, b–5]. Alkyl groups such as tris(trimethylsilyl)methyl (Tsi) and di(trimethylsilyl)methyl (Dis) are also used.

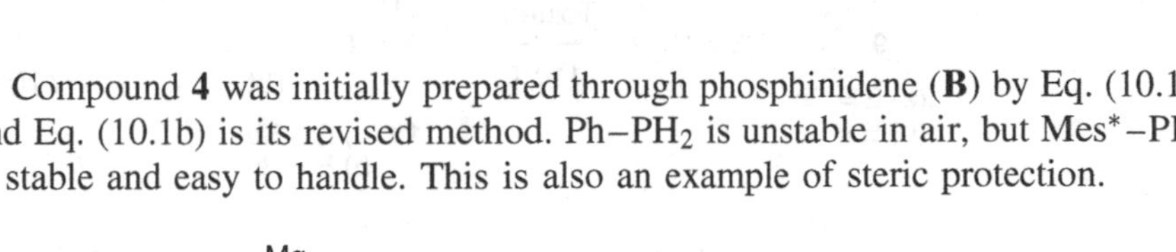

Me3Si, H, Me3Si, SiMe3, H, SiMe3, Me3Si, H, SiMe3 — Me3Si–C(SiMe3)2 — Me3Si–CH(SiMe3)

Mes* Tbt Tsi Dis

Compound **4** was initially prepared through phosphinidene (**B**) by Eq. (10.1a) and Eq. (10.1b) is its revised method. Ph–PH_2 is unstable in air, but Mes*–PH_2 is stable and easy to handle. This is also an example of steric protection.

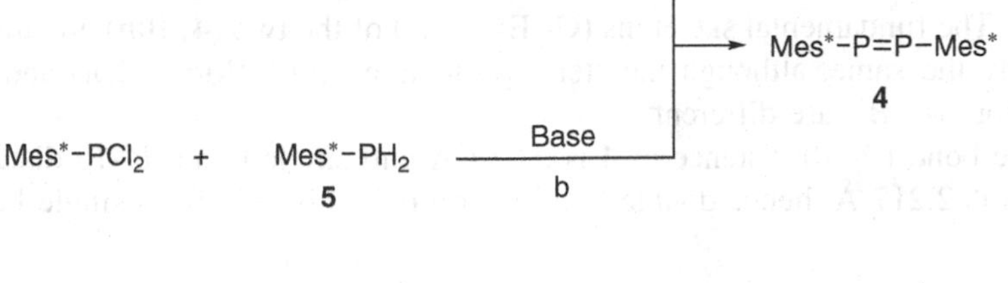

$$\text{Mes*–PCl}_2 \xrightarrow{\text{Mg}} [\text{Mes*–P:}]\ (\textbf{B}) \xrightarrow{\text{a}} \text{Mes*–P=P–Mes*}\ (\textbf{4})$$

$$\text{Mes*–PCl}_2 + \text{Mes*–PH}_2\ (\textbf{5}) \xrightarrow[\text{b}]{\text{Base}} \text{Mes*–P=P–Mes*}\ (\textbf{4}) \qquad (10.1)$$

The reductive coupling between different dichlorophosphines (Eq. 10.2) gives a mixed diphosphene (**6**). The reaction of the same type of dihydride and dichloride yields **7** in the presence of base, in which one of the trimethylsilyl groups is eliminated. This results from the nucleophilic attack of the chloride ion generated in situ to the silyl group (Eq. 10.3). It is shown that a variety of unsaturated compounds, such as Tsi–P=P–Tsi, Mes*–As=As–Dis, Tsi–As=As–Tsi, are prepared by the application of these methods.

$$\text{Mes*-PCl}_2 + \text{Tsi-PCl}_2 \xrightarrow{\text{Mg}} \underset{\mathbf{6}}{\text{Mes*-P=P-Tsi}} \tag{10.2}$$

$$\text{Mes*-PH}_2 + \text{Tsi-ECl}_2 \xrightarrow{\text{Base}} \underset{\mathbf{7}}{\text{Mes*-P=E-Dis}} \tag{10.3}$$

(**a**) E = P ; (**b**) E = As ; (**c**) E = Sb

Tbt is quite an efficient steric protective group because the two trimethylsilyl groups placed at 2,6-positions of the benzene ring surround and hide the reactive group at position 1; yet, there is a space made by one C–H group, allowing a reagent to approach (Eq. 10.4). Double bond compounds such as distibene (**10a**) and dibismuthene (**10b**) were synthesized by this method (Eq. 10.5). Stibinidene (Tbt–Sb:, **C**) was detected as an intermediate of the reaction of **9a** and $(Me_2N)_3P$; however, bismuthidene (Tbt–Bi:, **D**) could not be detected. Structures obtained by X-ray analysis are illustrated in Fig. 10.1.

$$\underset{\mathbf{8}}{3\ \text{TbtECl}_2} + 3\text{Li}_2\text{Se} \longrightarrow \underset{\mathbf{9}}{\text{cyclo-(TbtESe)}_3} \tag{10.4}$$

8: (**a**) E = Sb; (**b**) E = Bi — **9**: (**a**) E = Sb; (**b**) E = Bi

$$\mathbf{9} + (\text{Me}_2\text{N})_3\text{P} \xrightarrow[110^\circ\text{C}]{\text{Toluene}} \text{Tbt-E=E-Tbt} \tag{10.5}$$

9: (**a**) E = Sb; (**b**) E = Bi — **10** (**a**) E = Sb; (**b**) E = Bi

Compound **4** was first used to invalidate "the double bond rule," and **10b** bears a double bond of bismuth, which is the heaviest element before radioactive elements. The fundamental skeletons (C–E=E–C) of the two (**4, 10b**) are almost completely the same, although the steric protective group (Mes*, Tbt) and the central atom (P, Bi) are different.

Double bond (P=P) distance of **4** is 2.034 Å and single bond (P–P) distance of $(PhP)_5$ is 2.217 Å, hence double bond is contracted by 8% from single bond.

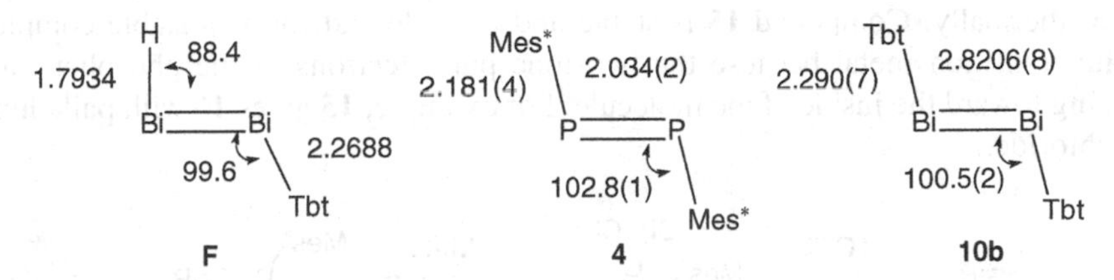

Figure 10.1 Structure of diphosphene (**4**) and dibismuthene (**10b**). Bond distance (Å), bond angle (°), **F** (calculated structure).

Double bond (Bi=Bi) distance of **10b** is 2.821 Å and single bond (Bi–Bi) distance of $Ph_2Bi-BiPh_2$ is 2.990 Å; therefore, 7% of bond contraction is observed [6]. For antimony (**10a**), bond contraction is 7%. Bond angle (∠C–E=E) of **4** is 102.8° and that of **10b** is 100.5°. They are essentially different from that of sp^2 hybridization (120°) and should become closer to the ideal angle of 90°. It is well known theoretically that the difference in size and energy between the s and p orbitals increases by descending the periodic table from nitrogen to bismuth. Thus hybridization becomes more difficult, and the electronic configuration is kept as $(ns)^2(np)^6$.

Especially, hybridization should be quite difficult for bismuth because the 6s orbital is considerably contracted by relativistic effect (Figs. 2.1 and 2.2).

Structure of **F** in Fig. 10.1 was obtained by theoretical calculation. On the basis of the result, bond angle (∠H–Bi=Bi) should be 90° in the ideal case and the angle (∠Tbt–Bi=Bi) is enlarged to 100°, probably because of steric hindrance between Tbt and Bi=Bi bond. Therefore, it is realized that the angle (∠C–E=E) should be 90° in the ideal case, and it is enlarged by steric hindrance between steric protection group and E=E bond. Group 15 elements differ considerably in their characters, especially in the reactivity of lone pair electrons. However, hybridization remains almost the same in building up these double bonds. Furthermore, compounds having a double bond between different elements are also stable, and their structures can be estimated based on **4** and **10b** [6, 7, 13].

$$\text{Mes}^*\text{–P} = \text{Bi–Bbt},\ \text{Bbt–Sb} = \text{Bi–Bbt}$$

$$\text{Bbt} = 2,6\text{-bis[bis(trimethylsilyl)methyl]-4}$$
$$\text{-[tris(trimethylsilyl)methyl]phenyl}$$

An allene-type compound bearing –P=C=P– bond (**11**) is obtained by treating the adduct of diphosphene (**4**) and dichlorocarbene with methyllithium (Eq. 10.6). Further, when the same type of reaction is repeated, cumulated double bond compound –P=C=C=P– (**12**) is prepared (Eq. 10.7).

A dimeric compound (**13**) of reaction product of Mes*P(H)Cl and lithium acetylide isomerizes (bond conversion) easily at room temperature to yield **14** in solution (Eq. 10.8), and **14** isomerizes further to diphosphinedenecyclobutene

(**15**) thermally. Compound **15** is stable and isolable and forms a stable complex with transition metal because the two lone pair electrons of the phosphene are facing toward the inside of the molecule. For example, **15** gives **16** with palladium dichloride.

Mes*P=PMes* (**4**) —:CCl_2→ [Mes*P–P(Mes*) bridged by CCl_2] —MeLi→ Mes*P=C=PMes* (**11**) (10.6)

Mes*P=C=PMes* (**11**) —:CCl_2→ [Mes*P–P(Mes*) bridged by C=CCl_2] —LiNapth→ Mes*P=C=C=PMes* (**12**)

Mes*–P(–C≡C–$SiMe_3$)–P(–C≡C–$SiMe_3$)–Mes* (**13**) —Room temp.→ Mes*–P=C=C(–$SiMe_3$)–C(–$SiMe_3$)=C=P–Mes* (**14**) (10.7)

—Δ→ **15** (cyclobutene with two exocyclic P=Mes* and two $SiMe_3$) —$PdCl_2$→ **16** (Cl_2Pd coordinated to both P of **15**) (10.8)

Distibene (**10a**) and dibismuthene (**10b**) protected by Tbt react to trap oxygen slowly in the air, and the trapped oxygen is cleaved to form a four-member ring (**17a,b**) (Eq. 10.9).

Tbt–E=E–Tbt —O_2, in the air→ Tbt–E(μ-O)$_2$E–Tbt (10.9)

10 (**a**) E = Sb (**b**) E = Bi; **17** (**a**) E = Sb (**b**) E = Bi

2,6-Dimesitylphenyl group (Dmp) is a kind of new protecting group and is designed to make a space for the reagent by rotation of the mesityl group. Dibismuthene (**19**) is easily prepared with Dmp. By reaction of **19** with zirconocene (**G**), generated by reduction of zirconocene dichloride with sodium, a three-member ring (**20**) is obtained (Eq. 10.10). Successful use of Dmp suggests new possibility for designing steric protection groups.

$$Cp_2ZrCl_2\ (\mathbf{18}) \xrightarrow{2Na} [Cp_2Zr]\ (\mathbf{G}) + R^1Bi{=}BiR^1\ (\mathbf{19}) \longrightarrow Cp_2Zr(BiR^1)_2\ (\mathbf{20}) \qquad (10.10)$$

R^1 = Dmp = di(2,6-mesityl)phenyl

10.3 UNSATURATED BONDS OF GROUP 14 ELEMENTS OF THIRD PERIOD AND HEAVIER

A carbene-type intermediate (**H**) of main group element generated in vapor phase is a singlet species of v letter structure, which dimerizes in solid to form E=E double bond (**21a,b**) (Eq. 10.11). Disilene (Si=Si) is a fundamental species of group 14 compounds of third period (Eq. 10.12). Disilene (**5**) was first reported in 1981, the same year as diphosphene (**4**) [2a,b,4a]. It was already mentioned that "the double bond rule" was completely denied by the synthesis and structure determination of **4** and **5**.

$$2\ Dis_2E{:}\ (\mathbf{H\ a,b}) \longrightarrow Dis_2E{=}EDis_2\ \ \mathbf{21}\ (\mathbf{a})\ E{=}Ge,\ (\mathbf{b})\ E{=}Sn \qquad (10.11)$$

Dis = di(trimethylsilyl)methyl : $(Me_3Si)_2CH-$

$$2\ Mes_2Si(SiMe_3)_2\ (\mathbf{22}) \xrightarrow{h\nu} (2\ Mes_2Si{:}\ (\mathbf{I})) \longrightarrow Mes_2Si{=}SiMes_2\ (\mathbf{5}) + 2\ Me_3Si{-}SiMe_3$$

Mes = mesityl = 2,4,6-trimethylphenyl

(10.12)

Structures of **5** and **21** are summarized in Table 10.1. The double bond consists of a dimer of carbenes; thus, the molecule is not planar, and there is angle θ between the plane of R–E–R and the axis of E=E. The angle θ increases in the order of silicon (18°), germanium (32°), and tin (41°). Double bond distance gradually becomes closer to single bond distance, and bond contraction decreases by 8, 4, and 2% for each element. The trend is contrary to double bonds of group 15 elements, and contraction stays almost constant at 7–9%.

A compound containing a double bond with different elements (**23**) is also prepared [8]. The structure can be estimated based on Table 10.1. A compound of boron–boron double bond (**24**) was recently reported, which is quite unique and coordinated with two carbenes [9]. Moreover, trisilaallene (**25**, Si=Si bond is bent) with a cumulated double bond [10] and disilyne (**26**) with a Si–Si triple bond [11] were synthesized recently, and they were shown to be stable. Compounds with a triple bond have also been prepared, such as Ge≡Ge, Sn≡Sn, Pb≡Pb.

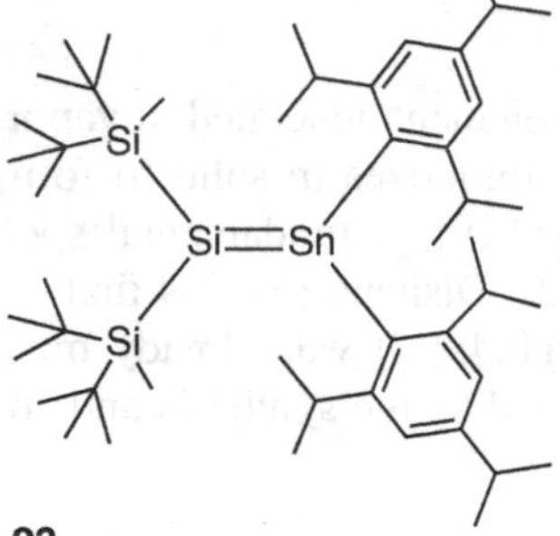

23 Si=Sn : 2.42 Å

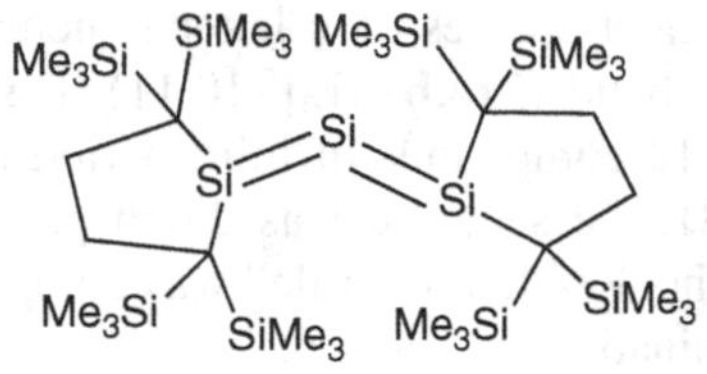

25

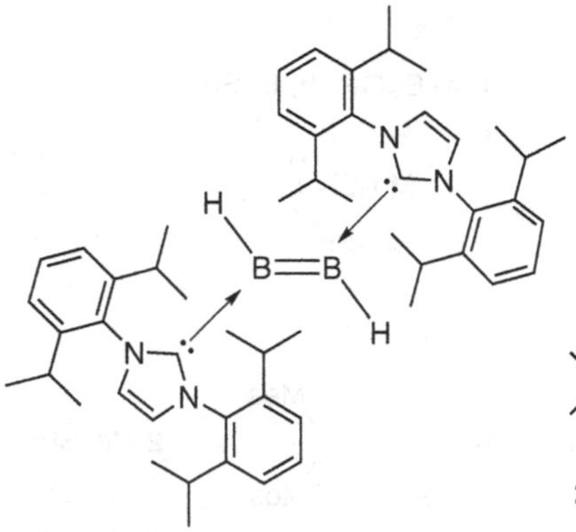

24 B=B : 1.560 Å

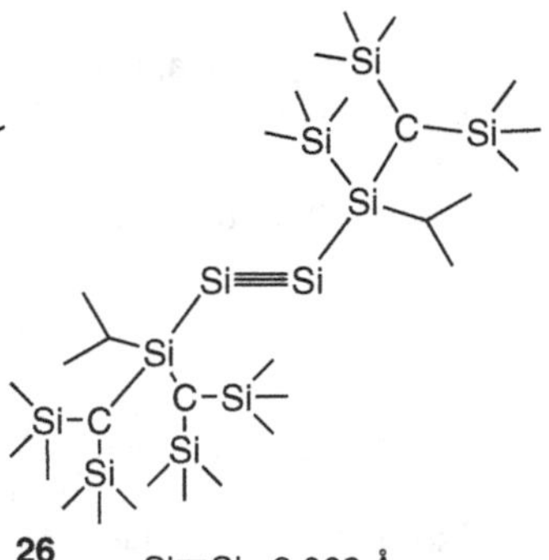

26 Si≡Si : 2.062 Å

Figure 10.2 Novel double and triple bond compounds. Hydrogen is written for B–H, methyl is abbreviated.

TABLE 10.1 Structure of Double Bond Compounds of Group 14 Elements

	$R_2E{=}ER_2$ (θ)	Bond angle (∠RER)	θ	Bond distance (E=E)	Bond distance* (E–E)	Bond contraction (%)
5	$(Mes)_2Si{=}Si(Mes)_2$	115.5	18	2.160	2.32	8
21a	$(Dis)_2Ge{=}Ge(Dis)_2$	112.5	32	2.347	2.44	4
21b	$(Dis)_2Sn{=}Sn(Dis)_2$	112	41	2.764	2.81	2

θ: angle between RER plane and E=E axis (°); bond angle (°) ; bond distance(Å) ; * literature.

Disilene (**5**) shows regular reactivity as a double bond, and halogen, hydrogen chloride, and alcohol add to the double bond (Eq. 10.13). The reaction of oxygen is the same type as in Eq. (10.9), and [2.2] type addition of acetylene affords useful compounds for further conversion.

$$
\mathbf{5}\ (Mes_2Si{=}SiMes_2) \begin{cases} \xrightarrow{\text{(i) } Cl_2} Mes_2SiCl{-}SiClMes_2 \\ \xrightarrow{\text{(ii) } HX} Mes_2SiH{-}SiXMes_2 \quad X{=}Cl, OR \\ \xrightarrow{\text{(iii) } O_2} \text{cyclo-}(Mes_2SiO)_2 \\ \xrightarrow{\text{(iv) } PhC{\equiv}CH} Mes_2Si{-}SiMes_2\ (\text{ring with } PhC{=}CH) \end{cases} \tag{10.13}
$$

Although a three-member ring compound of tin (**27**) does not contain any double bond, an interesting transformation of **27** is mentioned here. When **27** was heated in xylene at 200°C, pentastanna[1.1.1]propellane (**28**) was obtained. Compound **28** is a purple crystal and unstable in the air. Bond distance of Sn^1-Sn^2 is 2.856 Å and that of Sn^1-Sn^3 is 3.367 Å. If Sn^1-Sn^3 is supposed to be a bond, four bonds of Sn^1 (or Sn^3) should face the same direction in relation to a plane of $Sn^2-Sn^4-Sn^5$; this cannot be the case with tetrahedron structure and is an extremely unusual bond. At present, (Sn^1-Sn^3) is assumed not to be a bond but a diradical. This is a strange example of a bond (electronic state) between higher period main group elements [12a,b].

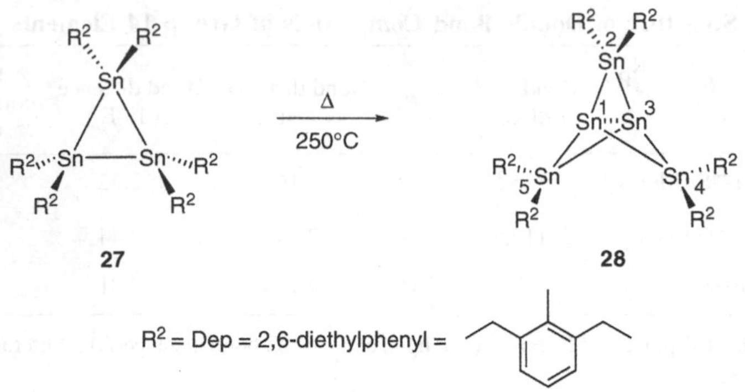

Δ, 250°C: 27 → 28 (10.14)

R^2 = Dep = 2,6-diethylphenyl =

10.4 AROMATIC COMPOUNDS OF GROUP 14 ELEMENTS

It is an interesting approach to try to synthesize aromatic compounds bearing main group elements of third period and heavier in the ring and investigate their aromaticity because compounds containing carbon–silicon and carbon–phosphorus double bonds have already been prepared. Silabenzene (**30a**) and germabenzene (**30b**) were prepared by treatment of the corresponding hydrochloride (**29**) bearing the skeleton with LDA (lithium diisopropylamide) (Eq. 10.15). Naphthalene and anthracene analogs (**32–34**) were similarly prepared (Fig. 10.3) [7,13]. These are isolable and thermally stable. It is clearly demonstrated that Tbt is an excellent steric protecting group, knowing that 1,4-di-*t*-butylsila and germabenzenes (**31a,b**) can be observed only at low temperatures.

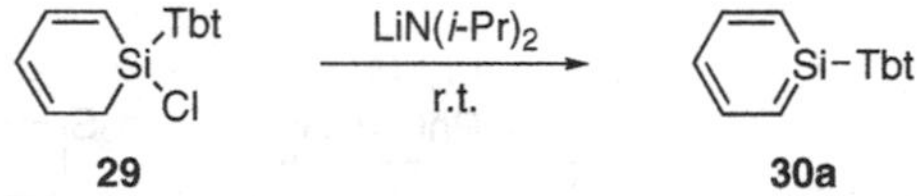

29 → $LiN(i\text{-}Pr)_2$, r.t. → **30a** (10.15)

It is confirmed by spectroscopy that compounds **30a** and **b** are planar and silicon and germanium are sp^2hybridized; consequently, they are aromatic and their π-electrons are delocalized. Based on theoretical calculation, aromatic stabilization of **30a** and **b** is 142.1 and 141.3 kJ/mol for each, and the values are close to the resonance energy (142.5 kJ/mol) of benzene. Heat of addition (DH) of water to Si=C and Ge=C double bond is −268.8 and −200.6 kJ/mol, respectively; however, it is only −29.7 kJ/mol to C=C double bond. Thus, the apparent contradiction that a heteroaromatic compound has aromaticity but high reactivity can be understood.

Actually, E=C double bond of **30** has remarkable reactivity, and reactions in Eq. (10.16) are quite similar to disilene (**5**) (Eq. 10.13).

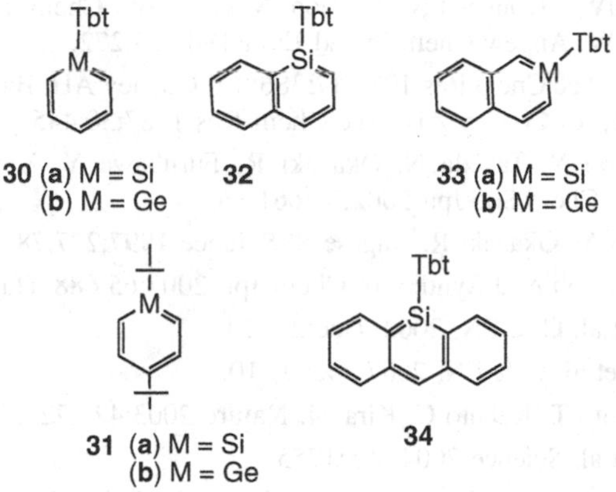
Tbt
M
30 (a) M = Si
(b) M = Ge
Tbt
Si
32
Tbt
M
33 (a) M = Si
(b) M = Ge
M
31 (a) M = Si
(b) M = Ge
Tbt
Si
34

Figure 10.3 Analogs of silabenzene.

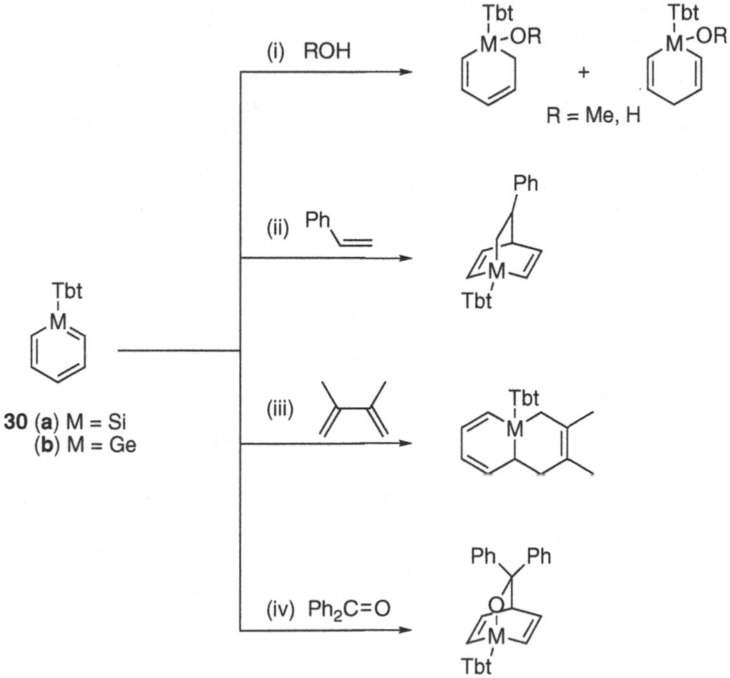
Tbt
M
30 (a) M = Si
(b) M = Ge
(i) ROH
Tbt
M-OR
+
Tbt
M-OR
R = Me, H
(ii) Ph
Ph
M
Tbt
(iii)
Tbt
M
(iv) Ph2C=O
Ph
Ph
O
M
Tbt

(10.16)

REFERENCES

1. (a) Yoshifuji M, Shima I, Inamoto N, Hirotsu K, Higuchi T. J Am Chem Soc 1981;103:4587; (b) Yoshifuji M, Shibayama K, Inamoto N, Matsushita T, Nishimoto K. J Am Chem Soc 1983;105:2495.
2. (a) West R, Fink MJ, Michl J. Science 1981;214:1343; (b) Michalczyk MJ, West R, Michl J. J Am Chem Soc 1984;106:821.

3. (a) Schmidt MW, Troung PN, Gordon MS. J Am Chem Soc 1987;109:5217; (b) Kutzelnigg W. Angew Chem Int Ed Engl 1984;23:272.
4. (a) Cowley AH. Acc Chem Res 1984;17:386; (b) Cowley AH, Barron AR. Acc Chem Res 1988;21:81; (c) Cowley AH. Acc Chem Res 1987;30:445.
5. Sasamori T, Arai Y, Takeda N, Okazaki R, Furukawa Y, Kimura M, Nagase S, Tokitoh N. Bull Chem Soc Jpn 2002;75:661.
6. Tokitoh N, Arai Y, Okazaki R, Nagase S. Science 1997;277:78.
7. Sasamori T, Tokitoh N. J Synth Org Chem Jpn 2007;65:688, (Japanese).
8. Sekiguchi A, et al. C & EN, 2002, Dec 23, 24.
9. Robinson GH, et al. C & EN, 2007, Oct 1, 10.
10. Ishida S, Iwamoto T, Kabuto C, Kira M. Nature 2003;421:725.
11. Sekiguchi A, et al. Science 2004;305:1755.
12. (a) Sita LR, Bickerstaff RD. J Am Chem Soc 1989;111:6454; (b) Sita LR, et al. C & EN, 1989, Aug 7, 29.
13. Tokitoh N. Acc Chem Res 2004;37:86.

CHAPTER 11

LIGAND COUPLING REACTION

11.1 INTRODUCTION

Ligand coupling reaction (LCR) is the intramolecular and concerted coupling (bond formation) of two ligands (substituents) which are bonded directly in σ-type bond to a hypervalent main group element. During the LCR, the valence of the central atom (M) decreases by 2, that is, $[(M^{(N+2)} \rightarrow M^{(N)}]$, and the stereochemistry of coupling substituents are retained. Therefore, the reaction does not proceed via any intermediate such as radicals.

There are three cases of hypervalent compounds undergoing ligand coupling: (i) stable compound; (ii) generated in situ by the substitution of σ-type ligand with nucleophile; (iii) generated in situ by the addition of a nucleophile to the double bond. These are illustrated as Eqs. (11.1–11.3). It is convenient to indicate two coupling ligands using an arc of a circle which are on the central atom $[M^{(N+2)}]$ of the hypervalent state [1].

$$\text{Nu–}M^{(N+2)}(X)L_3 \longrightarrow M^{(N)}L_3 + \text{NuX} \quad (11.1)$$

Organo Main Group Chemistry, First Edition. Kin-ya Akiba.

$$Nu^{\ominus} + L_nM^{(N+2)}\begin{smallmatrix}X\\Y\end{smallmatrix} \longrightarrow X^{\ominus} + L_nM^{(N+2)}\begin{smallmatrix}Nu\\Y\end{smallmatrix} \longrightarrow NuY + L_nM^{(N)} + X^{\ominus} \tag{11.2}$$

$$Nu^{\ominus} + L_nM^{(N+2)}(=X)Y \longrightarrow L_nM^{(N+2)}(X^{\ominus})\begin{smallmatrix}Nu\\Y\end{smallmatrix} \longrightarrow NuY + L_nM^{(N)}X^{\ominus} \tag{11.3}$$

In transition metal chemistry, technical terms of reductive elimination and oxidative addition are commonly used in order to express the apparent elimination and addition of ligands accompanied by the change of valence of the central metal. In these cases, reaction mechanism is often not precisely defined and can be ambiguous. Concerted reaction and also reactions proceeding through intermediate(s) such as radicals are involved. Moreover, in organic chemistry, reductive elimination is used to express a variety of reactions such as the elimination of hydrogen halide from a haloalkane with base and the elimination of halogen from 1,2-dihaloalkane with metals, and so on. Therefore, LCR is a new type of reaction and is different from the widely used reductive elimination, and is unique and specific for hypervalent compounds in general.

11.2 SELECTIVITY OF LIGAND COUPLING REACTION: THEORETICAL INVESTIGATION

One of the typical and fundamental examples of hypervalent compounds is 10-M-5 of group 15 elements and it was already explained in detail that the compound is trigonal bipyramid in general. The selectivity of the LCR of PR_5 (R = H; 10-P-5) was theoretically investigated on the basis of the conservation of orbital symmetry by Hoffmann. For a trigonal bipyramid (D_{3h} structure), the coupling between an apical group and an equatorial group is symmetry-forbidden. On the other hand, both equatorial–equatorial and apical–apical couplings are symmetry-allowed (Fig. 11.1) [2,3]. However, for PR_5 of C_{4v} square pyramid, coupling between an apical (L_1) group and basal (one of L_2–L_5) one is allowed. As for coupling between basal ligands, that of trans basal is symmetry-allowed and cis basal is symmetry-forbidden (Fig. 11.2).

For group 16 elements bearing lone pair electrons (C_{4v}), the most favorable transition structure is highly polarized with C_1 symmetry. This can be considered as a distorted trigonal bipyramid, in which an apical ligand and an equatorial ligand couple together and the third equatorial position is occupied by the lone pair of M (Fig. 11.3) [4]. For LCRs of MH_5 (M = P, As, Sb, Bi) and MH_4 (M = S, Se, Te), detailed theoretical investigations have been reported [5]. On the basis of the results, it is concluded that, for group 15 elements, equatorial–equatorial coupling is favored for three elements and apical–equatorial coupling is favored only for bismuth with a highly polarized transition structure.

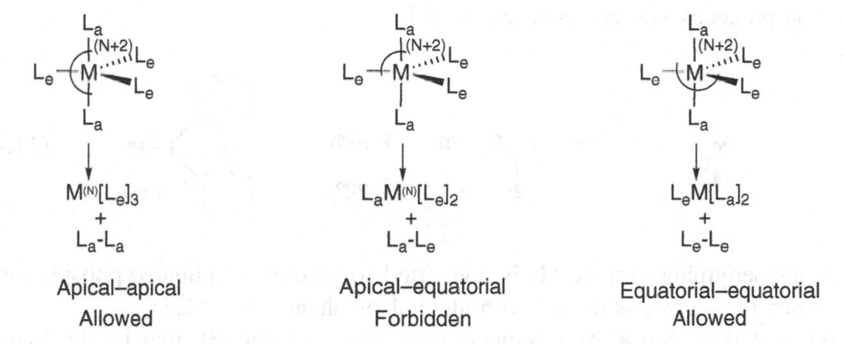

Figure 11.1 Selectivity of ligand coupling reaction of PR_5 (M = P, trigonal bipyramid) by Woodward–Hoffmann rule.

L_1-L_4 Apical–basal Allowed

L_2-L_4 Trans–basal Allowed

L_3-L_4 Cis–basal Forbidden

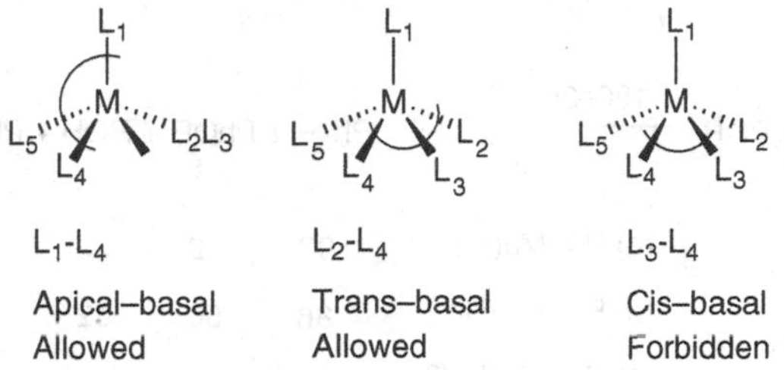

Figure 11.2 Selectivity of ligand coupling reaction of PR_5 (C_{4v} square pyramid) by Woodward–Hoffmann rule.

C_{4v} C_1 C_{2v}

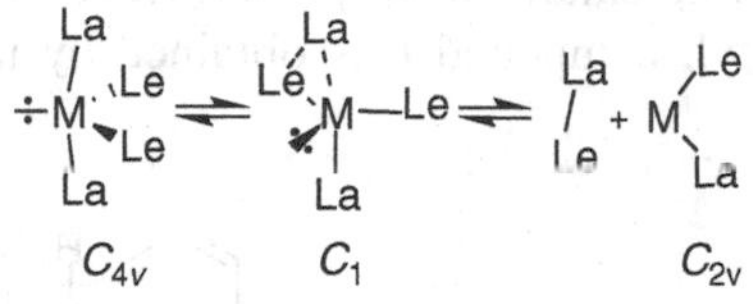

Figure 11.3 Selectivity of ligand coupling reaction of group 16 elements: apical–equatorial coupling is predicted through polar transition state.

11.3 LIGAND COUPLING REACTION OF ORGANIC COMPOUNDS OF PHOSPHORUS, ANTIMONY, AND BISMUTH

Pyrolysis of pentaphenylphosphorane (**1**) afforded the biphenyl, triphenylphosphine, and biphenylene derivative and benzene (Eq. 11.4). When the pyrolysis was carried out with styrene, polymerization took place to show the presence of radicals as intermediates, and the formation of a considerable amount of benzene should indicate the presence of the phenyl radical, thus suggesting that the main reaction proceeds via radical reactions [1].

$$\underset{\mathbf{1}}{Ph_5P} \xrightarrow{130} \underset{22\%}{Ph_3P} + Ph\text{-}H + \underset{22\%}{Ph\text{-}Ph} + \underset{10\%}{\text{(dibenzophosphole)P-Ph}} \qquad (11.4)$$

Pentaphenylphosphorane (**1**) is converted to alkoxy- or phenoxyphosphorane (**2**), when the former is heated with alcohol or phenol (Eq. 11.5).

When **2** is heated at high temperatures, phenyl ether (**3**), that is, the ligand coupling product, is obtained as the major product but radical reaction products such as alcohol and benzene are produced as byproducts (Eq. 11.6).

$$\underset{\mathbf{1}}{Ph_5P} + ROH \longrightarrow \underset{\mathbf{2}}{Ph_4POR} + PhH \qquad (11.5)$$

$$\underset{\mathbf{2}}{Ph_4POR} \xrightarrow{190\text{–}200\,°C} Ph_3P + \underset{\mathbf{3}}{PhOR} + ROH + PhH \qquad (11.6)$$

	Ph_3P	PhOR	ROH	PhH
(a) R = Me(%)	97	72	20	4
(b) R = *i*-Pr(%)	86	50	32	8
(c) R = $PhCH_2$(%)	95	91	–	11

Thermolysis of phosphoranes with two biphenylene ligands (**4**) resulted in trivalent phosphines (**5, 6**) (Eq. 11.7). When R is methyl or phenyl group, compound **5** is produced by ring enlargement (path a: LCR). With a bulky group such as 8-quinolyl or 9-anthryl, compound **6** is obtained by migration of R (path b: LCR).

(11.7)

4 → (a) 5; (b) 6

(a) R = Me, Ph ; (b) R = 8-Quinolyl, 9-anthranyl

Tetraphenylphosphonium bromide (**7**) reacted with *cis*- or *trans*-propenyllithium to yield *cis*- or *trans*-β-mehtylstyrene (**8**) in high yield (Eq. 11.8). It is remarkable that the LCR of the intermediate vinyltetraphenylphosphorane (**A, B**) proceeded with complete retention of stereochemistry. The proposed structures of **A** and **B**, however, were not investigated in detail.

$Ph_4\overset{\oplus}{P}\overset{\ominus}{Br}$ (**7**) — a: Me–CH=CH–Li (cis) → [A] → Ph–CH=CH–Me (*cis*-**8**); b: Me–CH=CH–Li (trans) → [B] → Ph–CH=CH–Me (*trans*-**8**) (11.8)

LCRs of bismuth compounds, among group 15 element compounds, are investigated the most and are valuable for application. This is based on the weak bond energy (143 kJ/mol) of bismuth–carbon, and the bismuth atom is large and easily takes two valence states of Bi(III) and Bi(V). Triphenyl(aryl)bismuth reacts with halogen easily to afford the stable pentavalent dihalotriphenyl(aryl)bismuth (**9**). Compound **9** is transformed with potassium carbonate to the cyclic carbonate (**10**) or with phenyllithium to pentaphenylbismuth (**11**). Compounds **10** and **11** are further converted to carboxylic acid derivative (**12**) or tetraphenylbismuthonium salt (**13**), and they are stable and used as reagents of pentavalent bismuth.

$Ph_3Bi + X_2 \longrightarrow Ph_3BiX_2$ (**9**) (**9** :(a)X = Cl; (b) X = Br)

9 → (c) PhLi → Ph_5Bi (**11**) → (e) HX → Ph_4BiX (**13**)

9 → (b) KF → Ph_3BiF_2

9 → (a) K_2CO_3 → Ph_3BiCO_3 (**10**) → (d) RCO_2H → $Ph_3Bi(OCOR)_2$ (**12**) (11.9)

Halobenzene is obtained according to the reaction of triphenylbismuth with mixed halogen, in which unstable pentavalent intermediates (**C, D**) are proposed to effect ligand coupling (Eq. 11.10). Stable and isolable derivatives of diazide (**14**) and dicyanide (**15**) give phenyl azide and benzonitrile thermally, and unstable dithiocyanate (**E**) yields phenylthiocyanate. These also proceed through LCRs of pentavalent bismuth compounds (Eq. 11.11).

$$Ph_3Bi \xrightarrow{(a)\ ICl} \left[Ph_3Bi(I)(Cl)\right]\ \mathbf{C} \longrightarrow PhI + Ph_2BiCl$$
$$Ph_3Bi \xrightarrow{(b)\ BrCN} \left[Ph_3Bi(Br)(CN)\right]\ \mathbf{D} \longrightarrow PhBr + Ph_2BiCN \qquad (11.10)$$

$$Ph_3BiCl_2\ (\mathbf{9a}) \xrightarrow{(a)\ NaN_3} Ph_3Bi(N_3)_2\ (\mathbf{14}) \xrightarrow{\Delta} PhN_3 + Ph_2BiN_3$$
$$Ph_3BiCl_2\ (\mathbf{9a}) \xrightarrow{(b)\ NaCN} Ph_3Bi(CN)_2\ (\mathbf{15}) \xrightarrow{\Delta} PhCN + Ph_2BiCN \qquad (11.11)$$
$$Ph_3BiCl_2\ (\mathbf{9a}) \xrightarrow{(c)\ NaSCN} \left[Ph_3Bi(SCN)_2\right]\ \mathbf{E} \longrightarrow PhSCN + Ph_2BiSCN$$

C-Phenylation of phenols is a difficult reaction, in general. This reaction was realized by the ligand coupling of the intermediate pentavalent bismuth compounds which are generated in situ by the reaction of phenols with **9** or **11** (Eq. 11.12). These proceed in high yield (75–85%). The yields increase when the reactions are carried out in the presence of a strong and bulky base (BTMG, *t*-butyltetramethylguanidine).

(a) 3,5-di-*t*-Bu-phenol $\xrightarrow{(a)\ Ph_3BiCl_2(\mathbf{9a}),\ BTMG}$ 2,6-diphenyl-3,5-di-*t*-Bu-phenol

(11.12)

(b) 2,6-dimethylphenol $\xrightarrow{(b)\ Ph_5Bi}$ 6-phenyl-2,6-dimethylcyclohexa-2,4-dienone

BTMG: *t*-butyltetramethylguanidine

Phenyl ether (O-arylation product) is usually obtained as a byproduct. The relative ratios of C-arylation and O-arylation were investigated by using 2-naphthol. The yields of the products are summarized in Table 11.1, with types of reagents (Ph_3BiXY) and conditions (neutral, acidic, and basic). Under neutral conditions with pentaphenylbismuth (**11**) or under basic conditions (with BTMG) with **13** or **9**, C-arylation is the major process (60–94%). Under acidic conditions,

TABLE 11.1 C-Phenylation and O-Phenylation of 2-Naphthol

2-Naphthol + Ph_3BiXY ⟶ 1-phenyl-2-naphthol + 2-phenoxynaphthalene

X	Y	Conditions	*C*-phenyl %	*O*-phenyl %
Ph	Ph	*a*	61	–
Ph	$OCOCF_3$	*a*	9	67
Ph	$OCOCF_3$	*b*	–	91
Ph	$OCOCF_3$	*c*	94	–
$OCOCF_3$	$OCOCF_3$	*c*	70	–
Cl	Cl	*c*	90	–

[a]Neutral conditions.
[b]Addition of trichloroacetic acid (0.6 eq.).
[c]Basic conditions using BTMG.

BTMG : $(Me_2N)_2C{=}N{-}t\text{-}Bu$ (Strong base)

O-arylation (67–91%) becomes the major route. The solvents are tetrahydrofuran (THF), benzene, and so on. [6].

C-Arylation takes place by ligand coupling at the 2-position of the phenoxy and 1-position of the phenyl groups of pentavalent bismuth. O-Arylation results from nucleophilic substitution at 1-position of the phenyl group with an external phenol, induced by the strong electron-withdrawing ability of an apical ligand (Fig. 11.4). In both cases, the high leaving ability of Ph_3Bi is supposed to play an important role (cf. Section 11.5 of the iodine compound).

On the basis of the mechanism of C-arylation, α-C-phenylation of a carbonyl group is expected to take place when an enolate generated from the carbonyl compound is trapped with Ph_3BiXY. α-C-Phenylation was investigated by using 2-ethoxycarbonylcyclohexanone **16**. The α-C-phenyl derivative **17** was obtained from **16** in good yield (57–91%) under neutral conditions with **10** (Ph_3BiCO_3) or **11** (Ph_5Bi) and under basic conditions with **9a** (Ph_3BiCl_2) or **13a** ($Ph_4BiOCOCF_3$). Under acidic conditions, the O-phenyl derivative (**18**) was the major product. These reactions proceed through intermediate (**F**) (Eqs. 11.13 and 11.14).

By employing 1,3-diketone (**19**), bismuth ylide (**21**) was obtained with **10** (Ph_3BiCO_3), and the diphenyl derivative (**20**) was afforded with **13b** (Ph_4BiOSO_2-*p*-Tol) under basic conditions. The reaction patterns of C-arylation and O-arylation are summarized in Fig. 11.5. It should be added here that **12** [$Ph_3Bi(OAc)_2$)] undergoes phenylation of heteroatoms of alcohol and amine at room temperature in high yield in the presence of copper powder or a copper catalyst [$Cu(OAc)_2$]. This is a useful synthetic method, although the detailed

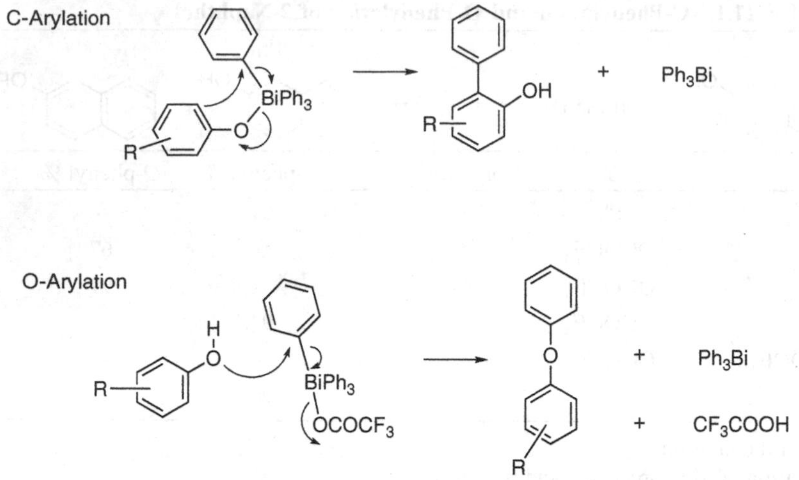

C-Arylation

$\text{Ar-O-BiPh}_4 \rightarrow$ 2-aryl phenol (R, OH) $+ \ Ph_3Bi$

O-Arylation

$\text{R-C}_6\text{H}_4\text{-OH} + \text{Ph-BiPh}_3\text{(OCOCF}_3) \rightarrow$ diaryl ether (R) $+ \ Ph_3Bi + CF_3COOH$

Figure 11.4 Mechanism of C-arylation and O-arylation.

reaction mechanism is not clear yet [6].

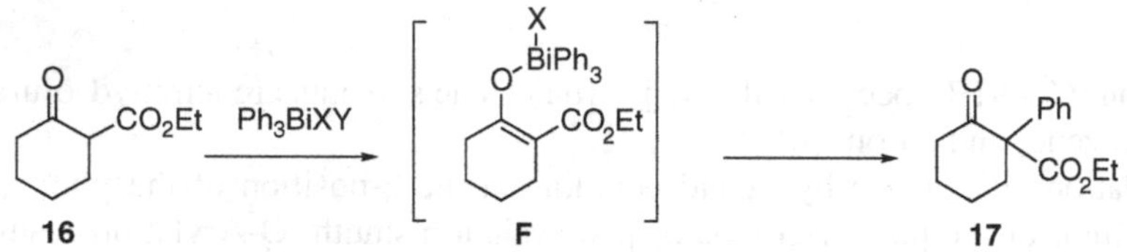

16 (2-CO_2Et cyclohexanone) $\xrightarrow{Ph_3BiXY}$ [**F** (enol-O-$Bi(X)Ph_3$, CO_2Et)] $\rightarrow$ **17** (2-Ph-2-CO_2Et cyclohexanone) (11.13)

Reagent	Yield
(a) Ph_3BiCO_3	73%
(b) Ph_3BiCl_2/BTMG	75%
(c) $Ph_4BiOCOCF_3$/BTMG	91%
(d) Ph_5Bi	57%

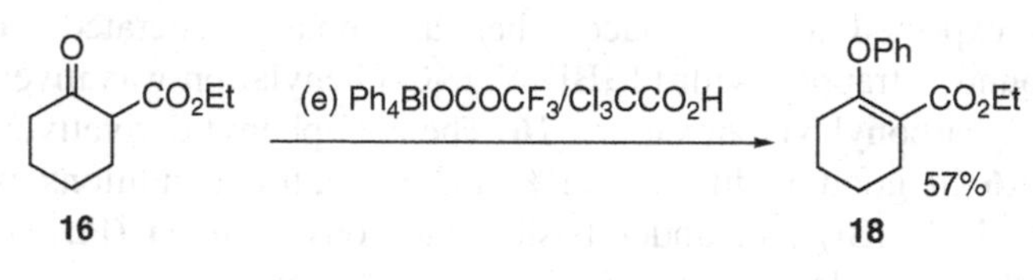

16 $\xrightarrow{\text{(e) } Ph_4BiOCOCF_3/Cl_3CCO_2H}$ **18** (OPh, CO_2Et) 57% (11.14)

19 (Me, Me) $\xrightarrow{\text{10 or 13b}}$ **20** (Ph, Ph; Me, Me) + **21** ($BiPh_3$; Me, Me) (11.15)

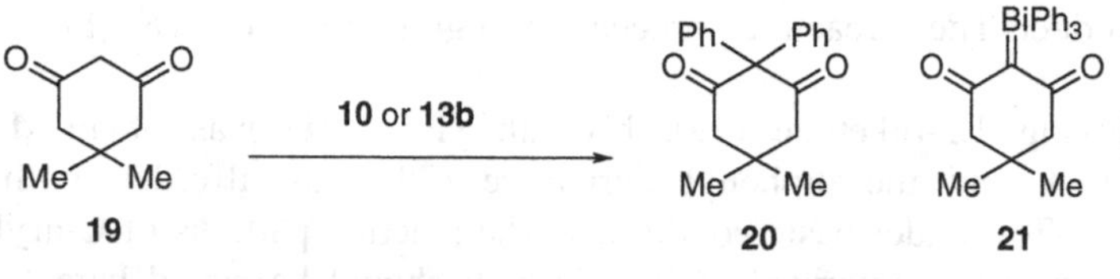

	20	21
10 Ph_3BiCO_3	–	75%
13b Ph_4BiOTs / BTMG	81%	–

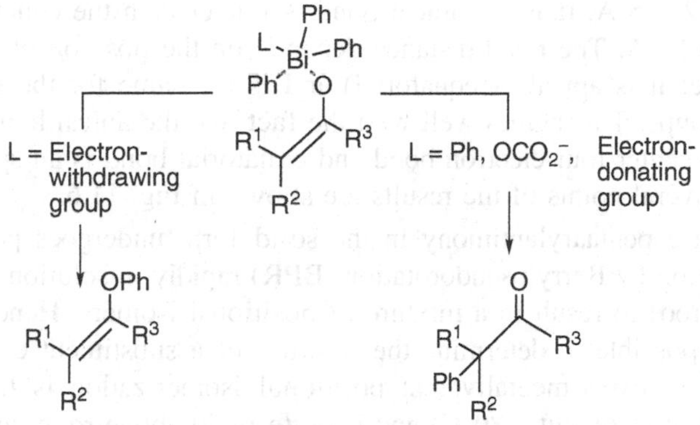

Figure 11.5 Selectivity of C-arylation and O-arylation.

The mechanism of LCR was investigated in detail by employing pentaarylantimony. Thermolysis of ^{14}C labeled pentaphenylantimony (**22**) in benzene afforded completely labeled biphenyl and triphenylantimony (Eq. 11.16). Styrene added to the solution did not polymerize at all. Hence, it is clear that the thermolysis is not a radical reaction but a concerted LCR (Eq. 11.16) [7].

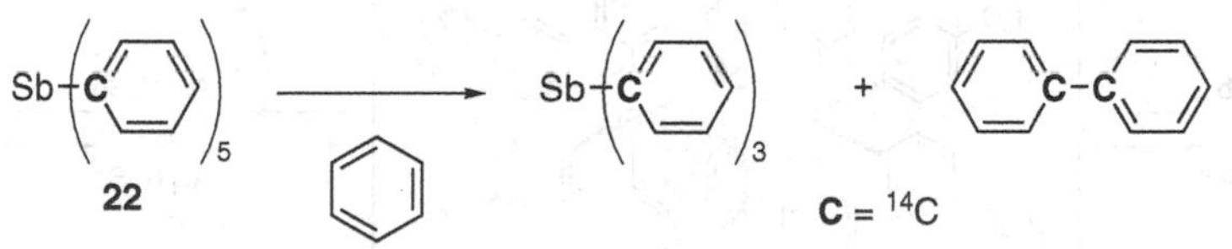

$$\mathrm{Sb(C_6H_5)_5}\ (\mathbf{22}) \xrightarrow{\text{benzene}} \mathrm{Sb(C_6H_5)_3} + \mathrm{C_6H_5{-}C_6H_5} \qquad \mathbf{C} = {}^{14}C \tag{11.16}$$

The selectivity of the LCR is determined by using substituted pentaarylantimony. The following experiments were performed.

1. All kinds of mixed pentaarylantimony (**23–28**) were synthesized as pure as possible.

Tol_5Sb	$ArTol_4Sb$	Ar_2Tol_3Sb	Ar_3Tol_2Sb
23	**24**	**25**	**26**
Ar_4TolSb	Ar_5Sb		
27	**28**		

Tol = p-$CH_3C_6H_4$, Ar = p-$CF_3C_6H_4$

2. Structure determination of all the mixed pentaarylantimony was carried out by X-ray analysis. It was clarified that all of them are almost perfect trigonal bipyramids. An electron-withdrawing group (Ar: p-$CF_3C_6H_4$) stays predominantly at the apical position. The bond distance of the apical bond is in the range of 2.242–2.263 Å and that of equatorial one is

2.140–2.155 Å; thus the apical bond is longer than the equatorial one by 0.10–0.15 Å. The bond distance depends on the position of a substituent (whether it is apical or equatorial) and is the same for the tolyl and the aryl groups. This agrees well with the fact that the apical bond consists of a three-center four-electron bond and equatorial bond is an sp^2 hybridized one. Several points of the results are shown in Fig. 11.6.

3. Any pure pentaarylantimony in the solid form undergoes positional isomerization by Berry pseudorotation (BPR) rapidly in solution (of course in vapor, too) to result in a mixture of positional isomers. Hence, it is actually impossible to determine the position of a substituent exactly. It was confirmed experimentally that positional isomerization is frozen at low temperatures (about −80°C) and is quite rapid above room temperature.
4. When pure **25** is heated in benzene (about 60°C), it results in a mixture of all kinds of pentaarylantimony (**23–28**), which should be due to the ligand exchange reaction (LER). LER is certainly unexpected because they (**23–28**) have no lone pair electrons. Then, LER should be specific for the hypervalent bond.

Tol_5Sb + Ar_5Sb ⇌ [dimeric Sb···Sb bridged intermediate; labels: CH_3, CF_3, H_3C, Sb, Sb] ⇌ $ArTol_4Sb$ + Ar_4TolSb ⇌ ...

Tol = p-$CH_3C_6H_4$, Ar = p-$CF_3C_6H_4$

(11.17)

5. LCR takes place at around 160°C in benzene in a sealed tube. Therefore, it is impossible to determine the selectivity of LCR of a certain pentaarylantimony in solution because equilibration among all kinds of pentaarylantimony takes place at much lower temperatures. In the search for a catalyst to accelerate LCR but not LER, LiTFPB [$(3,5\text{-}(CF_3)_2C_6H_3)_4B^-$ Li^+] and $Cu(acac)_2$ were found to be effective. In the presence of such a catalyst, the LCR, which can certainly be ascribed to pure pentaarylantimony, was obtained. But still there is uncertainty whether the selectivity really corresponds to a pure material or not.
6. The selectivity of LCR by flash vacuum thermolysis (FVT) has been determined. To a vacuum system, a stream of vapor of a pure sample of pentaarylantimony was introduced with argon under reduced pressure (about 10^{-3} mm Hg) by heating (about 40°C, conditions under which LER cannot be detected) in a furnace heated at 300°C. The products of the LCR

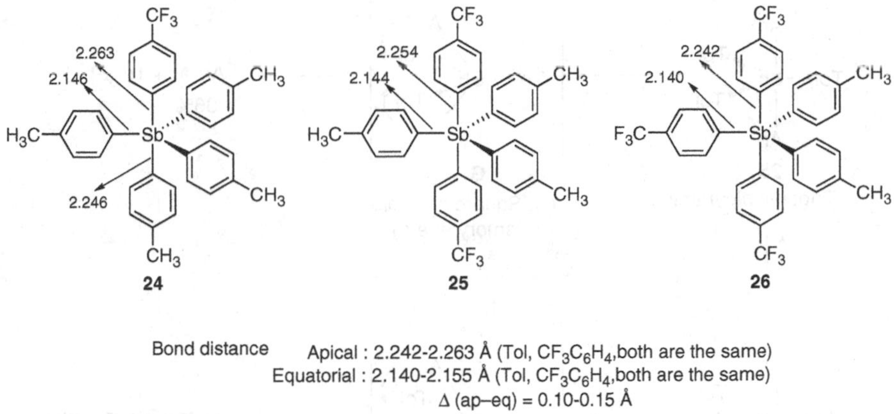

Figure 11.6 Structure of pentaarylantimony by X-ray analysis, which shows an almost perfect trigonal bipyramid (tbp).

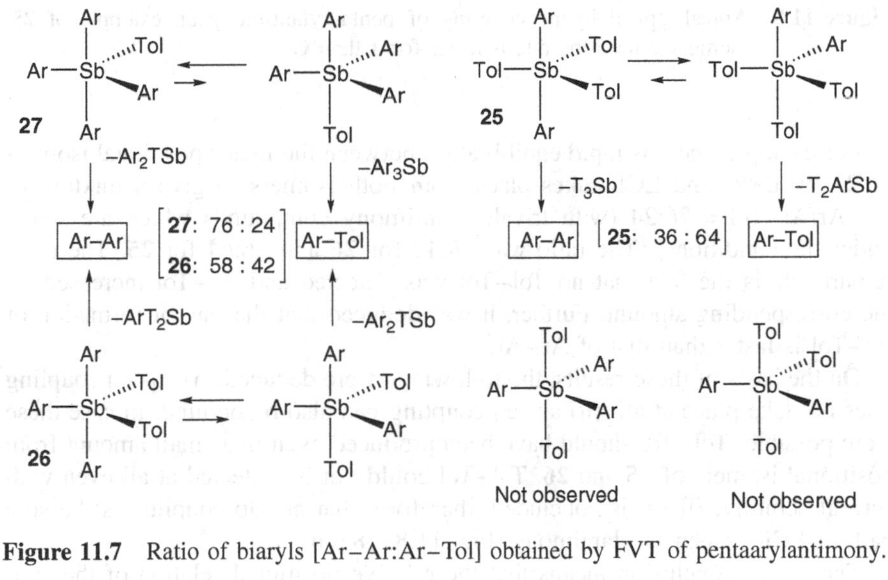

Figure 11.7 Ratio of biaryls [Ar–Ar:Ar–Tol] obtained by FVT of pentaarylantimony.

of a pure compound (of course under BPR) were determined quantitatively by trapping in a cold finger attached to the end of the vacuum system. The products were a mixture of Ar–Ar, Ar–Tol, and trivalent antimony. Surprisingly, Tol–Tol could not be detected at all. The results are illustrated in Fig. 11.7.

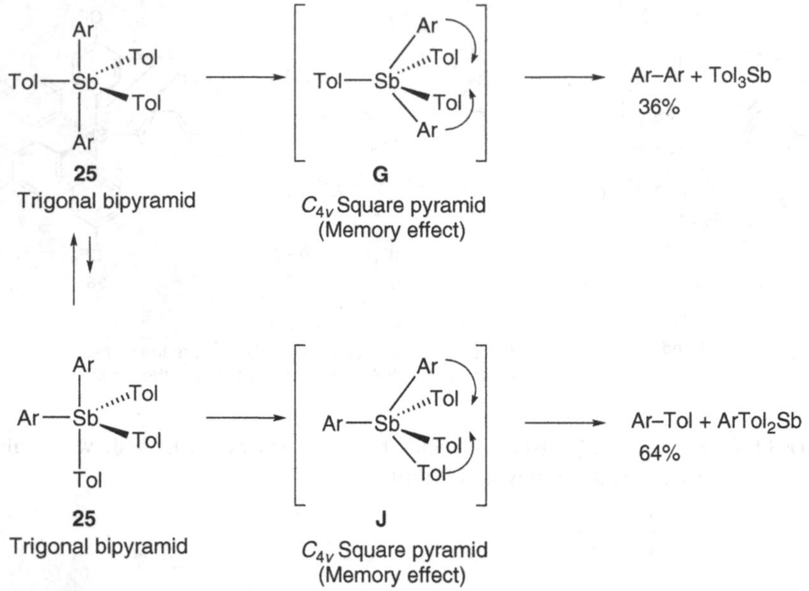

Figure 11.8 Apical–apical ligand coupling of pentaarylantimony an example of **25**: memory effect (the rate is faster for **J** than **G**).

For example, there is rapid equilibration between the major positional isomers of **27** by BPR, and LCR takes place from both isomers to give a mixture of Ar–Ar:Ar–Tol = 76:24 (with trivalent antimony compounds which are stable under the conditions). The ratio was 58:42 for **26** and 36:64 for **25**. The most remarkable is the fact that no Tol–Tol was detected and Ar–Tol increased by the corresponding amount. Further, it was deduced that the rate of formation of Ar–Tol is faster than that of Ar–Ar.

On the basis of these results, the followings were deduced: (i) eq–eq coupling does not take place at all; (ii) ap–eq coupling can also be omitted. In case these were possible, Tol–Tol should have been produced even in a small amount from positional isomers of **25** and **26**. Tol–Tol could not be detected at all even with careful scrutiny; (iii) it is concluded, therefore, that ap–ap coupling is the sole path of LCR of pentaarylantimony (Fig. 11.8) [8a, b].

The above conclusion means that the relative positional relation of the substituents is retained at the transition state (C_{4v} square pyramid structure) to proceed to the products: that is, once apical substituents start the bending motion to effect LCR, they go to the products. This can be called as a *memory effect*. This can be easily seen intuitively. Apical hypervalent bond (three-center four-electron bond) is longer and weaker than the sp^2-hybridized equatorial bonds. This is essentially the same as the kinetic isotope effect but appears much more effective because the difference in the character between apical and equatorial bonds is more essential than in the case of isotopes. The present conclusion on the selectivity of LCR of pentaarylantimony is different from the prediction of theoretical investigation on SbH_5 (eq–eq coupling was favored). This discrepancy would surely be resolved by future theoretical calculation using pentaarylantimony.

11.4 LIGAND COUPLING REACTION OF ORGANIC COMPOUNDS OF SULFUR, SELENIUM, AND TELLURIUM

Tetraarylchalcogens (10-M-4; Ar_4M: M = S, Se, Te) are rather unstable thermally compared to pentaarylnictogens (10-M-5; Ar_5M: M = P, As, Sb, Bi). Their stability increases according to the descending period (stability: Te > Se, S). Tetraarylsulfurane (**I**) was reported to be stable below 0°C and led to ligand coupling, as observed by ^{19}F NMR measurement. In order to stabilize **I**, the pentafluorophenyl group (electron-withdrawing group) was used (Eq. 11.18). In general, LCR proceeds almost quantitatively when sulfurane (**J**) is generated as an intermediate of the reaction of triarylsulfonium salt (**29**) and organolithium reagent (Eq. 11.19) [1,9].

$$C_6F_5SF_3 + 3C_6F_5Li \longrightarrow \quad SF_4 + 4C_6F_5Li \longrightarrow \left[(C_6F_5)_4S \right] (\mathbf{I}) \longrightarrow C_6F_5\text{-}C_6F_5 + (C_6F_5)_2S \tag{11.18}$$

$$\underset{\mathbf{29}}{Ar_3M^+X^-} + ArLi \longrightarrow \left[Ar_4M \right] (\mathbf{J}) \longrightarrow Ar\text{-}Ar + Ar_2M \tag{11.19}$$

(**a**) M = S, (**b**) M = Se

The fact that the intermediate is unstable means that a variety of reactions can proceed under mild conditions; consequently, many LCRs via sulfuranes have been investigated. A simple example is given in Eq. (11.20). But as in Eq. (11.21), it is apparent that LER occurs at **J** to yield different sulfuranes because three kinds of biaryls and two sulfides are obtained (Eq. 11.21). Selectivity of LCR of different kinds of aryl groups depends on the combination of aryls and the selectivity of LER. The selectivity of LER, however, is not the same as that of LCR, and therefore it is generally difficult to obtain a desired product in high yield by LCR. A ligand coupling of the biphenylene compound is an example (Eq. 11.22). As in Eq. (11.23), however, LCR is usefully applied as a method to introduce a phenyl group to heterocycles. Phenylation also proceeds for vinyllithiums and the stereochemistry of the vinyl group is retained

(Eq. 11.24). The result is the same as that of the phosphonium salt mentioned earlier (Eq. 11.8).

$$Tol_3S^+ BF_4^- + TolLi \longrightarrow \underset{100\%}{Tol\text{-}Tol} + \underset{100\%}{Tol\text{-}S\text{-}Tol} \quad (11.20)$$

$$Tol_3S^+ BF_4^- + PhLi \longrightarrow \underset{24\%}{Ph\text{-}Ph} + \underset{40\%}{Tol\text{-}Ph} + \underset{6\%}{Tol\text{-}Tol} + \underset{40\%}{Tol_2S} + \underset{3\%}{Tol\text{-}S\text{-}Ph} \quad (11.21)$$

(11.22)

(**a**) M = S, (**b**) M = Se

$Ph_3S^+ CF_3SO_3^-$ (11.23)

a: 70%, + Ph-S-Ph

b: 42%, + Ph-S-Ph

$Ph_3S^+ BF_4^-$ (11.24)

a: 63%, + Ph-Ph 38%

b: 67%, + Ph-Ph 6%

Sulfoxide can be prepared stably easily and its stereochemistry is stable and clarified well. The reaction of sulfoxide with organometallics is not so simple and there are several possibilities including BPR, LER, and LCR (Eq. 11.25). Optically active 1-phenylethyl-2-pyridyl sulfide (**30**) converts to sulfoxide (**31**) by oxidation with hydrogen peroxide. By addition of the Grignard reagent to **31**, 2-phenylethylpyridine (**32**) is obtained in high yield, keeping the optical activity of the phenylethyl group 100%. The mechanism of this reaction is explained as follows. Firstly, sulfurane (**K-a**) bearing a methyl group at an apical position is generated by the addition of methylmagnesium bromide to the sulfoxide (**31**) and then it is converted to **K-b** having 2-phenylethyl group at an apical position by BPR. LCR of apical–equatorial groups takes place at **K-b** to yield **32**, in which stereochemistry of the phenylethyl group is retained (Eq. 11.26) [9].

$R^1R^2S\rightarrow O + R^3M \rightarrow$ [$R^1R^2R^3S(OM)$ ⇌ φ ⇌ φ $R^3R^2R^1S(OM)$]

LER: $R^1R^3S\rightarrow O + R^3R^2S\rightarrow O + R^2M + R^1M$

LCR:
a: $R^1\text{-}R^2 + R^3\text{-S-OM}$
b: $R^1\text{-}R^3 + R^2\text{-S-OM}$
c: $R^2\text{-}R^3 + R^1\text{-S-OM}$

(11.25)

30 (S)-(−) $[\alpha]^{25}$D −375° —(H_2O_2/AcOH)→ **31** 91% S_SC_S ($+S_RC_S$) —(CH_3MgBr)→ [**K-a** ⇌ φ ⇌ φ **K-b**] **K** → (MeSOMgBr) + **32** 97%, *ee* = 100% (S)-(+) $[\alpha]^{25}$D +63°

(11.26)

Di(2-pyridyl) sulfoxide + EtMgBr, or 2-pyridyl methyl sulfoxide ($S-CH_3$) + 2-pyridyllithium (Li) → [**L**] → **33** + (RSOM)

(11.27)

According to a similar type of ligand coupling, 2,2′bipyridyl (**33**) is prepared through an intermediate (**L**) in about 70% yield. Stereochemistry of the vinyl group is completely retained by the reaction of 2-pyridyl vinyl sulfoxide with ethyl magnesium chloride (Eq. 11.28).

trans-vinyl —EtMgBr→ *trans*-vinyl

cis-vinyl —EtMgBr→ *cis*-vinyl

(11.28)

Thermolysis of tetramethyltellurium yields dimethyl telluride, ethane, and methane and the reaction is shown to involve radicals including methyl radical as an intermediate. On the other hand, tetraphenyltellurium undergoes LER with *t*-butyl thiol to generate intermediate **M** and LCR occurs at **M** to give diphenyl telluride and di-*t*-butyl disulfide [10a].

$$Me_4Te \longrightarrow Me_2Te + Me\text{–}Me + Me\text{–}H \quad (11.29)$$

$$Ph_4Te + 2t\text{-}BuSH \xrightarrow{-2PhH} [\mathbf{M}] \longrightarrow Ph_2Te + t\text{-}Bu\text{-}S\text{-}S\text{-}t\text{-}Bu \quad (11.30)$$

When tetraphenyltellurium is heated in toluene at around 80–140°C, biphenyl and diphenyl telluride are obtained quantitatively. The yield of the products does not decrease in the presence of furan or styrene, and styrene does not polymerize at all. Hence, it is confirmed that the reaction does not involve any radical reaction but is concerted LCR (cf. Eqs. 11.16, 11.26).

When a mixture of two kinds of tetraaryltelluriums (**34, 35**) is heated in solution, three kinds of biaryls including a mixed biaryl (Ar^1–Ar^2) are obtained quantitatively with three kinds of diaryl tellurides (Eq. 11.31). The formation of mixed tetraaryltellurium (**36, 37**, etc.) is observed in solution when the above reaction is carried out at lower temperatures (about 60°C). It is seen that all kinds of possible tetraaryltelluriums are formed by LER and then they are thermally decomposed by LCR, just like in the case of pentaarylantimony mentioned in

Section 11.3. It is believed that lone pair electrons play an important role in LER, but the electronic state of the intermediate (**N**) is not certain [10b]. There has been no investigation yet on the selectivity of LCR of tetraaryltellurium, that is, eq–eq, ap–ap, or ap–eq coupling. LCR of organoselenium compounds can be understood similar to sulfur and tellurium compounds; however, it is not as important as should deserve a mention here.

34 + **35** ⇌(LER) [**N**] ⇌

36 + **37** ⇌ ... —LCR→ Ar^1-Ar^1 + Ar^1-Ar^2 + Ar^2-Ar^2 + $(Ar)_2Te$

Ar = Ph, C_6D_5, p-$CH_3C_6H_4$

(11.31)

11.5 LIGAND COUPLING REACTION OF ORGANOIODINE COMPOUNDS

A mixture of iodoarene and halogenoarene results when diaryl iodonium salt is heated at high temperatures in solution or in the molten state. This is not useful as a synthetic reaction but is an example of LCR of apical–equatorial ligands at a highly polarized trivalent iodine intermediate (**O**). This is elucidated by determining the thermal decomposition products of the mixed diaryl iodonium salt [1].

$Ar_2I^+ X^-$ —Δ→ [**O**] ——→ ArX + ArI (11.32)

X = Cl, Br, I

The structure of diaryl iodonium salt is a distorted trigonal pyramid and it is called *iodane* with the valence of iodine of 3. (p-Chlorophenyl)(p-methylphenyl)(bromo)iodane (**38**) undergoes BPR rapidly in solution, and **38a** bearing the p-chlorophenyl group is favored over **38b** having the p-methylphenyl group at the apical position. But in the transition state of LCR,

P-b is favored than **P-a** because a partial negative charge appearing at bromine is more stabilized by equatorial *p*-chlorophenyl group and a partial positive charge appearing at iodine is better stabilized with the *p*-methylphenyl group. Therefore, the reaction path through **P-b** plays the major role and a combination of *p*-bromochlorobenzene and *p*-methyliodobenzene becomes the major product [11].

38a P-a Minor path 38b P-b Major path ψ Δ

(11.33)

LCR is accelerated when a benzene ring has ortho substituents (**39**), and it is realized that steric factor is also important along with electronic factor at a transition state (**O**) (Eq. 11.34).

39 X = Cl 95 : 5; Br 96 : 4; I 84 : 16 + (PhI + PhX)

(11.34)

Bromide can substitute tetrafluoroborate and there are two kinds of isomers (**41a, b**) that give vinyl bromide and vinyl iodide, respectively (Eq. 11.35). Vinyliodane (**42**) reacts with lithium cuprate to afford the substituted vinyl compound (Eq. 11.36). Stereochemistry of vinyliodane is retained by the substitution (Eq. 11.37).

C-Phenylation using diphenyliodonium salts are reported with enolates (Eqs. 11.38–11.40). The vinyl group and the phenyl group compete for the substitution, and the vinyl group is favored over the phenyl group (Eq. 11.40).

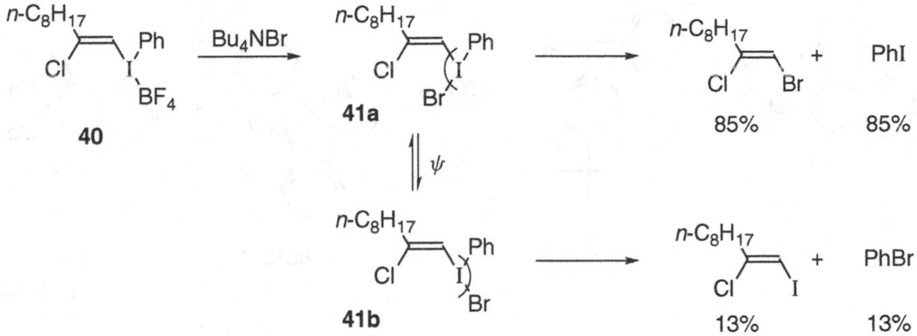

(11.35)

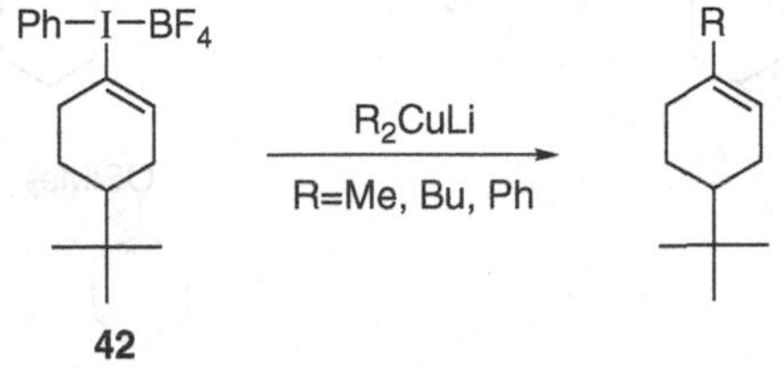

(11.36)

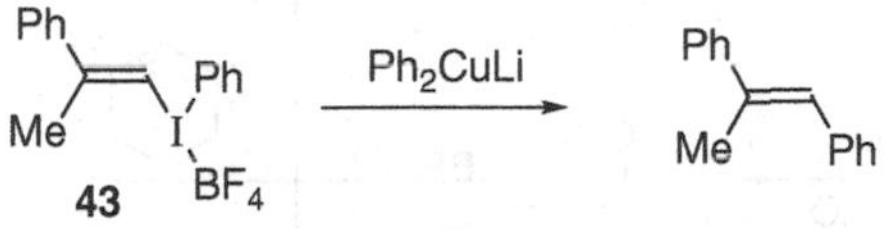

(11.37)

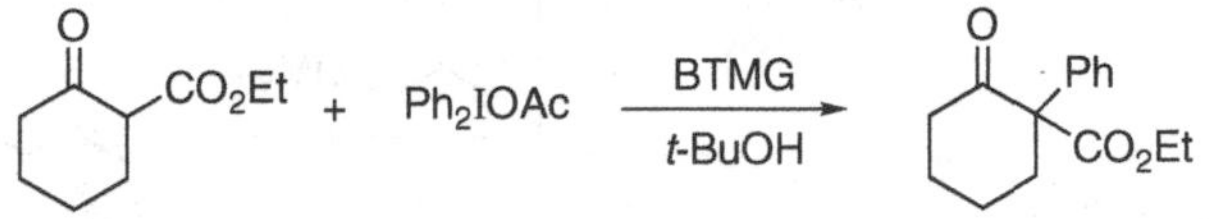

(11.38)

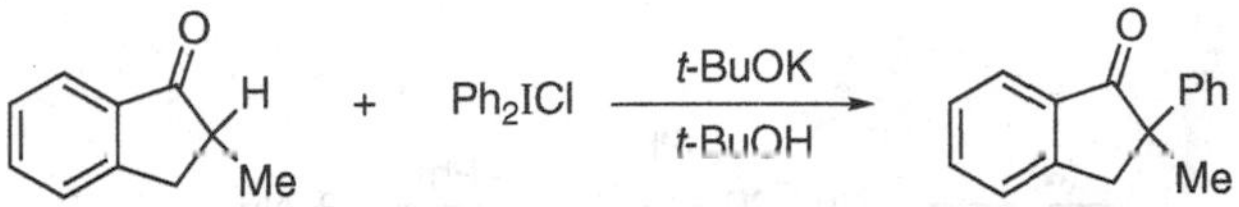

(11.39)

Silyl enol ether is also phenylated with Ph_2IF, and silyl enol ether reacts with iodosobenzene to yield phenacyliodane (**Q**) and **Q** is used for further conversions (Eq. 11.42). These are examples of LCRs through the diphenyliodane intermediate. Similar arylation using bismuth compounds have already been mentioned (Eqs. 11.13–11.15, Fig. 11.5).

Thus, it is apparent that the reactivity of bismuth (V) and iodine (III) resembles well. This is due to the extremely large leaving group ability of bismuth (III) and iodine (I) and also due to the easy conversion of Bi(V) to Bi(III) and I(III) to I(I).

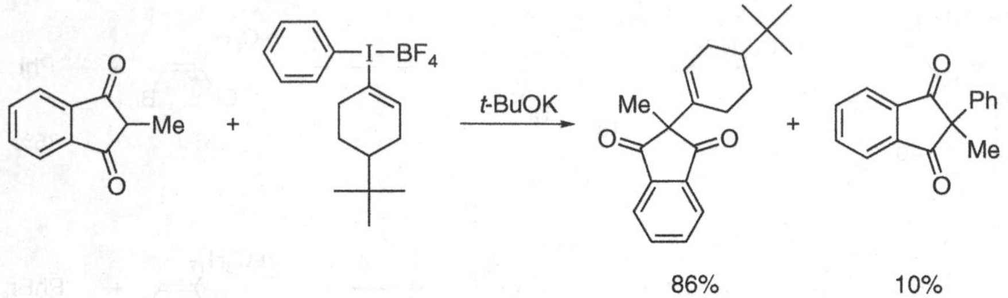

86% 10%

(11.40)

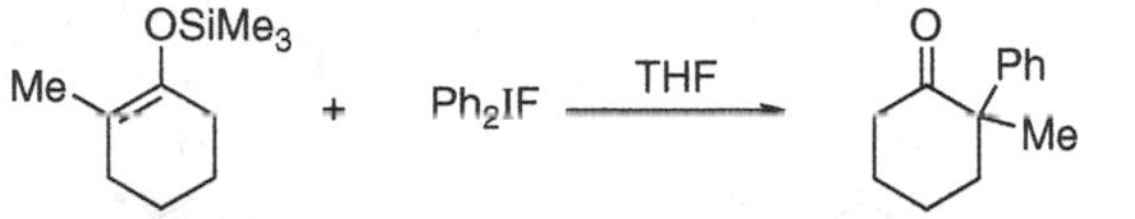

(11.41)

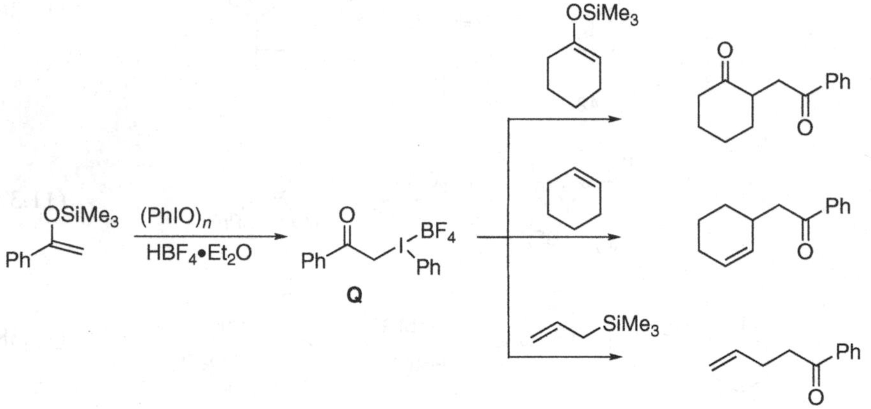

(11.42)

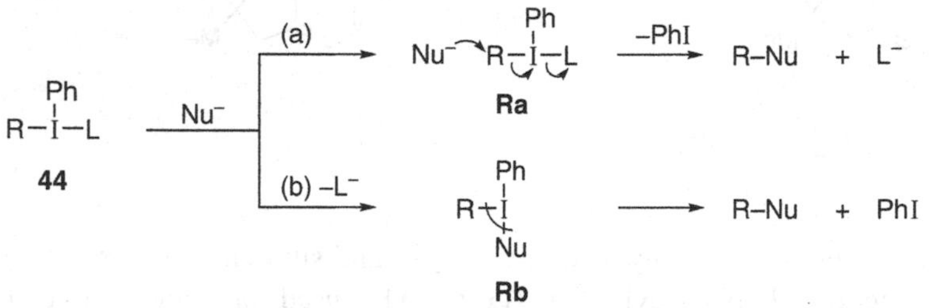

(11.43)

Finally, we add here as a comment that it is difficult to determine and scrutinize the mechanism of the reaction of iodane (**44**) with a nucleophile and whether it proceeds through S_N2 type substitution (**Ra**) or ligand coupling at the iodane (**Rb**) (Eq. 11.43). As an example, oxidative coupling of butyllithium to afford octane is mentioned, which is effected by (diacetoxy) phenyl iodane (**45**) (Eq. 11.44). It is difficult to see whether the first intermediate (**S**) results in the product directly by S_N2 type substitution or through the second intermediate (**T**) by LCR. Anyway, it is apparent that LCR is interesting, useful, and unique for hypervalent compounds.

OAc Ph—I OAc **45** —BuLi→ Ph—I(OAc)(BuLi) **S** —(−LiOAc, −PhI)→ octane; via **T** (Ph—I with two butyl groups) (11.44)

REFERENCES

1. Finet J-P. Volume 18, Ligand Coupling Reactions with Heteroatomic compounds, Tetrahedron organic chemistry series. Pergamon, Oxford, UK (Elsevier Science Ltd.); 1998.
2. Hoffmann R, Howell JM, Mutterties EL. J Am Chem Soc 1972;94:3047.
3. Kutzelnigg W, Wasilewski J. J Am Chem Soc 1982;104:953.
4. Moc J, Dorigo AE, Morokuma K. Chem Phys Lett 1993;204:65.
5. Moc J, Morokuma K. J Am Chem Soc 1995;117:11790.
6. Finet J-P. Chem Rev 1989;89:1487.
7. Shen K, McEwen WE, Wolf A. J Am Chem Soc 1969;91:1283.
8. (a) Akiba K.-y. Pure Appl Chem 1996;68:837; (b) Schröder G, Okinaka T, Mimura Y, Watanabe M, Matsuzaki T, Hasuoka A, Yamamoto Y, Matsukawa S, Akiba K.-y. Chem Eur J 2007;13:2517.
9. Oae S. Acc Chem Res 1991;24:202.
10. (a) Barton DHR, Glover SA, Ley SV. J Chem Soc Chem Commun 1977:266; (b) Glover SA. J Chem Soc [Perkin 2] 1980:1338.
11. Ochiai M. Top Curr Chem 2003;224:5.

[illegible] directly by [illegible] substitution or through the second intermediate (T) by LCR. Anyway, [illegible] that LCR is interesting [illegible] useful [illegible] technique for hypervalent [illegible] compounds.

(11.4.1)

REFERENCES

[illegible] Akiba KY, [illegible] Hypervalent [illegible] with [illegible] compounds [illegible] organic chemistry [illegible] Pergamon (Oxford) Ltd. (Elsevier Science Ltd.) [illegible]

2. Hoffmann R, [illegible] Am Chem Soc 1972;94:4047 [illegible]

3. Kutzelnigg W. [illegible] Chem [illegible]

4. [illegible] J Chem Phys [illegible]

5. Moc [illegible] J Am Chem Soc 1995;117:11790.

6. [illegible] Chem Rev [illegible]

7. Seu[illegible] K, McBean [illegible] A. [illegible] J Am Chem Soc [illegible]

8. [illegible] Kaw[illegible] Pure Appl Chem [illegible] Okuda T, [illegible] Matsukawa S, Akiba K-y. [illegible]

9. [illegible] Acc Chem Res [illegible]

10. [illegible] Martin [illegible] J Chem Soc Chem Commun 1977;269 [illegible]

11. [illegible] SA [illegible]

12. [illegible] Chem [illegible]

NOTES 9

HEXAVALENT ORGANOTELLURIUM COMPOUNDS

Tetraphenyltellurium (**1a**, Ph_4Te, tellurane) undergoes ligand coupling reaction (LCR) at 80–150°C and yields diphenyl telluride (Ph_2Te) and biphenyl (Ph_2) (cf. Eq. 11.31) [1]. Thermal instability of compound **1** would be ascribed to the presence of unshared electron pair and a slightly distorted trigonal bipyramidal structure. Then, can hexavalent organotellurium compounds (pertellurane) be prepared by oxidation of unshared electron pair? Actually, tetramethyltellurium (tellurane) and hexamethyltellurium (pertellurane) were prepared and their structures have been determined already (Eq. N9.1) [2,3].

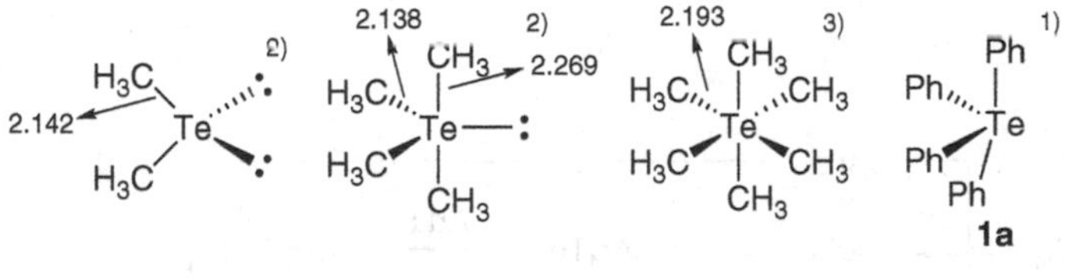

2.142 2) 2.138 2) 2.269 2.193 3) 1) **1a** (N9.1)

Hexaphenyltellurium (**5a**) and hexa(p-trifluoromethylphenyl)tellurium (**5b**) were prepared recently and their structures were determined by X-ray analysis [4]. Syntheses of aromatic organotellurium compounds are summarized in Figure N9.1. Pentaaryltellurium anion **2** is in equilibrium between tetraaryltellurium **1** and aryllithium in solution; however, its halide **3** is stable [5]. Stable pentaaryltellurium cation **4** was obtained as a result of abstraction of a halide ion with silver salt [6]. Hexaaryltellurium was obtained as a stable compound

Organo Main Group Chemistry, First Edition. Kin-ya Akiba.

by reaction of **4** with aryllithium. Although pentaaryltellurium anion **2** is quite unstable thermally, it was crystallized at low temperature for X-ray analysis.

$$[Ar_5Te]^+TfO^- + RLi \longrightarrow Ar_5TeR \quad (N9.2)$$

4a, b → **7**

Ar: **a** = Ph; **b** = *p*-$CF_3C_6H_4$

R: **a** = Ar; **b** = CH_2=–CH; **c** = Me; **d** = Bu

Although hexaphenyltellurium (**5a**) reacted with sulfuryl chloride (SO_2Cl_2) easily at room temperature to give the corresponding halide (**3a**), hexa(*p*-trifluoromethylphenyl)tellurium (**5b**) did not react with SO_2Cl_2, Br_2, XeF_2, etc., at all and was recovered intact. Moreover, **5b** was inert for organolithiums and reducing agents. Carbon-tellurium bond of **5b** was cleaved with potassium graphite (KC_8) to give pentaaryltellurium anion (**6b**). The anion (**6b**) reacted with a variety of electrophiles and afforded monomethyl derivative (**7c**) quantitatively [7].

$$Ph_6Te + SO_2Cl_2 \longrightarrow Ph_5TeCl^{5)} \quad (N9.3)$$

5a → **3a**

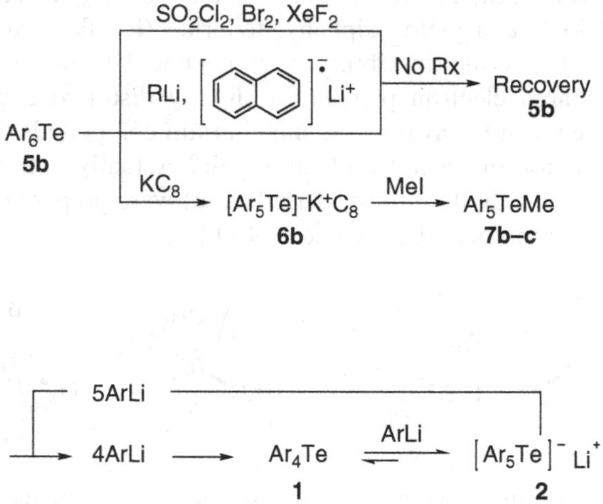

$$Ar_6Te\ (\mathbf{5b}) \xrightarrow{SO_2Cl_2,\ Br_2,\ XeF_2;\ RLi,\ [C_{10}H_8]^{\bullet-}Li^+} \text{No Rx} \longrightarrow \text{Recovery } \mathbf{5b} \quad (N9.4)$$

$$Ar_6Te\ (\mathbf{5b}) \xrightarrow{KC_8} [Ar_5Te]^-K^+C_8\ (\mathbf{6b}) \xrightarrow{MeI} Ar_5TeMe\ (\mathbf{7b\text{–}c})$$

$$TeCl_4 \xrightarrow{4ArLi} Ar_4Te\ (\mathbf{1}) \underset{}{\overset{ArLi}{\rightleftharpoons}} [Ar_5Te]^-Li^+\ (\mathbf{2})$$

$$TeCl_4 \xrightarrow{5ArLi} [Ar_5Te]^-Li^+\ (\mathbf{2})$$

$$[Ar_5Te]^-Li^+\ (\mathbf{2}) \xrightarrow{SO_2Cl_2 \text{ or } Br_2} Ar_5TeX\ (\mathbf{3}) \xrightarrow{Ag^+Y^-} [Ar_5Te]^+Y^{-\ 6)}\ (\mathbf{4}) \xrightarrow{ArLi} Ar_6Te^{\ 4)}\ (\mathbf{5})$$

3 X = Cl; Br; **4** Y = ClO_4; TfO

1 Ar: **a** = Ph; **b** = *p*-$CF_3C_6H_4$; **c** = *p*-MeC_6H_4; **d** = *p*-$MeOC_6H_4$
2, 3, 4, 5 Ar: **a** = Ph; **b** = *p*-$CF_3C_6H_4$

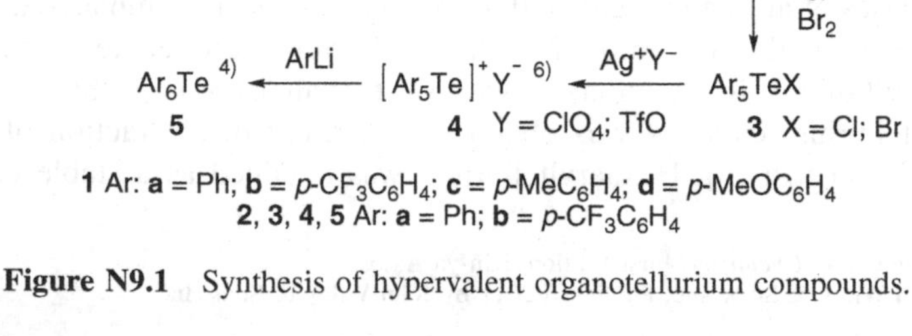

Figure N9.1 Synthesis of hypervalent organotellurium compounds.

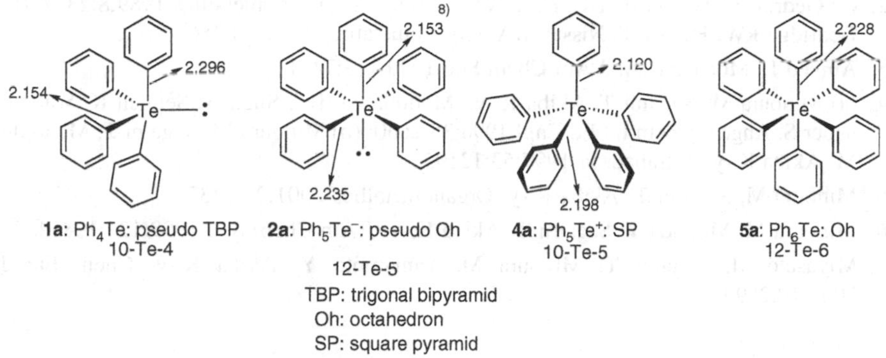

1a: Ph_4Te: pseudo TBP 10-Te-4 **2a**: Ph_5Te^-: pseudo Oh 12-Te-5 **4a**: Ph_5Te^+: SP 10-Te-5 **5a**: Ph_6Te: Oh 12-Te-6

TBP: trigonal bipyramid
Oh: octahedron
SP: square pyramid

Figure N9.2 Structures of hypervalent aromatic tellurium compounds (bond length in angstrom).

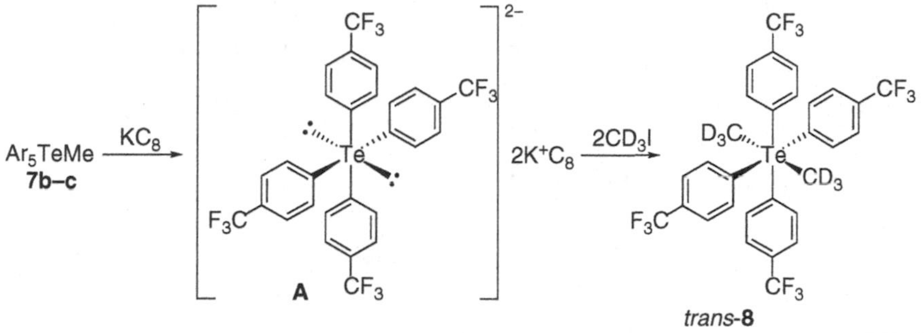

Ar_5TeMe **7b–c** $\xrightarrow{KC_8}$ [**A**]$^{2-}$ $2K^+C_8$ $\xrightarrow{2CD_3I}$ *trans*-**8**

(N9.5)

When **7c** was treated with KC_8, dianion (**A**) is generated in situ and the dianion was methylated to afford *trans*-dimethyl derivative (*trans*-**8**).

The structures of several aromatic tellurium compounds are summarized in Figure N9.2. The structure of tetraphenyltellurium (**1a**) is a pseudo trigonal bipyramid as was already mentioned and that of pentaphenyltellurium anion (**2a**) is a pseudo octahedron probably due to the repulsion of unshared electron pair and four phenyl groups. Pentaphenyltellurium cation (**4a**) is a square pyramid. It is noted here by X-ray analysis that the structures of both anion and cation of the same component (i.e., **2a** and **4a**) are different, which illustrates the steric effect of unshared electron pair clearly.

REFERENCES

1. (a) Wittig G, Fritz H. Justus Liebigs Ann Chem 1952;577:39; (b) Barton DHR, Glover SA, Ley SV. J Chem Soc Chem Comm 1977:266; (c) Glover SA. J Chem Soc [Perkin 1] 1980:1338.

2. (a) Gedridge RW, Harris DC, Higa KT, Nissan RA. Organometallics 1989;8:2817; (b) Gedridge RW, Higa KT, Nissan RA. Organometallics 1991;10:286.
3. Ahmed L, Morrison JA. J Am Chem Soc 1990;112:7411.
4. (a) Minoura M, Sagami T, Akiba K.-y., Modrakowski C, Suda A, Seppelt K, Wallenhauer S. Angew Chem Int Ed Engl 1996;35:2660; (b) Minoura M, Sagami T, Miyasato M, Akiba K.-y. Tetrahedron 1997;53:12195.
5. Minoura M, Sagami T, Akiba K.-y. Organometallics 2001;20:2437.
6. Minoura M, Mukuda T, Sagami T, Akiba K.-y. J Am Chem Soc 1999;121:10852.
7. Miyasato M, Sagami T, Minoura M, Yamamoto Y, Akiba K.-y. Chem Eur J 2004;10:2590.

CHAPTER 12

HYPERVALENT CARBON COMPOUNDS: CAN HEXAVALENT CARBON EXIST?

12.1 INTRODUCTION

Methyl bromide is converted to methanol by sodium hydroxide when they are heated in a water–ethanol solution. Let us consider the mechanism of the fundamental reaction generally and in detail.

As the valence of carbon is 4 and saturated carbon is sp^3 hybridized, carbon bears four σ-bonds with different substituents. It is well established experimentally for the reaction of the saturated carbon compound (**1**) and a nucleophile ($Nu{:}^-$) that (i) the rate of the reaction is second order and (ii) the configuration of the central carbon is inverted to yield **1′**. On the basis of these facts, **A** should be invoked as the transition state in which three substituents (a, b, c) stay in a plane perpendicular to a line of $[Nu\cdots C\cdots Y]^-$ (Y: leaving group). The configuration of the central carbon has changed to sp^2 from sp^3 hybridization (Eq. 12.1).

$$HO^- + CH_3Br \longrightarrow CH_3OH + Br^-$$

$$Nu{:}^- + \underset{\mathbf{1}}{abcC{-}Y} \longrightarrow \left[\underset{\mathbf{A}}{\overset{\delta-}{Nu}\cdots C(abc)\cdots\overset{\delta-}{Y}}\right] \longrightarrow \underset{\mathbf{1'}}{Nu{-}C(abc)} + Y^- \qquad (12.1)$$

Organo Main Group Chemistry, First Edition. Kin-ya Akiba.

In **A**, the central carbon bears five bonds to connect substituents, and hence it has 10 valence electrons formally. The two electrons in excess of the octet are used to form a linear bond holding a nucleophile (Nu:$^-$) and a leaving group (Y). The unique bond, therefore, bears four electrons and consists of a three-center four-electron bond (3c–4e bond). The bond is also called a *hypervalent bond*, and as stated several times in this book, it is electron-rich, longer, and weaker than a regular σ-bond. Then, substituent (Y) leaves as a negative ion (Y$^-$) accompanied by inversion of the carbon. This is the S_N2 reaction.

It has been a long-standing dream and objective to stabilize the transition state or to synthesize a hypervalent compound of carbon. Chemistry of hypervalent compounds of the third period such as phosphorus, sulfur, and silicon and heavier has been established by the effort of many groups during several decades [1a]. It started from the dream of Staudinger (1919) followed by Wittig (1947) (cf. Notes 6) and then was expanded by Breslow, Martin, Corriu, Barton, and the author, among others. Synthesis of hypervalent carbon compounds, which is the central element of organic chemistry, had been the target for about 40 years, but in vain. Four conceptual models for the synthesis, each leading to the formation of the 3c–4e bond, are summarized in Fig. 12.1. They are (i) two radicals combine with two unpaired electrons of the p orbital of X; (ii) two nucleophiles (lone pair electrons) combine with each vacant lobes of the p orbital of X; (iii) a nucleophile (lone pair electrons) donates electrons to the σ^*orbital of the Z–X bond and combines with X; and (iv) a nucleophile (lone pair electrons) attacks an onium salt to be bonded with X [1b].

Hypervalent compounds of main group elements of the third period and heavier, such as phosphorane, sulfurane, and so on, were prepared by method (iv). As the onium salt of carbon cannot exist, method (ii) was employed for attempted synthesis.

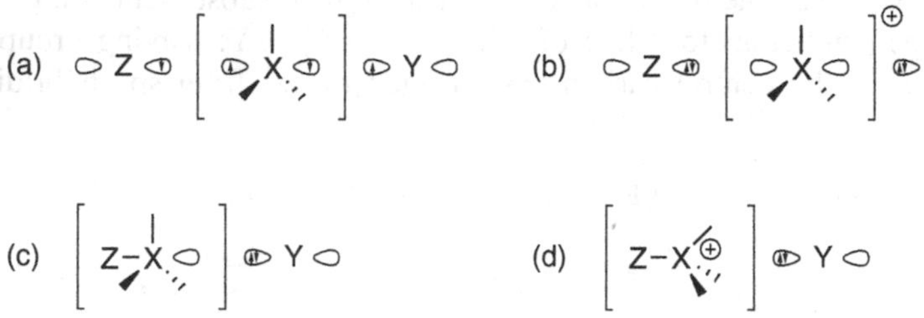

Figure 12.1 Conceptual models to form pentacoordinate hypervalent species. (a) Two radicals combine with two unpaired electrons of p orbital. (b) Two nucleophiles combine with each vacant lobe of p orbital. (c) A nucleophile donates electrons to σ^*orbital of Z–X bond (transition state of S_N2). (d) A nucleophile attacks onium salt to be bonded (e.g., phosphorane, sulfurane, etc.)

12.2 ATTEMPTS FOR PENTACOORDINATE HYPERVALENT CARBON SPECIES

In 1968, Breslow prepared a triphenylmethyl cation (**2**) in which each phenyl group bears a 2-methylthiomethyl group, expecting the formation of 3c–4e bond by the interaction of two sulfur atoms with a vacant 2p orbital of the central carbocation (method ii). But **2** turned out to be a sulfonium salt (**3**) and the objective was not attained (Eq. 12.2) [2]. In 1985, Hojo prepared 2-isopropyl cation attached to 1-position of fluorene skeleton, but a sulfonium salt (**4**) of five-member ring was obtained [3].

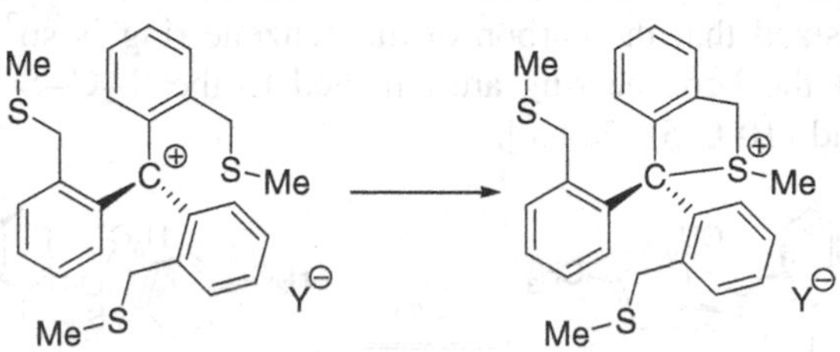

Me S C S Me Me S Y **2** → Me S C S Me Me S Y **3** (12.2)

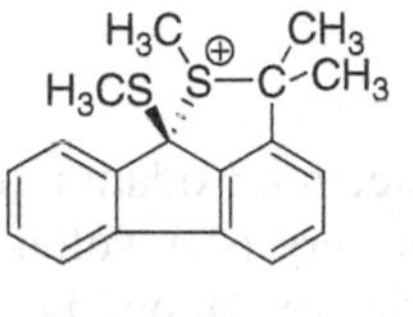

H_3C CH_3 H_3CS S C CH_3 **4**

In 1973, Martin tried to prepare pentacoordinate hypervalent carbon species (**B**) with the S–C–S hypervalent bond. It was designed so that the 2-isopropyl cation attached at the 9-position of the anthracene skeleton should accept two lone pair electrons of the sulfur of the phenylthio groups at 1,8-positions (Eq. 12.3). Again, sulfonium salt (**5**) was obtained, and equilibration was observed between **5** and **5′**. Kinetic parameters of the equilibration were $\Delta H^{\ddagger} = 13$ kcal/mol and $\Delta S^{\ddagger} = -5$ eu. Martin called the equilibration as *bell-clapper rearrangement* to express the movement and inversion of the central carbon [4].

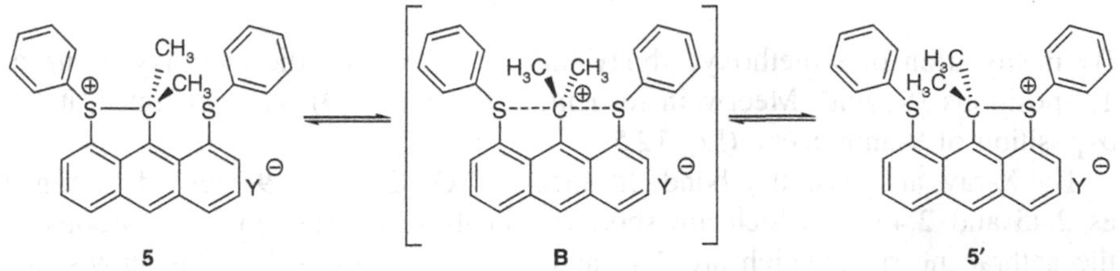

CH_3 CH_3 S C S Y **5** ⇌ [H_3C CH_3 S----C----S Y] **B** ⇌ H_3C H_3C S C S Y **5′**

(12.3)

Martin further prepared a formal dication with anthracene skeleton (**6**) having two *p*-methylphenylthio groups at 1,8-positions in which a 2′,6′-(dimethoxy)phenyl group is attached at the 9-position. He expected that a formal carbocation at 1′-position of the benzene ring should attract two phenylthio groups to build a pentacoordinate S–C–S hypervalent bond. The benzene ring is perpendicular to the anthracene skeleton and the two methoxy groups at 2′,6′-positions of the benzene ring should contribute to stabilize the molecule (**6**). In the formal dication (**6**), chemical shift (δ) of 1′-carbon (^{13}C) of the benzene is 109.3. Coupling constant of 1′-carbon with 2′,6′-carbons is 61.9 Hz and that with 9-carbon of the anthracene is 56.3 Hz. The corresponding values of neutral compound (**7**) are 118.0 (δ), 74.3, and 40.3 Hz. Based on the result, Martin hypothesized that the carbon of the benzene ring is sp^2 hybridized and the electrons of the benzene ring are attracted to the S–C–S bond to form a hypervalnet bond (10-C-5) [5a,b,c].

6 ⇌ 7 (+2e / −2e)

(12.4)

In addition, reversible two-electron oxidation (cyclic voltammetry) of a neutral compound (**7**) occurs easily at very near voltages (0.98, 1.02 V). Without the 1,8-phenylthio groups, electric oxidation is quite unlikely but reversible one-electron oxidation takes place at 1.27 V. This shows clearly that the 1,8-phenylthio groups stabilize the resulting dication (**6**) [5d]. This is also supportive of Martin's hypothesis. He could not, however, isolate crystals of **6**, unfortunately. Hence at present, the possibility cannot be ruled out that **6** is considerably closer to a model of the transition state of S_N2 than **5** but is still in rapid equilibration like in Eq. (12.3) (bond switching).

12.3 SYNTHESIS OF PENTACOORDINATE HYPERVALENT CARBON SPECIES (10-C-5) AND BOND SWITCHING AT CARBON AND BORON

By methylation of 9-methoxycarbonylanthracene bearing the methoxy group at 1,8-positions (**8**) with Meerwein reagent, carbocation (**9**) was generated at the α-position of 9-anthracene (Eq. 12.5).

By X-ray analysis, the bond distances of O–C–O of **9** were determined as 2.45 and 2.43 Å, which are shorter than those of 1,9- and 9,8-carbons of the anthracene ring, which are 2.49 and 2.52 Å, respectively. This shows that the α-carbocation of the anthracene ring withdraws the lone pair electrons of

1,8-methoxy groups to form a hypervalent bond. The chemical shift (δ) of the carbonyl carbon (^{13}C) shifts downfield from 172.3 (**8**) to 192.6 (**9**). Compound **9** is the first example of hypervalent carbon species (10-C-5) whose structure has been unequivocally determined. The basic reason for this success is the instability of an oxonium salt compared to a sulfonium salt, thus O–C–O bond is kept balanced [6a,b].

2.45 Å, 2.43 Å

$Me_3O^+BF_4^-$, CH_2Cl_2

$B_2F_7^-$

8 → **9** (12.5)

^{13}CNMR: δ 172.3 → δ 192.6

Until here, the hypervalent bond was tried to form at a carbocation by attracting electrons of the lone pair of sulfur and oxygen (Fig. 12.1b). The transition state of S_N2 bears a formal negative charge in the 3c–4e bond. In order to realize this, an attempt was made to attack a C–O bond having two trifluoromethyl groups with an oxide anion from the rear side in **10** to yield **C** (Fig. 12.2**c**). The structure of the resulting compound (**11**) was analyzed by X-ray to show distances of O–C–O to be $O^1–C^{15} = 1.417$ and $O^2–C^{15} = 2.896$ Å, and thus the oxide anion faces the potassium cation of 18-crown-6. Unfortunately, the 3c–4e bond was not formed (Fig. 12.2)[6b].

In a hypervalent carbon compound **9** (10-C-5), two methoxy groups at the 1,8-positions are forced to stay close to the α-carbocation of the 9-position by the anthracene skeleton. Next, it was investigated whether the steric constraint is really necessary. Accordingly, benzene derivatives were prepared that have 2,6-substituents bearing heteroatoms (Eq. 12.6).

Triphenylmethyl alcohol (**12**) bearing the methylthiomethyl group at the 2,6-positions was prepared and treated with perchloric acid to yield a sulfonium salt (**13**). The structure of **13** was determined by X-ray analysis. When ^{1}H NMR of **13** was measured at room temperature, the methyl ($\delta = 1.6$) and the methylene groups appeared as broad peaks. When measured at 100°C, the methyl and

KH, 18-crown-6, THF

K^+-18C6

10 **C** **11**

$O^1–C^{15}$: 1.471 Å
$O^2–C^{15}$: 2.986 Å

Figure 12.2 Attempt to prepare 3c–4e bond bearing a negative charge.

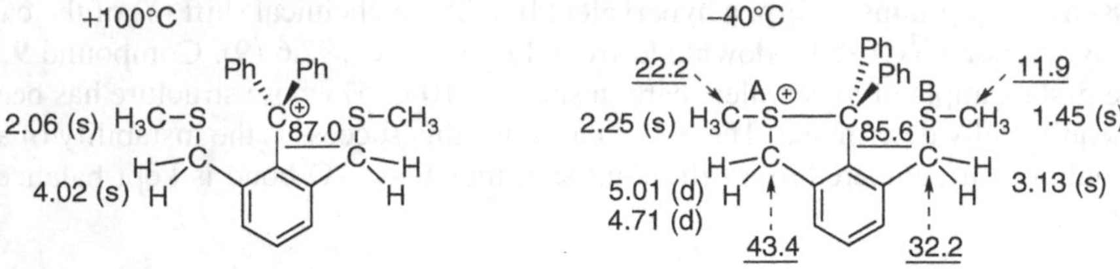

Figure 12.3 Influence of temperature on the ^{1}H and ^{13}C spectra of **13** (CD_3CN, ppm).

the methylene groups each gave a sharp singlet peak ($\delta = 2.06, 4.02$) (**13D**). At low temperatures (−40°C), the following peaks were observed: sharp singlets of two methyl groups ($\delta = 1.45, 2.25$), a sharp singlet of one methylene group ($\delta = 3.13$), and two doublets ($\delta = 4.71, 5.01$) of the other methylene group (**13E**). The low-temperature spectrum corresponds well to the structure obtained by X-ray analysis. In Fig. 12.3, the observed chemical shifts of ^{13}C are underlined.

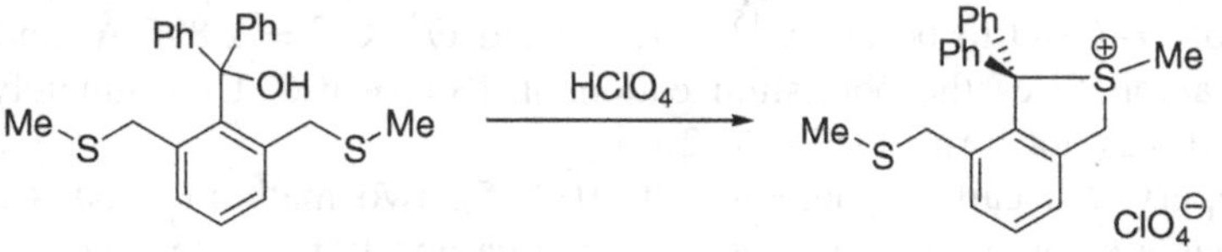

(12.6)

These results illustrate the presence of rapid bond switching equilibration between **13** and **13′** in solution, which is similar to that in compound **5**. A selenium compound substituted for sulfur behaves just like **13** (Eq. 12.7).

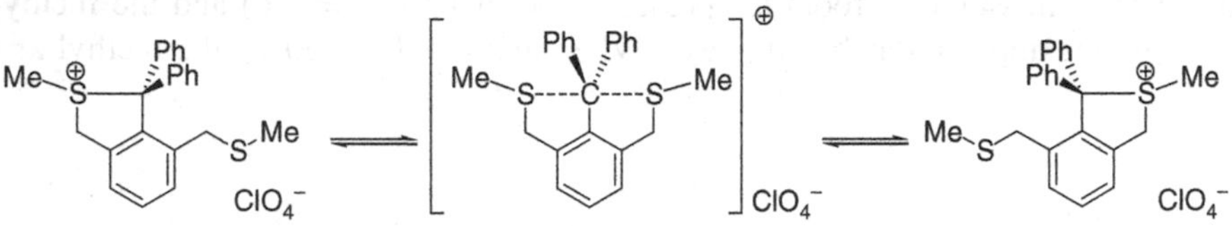

Bonds witching equilibration (piston-rod,or bell-clapper,mechanism)

(12.7)

Triphenylmethyl alcohol (**14**) with *p*-methylphenyloxymethyl groups at the 2,6-positions was prepared and treated with perchloric acid to yield a dark green solid (**15a, b**), from which bright crystals were obtained (Eq. 12.8).

14

HClO$_4$or HBF$_4$

15
(a) X = Cl Dark green solid
(b) X = F

ClO_4^-or BF_4^-

(12.8)

The bond angle of O–C–O in compound **15** is about 160° and is nearly linear although it deviates from the ideal angle of 180°. The structure of compound **15** is symmetric. The C–O bond distance in **15a** and **b** bearing the *p*-methylphenyloxymethyl group is 2.67–2.69 Å but that in **15c–e** having the *p*-methoxyphenyloxymethyl group is slightly elongated as 2.70–2.78 Å. In compounds with the *p*-methoxyphenyloxymethyl group, the C–O distance (2.705, 2.718, 2.77 Å) increases gradually according to the electron-donating ability of each 9-phenyl groups with para substituents such as hydrogen, fluorine, and methoxy groups. The trend is the same for **15a** and **b** (Table 12.1). This is expected, considering that the electron-withdrawing ability of the carbocation decreases according to the increase in the electron-donating ability of the substituents. These facts are coincident with the expected character of the 3c–4e bond. It is clearly demonstrated that the 3c–4e bond of carbon, that is, the 10-C-5 hypervalent carbon compound, can be formed even in the absence of a steric constraint [7].

TABLE 12.1 Penta Coordinate Hypervalent Carbon Compounds (10-C-5) with Flexible Ligands

ClO_4^- PF_6^- or ClO_4^-

Bond distance (Å)	X = Cl **15a**	X = F **15b**	X = H **15c**	X = F **15d**	X = OMe **15e**
C^1–O^1	2.671(4)	2.690(4)	2.705(2)	2.718(5)	2.77(1)
C^1–O^2	2.682(4)	2.690(4)	2.705(2)	2.718(5)	2.78(1)
Bond angle (°) O^1–C^1–O^2	161.9(1)	162.3(1)	158.9(2)	160.4(4)	159.1(4)

Boron compounds are trivalent (6-B-3) and are easily prepared as stable, because they are neutral, although they are isoelectronic with the carbocation (6-C-3). Penta- and hexavalent hypervalent boron compounds (**16, 17**) were already prepared [8].

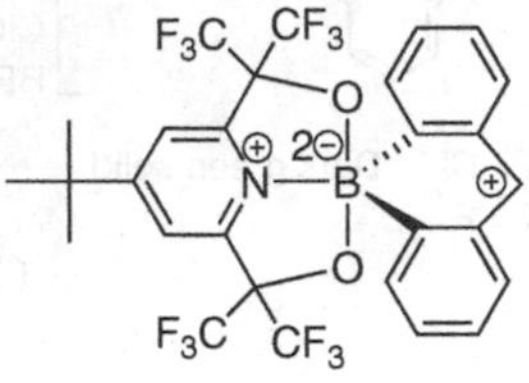

16 10–B–5

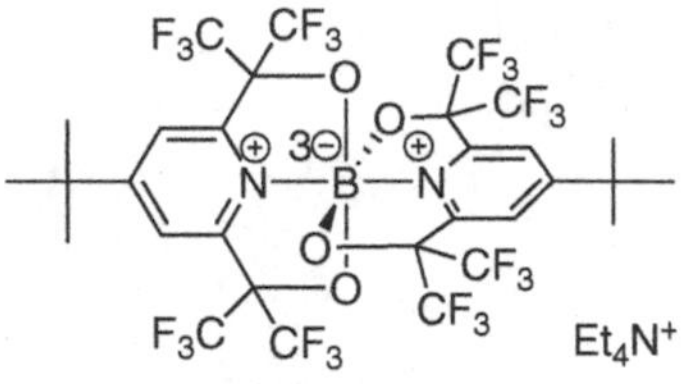

17 12–B–6

In order to see the interaction between boron and nitrogen, anthracenes bearing boron at the 9-position and dimethylamino groups at the 1,8-positions were prepared (**18**–**20**) [6b,c,d]. On the basis of X-ray analysis, it was shown that a boron–nitrogen bond was formed with one nitrogen (1.809, 1.739, 1.664 Å) and the other nitrogen was not bonded (2.94, 3.12, 3.12 Å). The boron–nitrogen distance decreases according to the increase of the electron-withdrawing ability of the substituents on the boron, but the distance between the 1,8-nitrogens is kept constant (4.80 Å) [6b]. The situation is the same as in sulfonium salts (**5**).

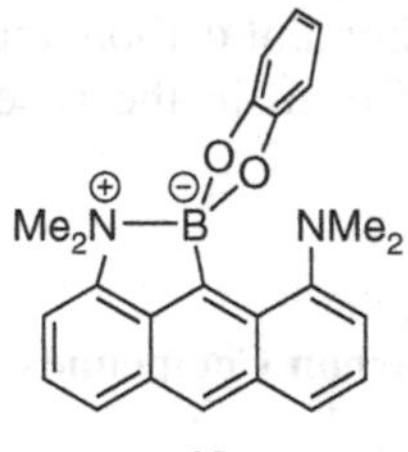

	18	**19**	**20**
Bond distance of B-N (Å)	1.809(2), 2.941(2)	1.739(2), 3.124(3)	1.664(3), 3.129(3)
Distance of N-B-N (Å)	4.75	4.86	4.79

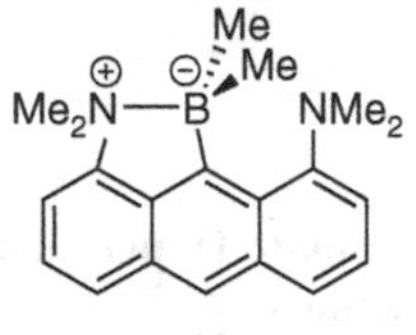

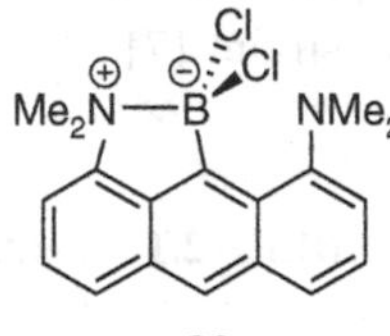

When the ^{1}H NMR spectra of these compounds were measured at room temperature, only one sharp singlet peak appeared for the four methyl groups, and the anthracene ring was symmetric (two kinds of doublets, one triplet, one singlet). The spectrum corresponded to that of a sulfonium salt (**13**) at high temperature (100°C). But even at as low as −80°C, the ^{1}H NMR of these compounds did not change and stayed the same as they were at room temperature. These facts imply that even the most unsymmetric compound of **20** undergoes quite rapid bond switching at −80°C. The bond switching proceeds along (almost) the line of N–B–N, a 3c–4e bond is formed at the transition state, the central boron becomes pentacoordinate (hypervalent), and the configuration is inverted (Eq. 12.9).

20 ⇌ [**20**‡] ⇌ **20′** (12.9)

5-Coordinate hypervalent boron

The bond switching of Eq. (12.9) is schematically illustrated in Fig. 12.4a. The distance between two nitrogens (N^1, N^2) of dimethylamino groups at the 1,8-positions of anthreacene is kept constant (4.80 Å) and that of the bonded boron (B^1) and nitrogen (N^1) is 1.74 Å. The distance corresponds to 110% of the sum of the covalent radii of boron and nitrogen. The distance of the nonbonded boron (B^1) and nitrogen (N^2) is 3.06 Å. Because of bond switching, $\mathbf{B}^1$ shifts to $\mathbf{B}^{1'}$ on the line to form a new bond between boron (B^1) and nitrogen (N^2), the B^1–N^1 distance changes to 3.06 Å, and simultaneously the configuration of the boron is inverted. Bond switching occurs rapidly even at low temperature.

Then why is the bond between the boron and nitrogen elongated by 110% of the sum of covalent radii? This happens certainly by the donation of the lone pair electron of the nonbonded nitrogen (N^2) to the σ^* orbital of the bonded boron (B^1) and nitrogen (N^1), and hence the bond is weakened and elongated to a certain degree (10% here). That is, Fig. 12.4a corresponds to an early transition state of S_N2, in general. Hence these compounds happened to be suitable models to undergo rapid and eternal bond-switching.

The essential idea mentioned in Fig. 12.4a can be generalized as in Fig. 12.4b. When a nucleophile approaches the central atom (boron here) to 110% of its van der Waals radius (for boron: 2.29 Å) from the rear side, interaction between the two atoms starts to take place and the lone pair electrons are donated to the σ^*orbital of the covalent bond [boron (B^1)–nitrogen (N^1) here] and the bond is gradually elongated and finally broken. The estimated value of 110% would have a real meaning. It is hoped that the idea expressed here will be investigated theoretically as a model of S_N2 employing a variety of combination of elements and nucleophiles [9].

12.4 ATTEMPTS TO HEXACOORDINATE HYPERVALENT CARBON SPECIES (12-C-6)

Is there any possibility for hexacoordinate hypervalent carbon species (12-C-6) to exist? In principle, it can be realized in case four nucleophiles coordinate to two vacant 2p orbitals of the sp carbon. It may certainly be better to place two positive charges on the central carbon to let the coordination become

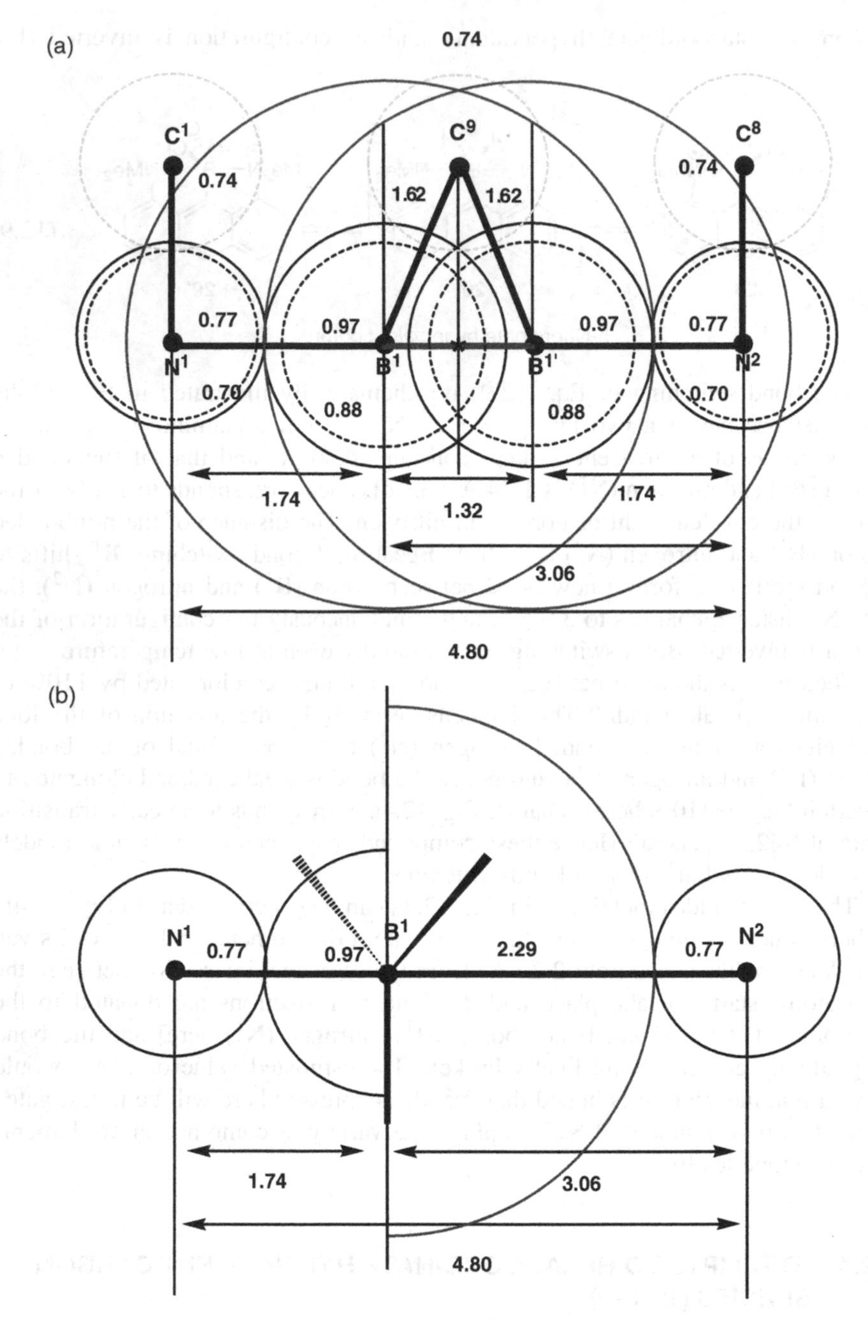

Figure 12.4 Bond switching diagram of N–B · · · N on the anthracene skeleton (C^1, C^9, C^8) **(18, 19, 20)**. (a) Dotted circle represents covalent radius: N, 0.70; B, 0.88; C, 0.74 Å. Solid circle represents 110% of covalent radius: N, 0.77; B, 0.97 Å. (b) Large half circle represents 110% of van der Waals radius of B: 2.29 Å. Others represent 110% of covalent radius of N and B.

stronger. Imaginary models can be illustrated as **21, 22**, and **23**. Two 3c–4e bonds should be formed to be coordinated by two waters or dimethyl ethers for each one, in which there are two positive charges and two orthogonal vacant 2p orbitals on the central carbon. The two hydrogen (**21**), methyl (**22**), or phenyl (**23**) groups are connected to the sp carbon. They are quasi stable by theoretical calculation and the structure and the carbon–oxygen bond energy are obtained.

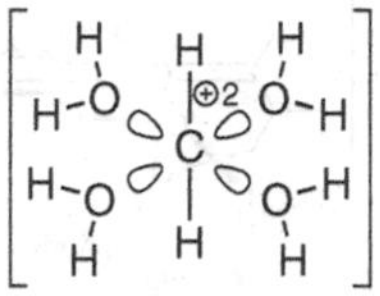

21 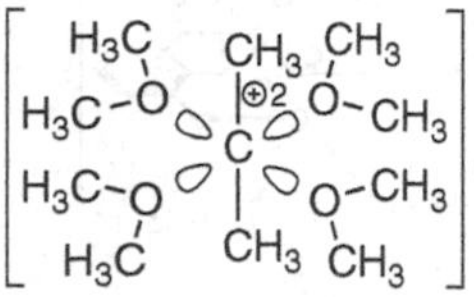 **22** 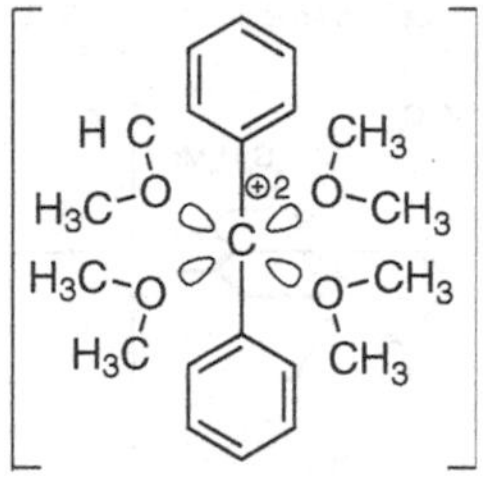 **23**

As model compounds for experimental synthesis, compound **24** having two orthogonal anthracene skeletons is derived from **9** and compound **25** bearing phenyloxymethyl arms at the 2,6-positions of a benzene ring (pincer ligand) is deduced from **15**. Moreover, by dimethylation of thioxanthene (**26**) or xanthene allenes (**28**), it is expected to generate dicationic carbon species resembling **24b** or **25b** to build up hypervalent carbon compounds (12-C-6) of **27** or **29**.

Actually, **26** and **27** have been synthesized and their structures determined by X-ray analysis. The thioxanthene skeleton of **27** is considerably strained, and the distances of the four C(allene)–O bonds are 2.641, 2.706, 2.750, and 2.673 Å (av. = 2.693 Å). These distances are slightly larger ($\Delta = 0.25$ Å) than the C–O bond of **9** (2.43, 2.45 Å) but are not much longer than that of **15** (av. = 2.711 Å) and are almost equal to it. In conclusion, it can be believed that there are two 3c–4e bonds in the C–O of **27**, and **27** serves as the first example of a hexacoordinate hypervalent carbon compound (12-C-6) [10].

On the basis of theoretical calculations, the 3c–4e bond of **27** is the weakest among **24, 25, 27**, and **29** and the order of their bond energy is **25** > **24** > **29** > **27**. The distance of the four C(allene)–O bonds of **25** is calculated as 2.484 Å and the bond energy is 26.9 kcal/mol; the presence of a typical 3c–4e bond was clearly demonstrated by molecular orbital calculations [11]. On the basis of these results, there is no doubt as to the presence of a hexacoordinate hypervalent carbon compound (12-C-6). On the other hand, the question still is: in compound **27** a carbodication is surrounded by and floating in the sea of electrons of the four oxygens and what kind of bonds are they? In case the fundamental ideas mentioned here to expand the valence of the main group elements are applied to elements of the third period and heavier, interesting development of chemistry can be expected.

24a
Hexa coordinate hypervalent carbon species

24b
Two vacant orbitals lie on the central sp carbon

24c
Positive charge lies in each benzene ring

25a
Hexa coordinate hypervalent carbon species

25b
Two vacant orbitals lie on the central sp carbon

25c
Positive charge lies in each benzene ring

26 —2 MeX→ **27**

28 —2 MeX→ **29**

REFERENCES

1. (a) Musher JI. Angew Chem Int Ed Engl 1969;8:54; (b) Akiba K.-y., editor. Chemistry of hypervalent compounds. Weinheim, Wiley-VCH; 1999.
2. (a) Breslow R, Garratt S, Kaplan L, LaFollette D. J Am Chem Soc 1968;90:4051; (b) Breslow R, Kaplan L, LaFollette D. J Am Chem Soc 1968;90:4056.
3. Hojo M, Ichi T, Shibato K. J Org Chem 1985;50:1478.

4. (a) Basalay RJ, Martin JC. J Am Chem Soc 1973;95:2565; (b) Martin JC, Basalay RJ. J Am Chem Soc 1973;95:2572.
5. (a) Martin JC. Science 1983;221(4610): 509; (b) Forbus TR Jr, Martin JC. J Am Chem Soc 1979;101:5057; (c) Forbus TR Jr, Martin JC. Heteroatom Chem 1993;4: 113–128, 129; (d) Forbus TR Jr, Kahl JL, Faulkner LR, Martin JC. Heteroatom Chem 1993;4:137.
6. (a) Akiba K.-y,, Yamashita M, Yamamoto Y, Nagase S. J Am Chem Soc 1999;121:10644; (b) Yamashita M, Yamamoto Y, Akiba K.-y., Hashizume D, Iwasaki F, Takagi N, Nagase S. J Am Chem Soc 2005;127:4354; (c) Yamashita M, Yamamoto Y, Akiba K.-y., Nagase S. Angew Chem Int Ed 2000;39:4055; (d) Yamashita M, Kamura K, Yamamoto Y, Akiba K.-y. Chem Eur J 2002;8:2976.
7. Akiba K.-y., Moriyama Y, Mizozoe M, Inohara H, Nishii T, Yamamoto Y, Minoura M, Hashizume D, Iwasaki F, Takagi N, Ishimura K, Nagase S. J Am Chem Soc 2005;127:5893.
8. Lee DY, Martin JC. J Am Chem Soc 1984;106:5745.
9. (a) Yamamoto Y, Akiba K.-y. J Synth Org Chem Jpn 2004;62:1128; (b) Akiba K.-y., Yamamoto Y. Heteroatom Chem 2007;18:161.
10. Yamaguchi T, Yamamoto Y, Kinoshita D, Akiba K.-y., Zhang Y, Reed CA, Hashizume D, Iwasaki F. J Am Chem Soc 2008;130:6894.
11. (a) Kikuchi Y, Ishii M, Akiba K.-y., Nakai H. Chem Phys Lett 2008;450:37. (b) Nakai H, Okoshi M, Atsumi T, Kikuchi Y, Akiba K.-y. Bull Chem Soc Jpn 2011;84:505.

NOTES 10

MAIN GROUP ELEMENT PORPHYRINS

Porphyrin and its related compounds, such as hemoglobin (Fe(II) complex), chlorophyll (Mg(II) complex), and vitamin B_{12} (Co(III) complex), are important and widely present as biomolecules. Central elements of these are transition metals; however, porphyrins bearing main group elements as the central atom are also known to have unique structures and characteristics [1].

Porphyrin (**1**) is a planar 20-member ring, in which four pyrrole rings are connected with four one-carbon groups (–CH=). Porphyrin consists of 18π $(4n + 2)$ aromatic system because carbon–carbon double bonds in two pyrrole rings (**B, D**) are not counted in the cyclic conjugated system. Skeleton of porphyrin and its representative derivatives are illustrated in Fig. N10.1. When two hydrogens in the core are substituted for another element, a variety of element porphyrins are obtained.

As main group element porphyrins, groups 1, 2, 13, 14, and 15 element porphyrins are known but groups 16, 17, 18 element porphyrins are not known [2].

Group 1 Derivatives: A lithium ion sits in the dianion core of porphyrin, and another one is coordinated with four tetrahydrofuran (THF) molecules and located outside the core as a counter ion. Na^+ and K^+ are coordinated with two THF molecules, and they coordinate with two nitrogens of the core from above and below the molecular plane.

Organo Main Group Chemistry, First Edition. Kin-ya Akiba.

1 H_2(Porphyrin)

TTP: Tetraphenylporphyrin (phenyl groups: 5, 10, 15, 20)
OEP: Octaethylporphyrin (ethyl groups: 2, 3, 7, 8, 12, 13, 17, 18)
OETPP: Octaethyltetraphenylporphyrin
(Ethyl groups: 2, 3, 7, 8, 12, 13, 17, 18; phenyl groups: 5, 10, 15, 20)

Figure N10.1 Structure of porphyrin and its derivatives.

Group 2 Derivatives: Mg porphyrin is stable and has been investigated in detail in relation to chlorophyll. Be, Ca, Sr, Ba porphyrins are unstable and are not well investigated.

Group 13 Derivatives: Boron is not contained in the core but binds with a nitrogen of two porphyrins and forms an–N–B(F)–O–B(F)–N–bridge. Al, Ga, In, Tl porpyrins have been prepared. Among them, Al porpyrins are effectively used as polymerization and carbon dioxide fixing catalysts by the Inoue and Aida group in Japan [3a,b].

Group 14 Derivatives: Two oxidation states of two and four valences are known for each element. Elements of higher oxidation states bear two axial

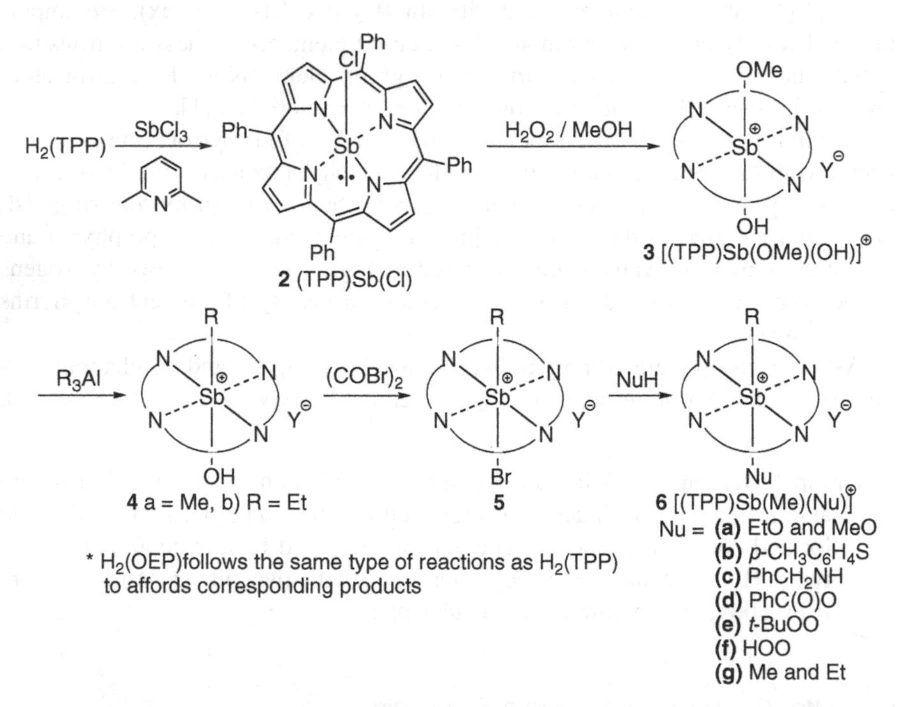

Figure N10.2 Synthesis of hypervalent TPP and OEP derivatives of antimony.

ligands. Derivatives of Si(IV), Ge(IV), Sn(II), Sn(IV), Pb(II) are stable, but those of Si(II), Ge(II), Pb(IV) are unstable or unknown.

Group 15 Derivatives: Two oxidation states of three and five are known for each element. Derivatives of P(III) are easily oxidized to those of P(V), and Bi(III) derivatives are known, but Bi(V) are not. Derivatives of As were reported earlier; they were initially not examined properly but were properly prepared recently. Their structures were determined by X-ray analysis [4]. For antimony and phosphorus derivatives, those bearing carbon substituent(s) as axial ligand(s) were synthesized recently, and some of them are briefly mentioned here.

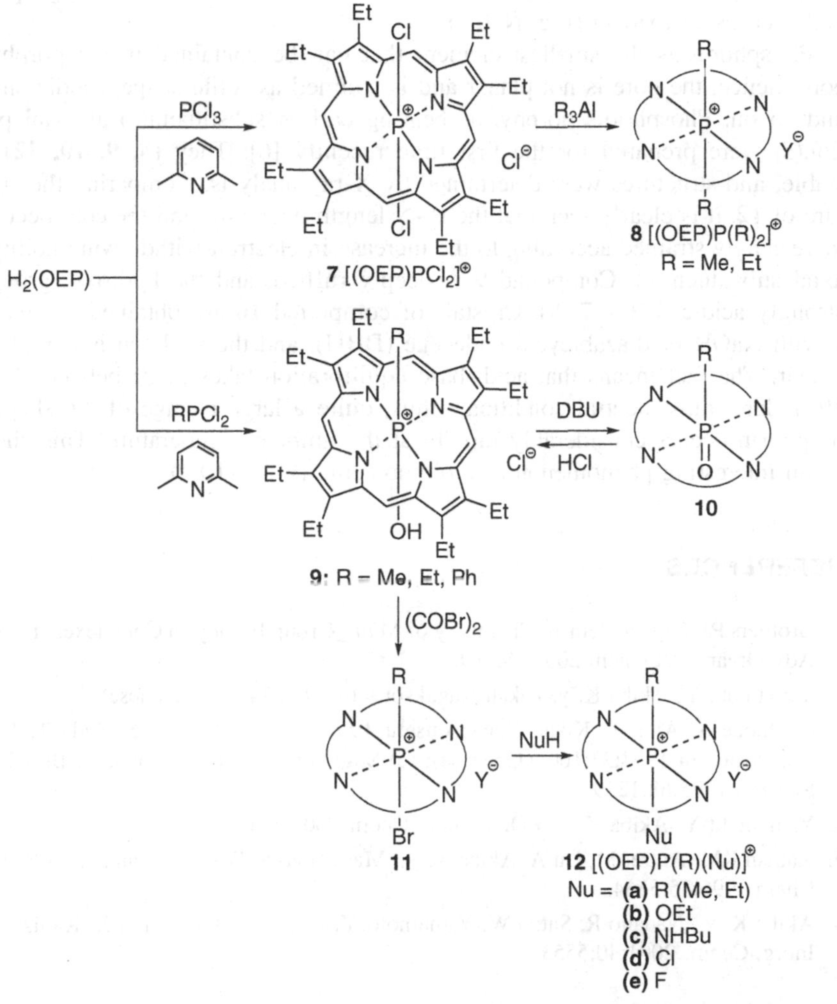

Figure N10.3 Synthesis of hypervalent OEP derivatives of phosphorus.

Antimony(III) porphyrin (**2**) is obtained in high yield by the reaction of H_2(TPP) and antimony trichloride in the presence of 2,6-dimethylpyridine. Unshared electron pair of **2** is considerably reactive and gives Sb(V) derivative of $[(TPP)Sb(OMe)(OH)]^+ Y^-$ (**3**) by treatment with hydrogen peroxide in methanol. The methoxy group of **3** is converted to alkyl groups with trialkylaluminums to give **4**. The hydroxyl group of **4** is converted to the bromide (**5**) with oxalyl dibromide, and **5** yields a variety of **6** by nucleophilic substitution. Compounds **3, 4, 6** are stable thermally, and the structures determined by X-ray analysis to show their porphyrin skeletons are planar. Their axial bonds are hypervalent, and compounds **4** and **6** are the first ones bearing carbon substituent(s) at axial position(s) [5]. Compounds **6e** and **f** are peroxides, and **6g** is a dialkyl derivative. Hydroperoxides of transition metal porphyrins are reactive and unstable to be isolated but compound **6f** is isolable in high purity and used as an oxidant (Fig. N10.2).

Phosphorus is the smallest element that can be contained in the porphyrin core; hence, the core is not planar and is strained as ruffle shape, saddle shape, and so on. Phosphorus porphyrins bearing carbon substituent(s) at axial position(s) were prepared for the first time recently [6]. They (**8, 9, 10, 12**) are stable, and structures were determined by X-ray analysis. Comparing the structure of **12**, it is clearly seen that the P–N length decreases and the core becomes increasingly strained according to the increase in electron-withdrawing ability of axial substituent(s). Compound **9** is deeply ruffled, and the hydroxyl group is strongly acidic (pH $\approx$ 7–8). Crystals of compound **10** are obtained by treating **9** with NaOH or diazabicyclo undecene (DBU), and the skeleton is completely planar. The fact means that acid–base equilibration takes place between **9** and **10** under almost neutral conditions; thus, quite a large change of the shape of porphyrin occurs considerably rapidly at the ambient temperature. This should be an interesting phenomenon as a biomolecule (Fig. N10.3).

REFERENCES

1. Brothers PJ, Organoelement Chemistry of Main_Group Porphyrin Complexes (review). Adv Organomet Chem 2001;48:289.
2. Yamamoto Y, Akiba K.-y. Kikan Kagaksousetu 1998;34:93. (Japanese).
3. (a) Inoue S, Aida T. Kikan Kagaksousetu 1993;18:55; (b) Inoue S, Aida T. Kikan Kagaksousetu 1994;23:160, (Japanese); (c) Sugimoto H, Aida T, Inoue S. Bull Chem Soc Jpn 1995;68:1239.
4. Yamamoto Y, Akiba K.-y. J Organomet Chem 2000;611:200.
5. Kadish KM, Autret M, Ou A, Akiba K.-y., Masumoto S, Wada R, Yamamoto Y. Inorg Chem 1996;35:5564.
6. Akiba K.-y., Nadano R, Satoh W, Yamamoto Y, Nagase S, Ou Z, Tan X, Kadish KM. Inorg. Cehm. 2001;40:5553.

INDEX

Printed in the United States
By Bookmasters